T0135145

Lecture Notes in Computer Science   **14720**

Founding Editors

Gerhard Goos
Juris Hartmanis

The series Lecture Notes in Computer Science (LNCS), including its subseries Lecture Notes in Artificial Intelligence (LNAI) and Lecture Notes in Bioinformatics (LNBI), has established itself as a medium for the publication of new developments in computer science and information technology research, teaching, and education.

LNCS enjoys close cooperation with the computer science R & D community, the series counts many renowned academics among its volume editors and paper authors, and collaborates with prestigious societies. Its mission is to serve this international community by providing an invaluable service, mainly focused on the publication of conference and workshop proceedings and postproceedings. LNCS commenced publication in 1973.

Fiona Fui-Hoon Nah · Keng Leng Siau
Editors

HCI in Business, Government and Organizations

11th International Conference, HCIBGO 2024
Held as Part of the 26th HCI International Conference, HCII 2024
Washington, DC, USA, June 29 – July 4, 2024
Proceedings, Part I

 Springer

Editors
Fiona Fui-Hoon Nah
Singapore Management University
Singapore, Singapore

Keng Leng Siau
Singapore Management University
Singapore, Singapore

ISSN 0302-9743   ISSN 1611-3349 (electronic)
Lecture Notes in Computer Science
ISBN 978-3-031-61314-2   ISBN 978-3-031-61315-9 (eBook)
https://doi.org/10.1007/978-3-031-61315-9

Foreword

This year we celebrate 40 years since the establishment of the HCI International (HCII) Conference, which has been a hub for presenting groundbreaking research and novel ideas and collaboration for people from all over the world.

The HCII conference was founded in 1984 by Prof. Gavriel Salvendy (Purdue University, USA, Tsinghua University, P.R. China, and University of Central Florida, USA) and the first event of the series, "1st USA-Japan Conference on Human-Computer Interaction", was held in Honolulu, Hawaii, USA, 18–20 August. Since then, HCI International is held jointly with several Thematic Areas and Affiliated Conferences, with each one under the auspices of a distinguished international Program Board and under one management and one registration. Twenty-six HCI International Conferences have been organized so far (every two years until 2013, and annually thereafter).

Over the years, this conference has served as a platform for scholars, researchers, industry experts and students to exchange ideas, connect, and address challenges in the ever-evolving HCI field. Throughout these 40 years, the conference has evolved itself, adapting to new technologies and emerging trends, while staying committed to its core mission of advancing knowledge and driving change.

As we celebrate this milestone anniversary, we reflect on the contributions of its founding members and appreciate the commitment of its current and past Affiliated Conference Program Board Chairs and members. We are also thankful to all past conference attendees who have shaped this community into what it is today.

The 26th International Conference on Human-Computer Interaction, HCI International 2024 (HCII 2024), was held as a 'hybrid' event at the Washington Hilton Hotel, Washington, DC, USA, during 29 June – 4 July 2024. It incorporated the 21 thematic areas and affiliated conferences listed below.

A total of 5108 individuals from academia, research institutes, industry, and government agencies from 85 countries submitted contributions, and 1271 papers and 309 posters were included in the volumes of the proceedings that were published just before the start of the conference, these are listed below. The contributions thoroughly cover the entire field of human-computer interaction, addressing major advances in knowledge and effective use of computers in a variety of application areas. These papers provide academics, researchers, engineers, scientists, practitioners and students with state-of-the-art information on the most recent advances in HCI.

The HCI International (HCII) conference also offers the option of presenting 'Late Breaking Work', and this applies both for papers and posters, with corresponding volumes of proceedings that will be published after the conference. Full papers will be included in the 'HCII 2024 - Late Breaking Papers' volumes of the proceedings to be published in the Springer LNCS series, while 'Poster Extended Abstracts' will be included as short research papers in the 'HCII 2024 - Late Breaking Posters' volumes to be published in the Springer CCIS series.

I would like to thank the Program Board Chairs and the members of the Program Boards of all thematic areas and affiliated conferences for their contribution towards the high scientific quality and overall success of the HCI International 2024 conference. Their manifold support in terms of paper reviewing (single-blind review process, with a minimum of two reviews per submission), session organization and their willingness to act as goodwill ambassadors for the conference is most highly appreciated.

This conference would not have been possible without the continuous and unwavering support and advice of Gavriel Salvendy, founder, General Chair Emeritus, and Scientific Advisor. For his outstanding efforts, I would like to express my sincere appreciation to Abbas Moallem, Communications Chair and Editor of HCI International News.

July 2024   Constantine Stephanidis

HCI International 2024 Thematic Areas
and Affiliated Conferences

- HCI: Human-Computer Interaction Thematic Area
- HIMI: Human Interface and the Management of Information Thematic Area
- EPCE: 21st International Conference on Engineering Psychology and Cognitive Ergonomics
- AC: 18th International Conference on Augmented Cognition
- UAHCI: 18th International Conference on Universal Access in Human-Computer Interaction
- CCD: 16th International Conference on Cross-Cultural Design
- SCSM: 16th International Conference on Social Computing and Social Media
- VAMR: 16th International Conference on Virtual, Augmented and Mixed Reality
- DHM: 15th International Conference on Digital Human Modeling & Applications in Health, Safety, Ergonomics & Risk Management
- DUXU: 13th International Conference on Design, User Experience and Usability
- C&C: 12th International Conference on Culture and Computing
- DAPI: 12th International Conference on Distributed, Ambient and Pervasive Interactions
- HCIBGO: 11th International Conference on HCI in Business, Government and Organizations
- LCT: 11th International Conference on Learning and Collaboration Technologies
- ITAP: 10th International Conference on Human Aspects of IT for the Aged Population
- AIS: 6th International Conference on Adaptive Instructional Systems
- HCI-CPT: 6th International Conference on HCI for Cybersecurity, Privacy and Trust
- HCI-Games: 6th International Conference on HCI in Games
- MobiTAS: 6th International Conference on HCI in Mobility, Transport and Automotive Systems
- AI-HCI: 5th International Conference on Artificial Intelligence in HCI
- MOBILE: 5th International Conference on Human-Centered Design, Operation and Evaluation of Mobile Communications

List of Conference Proceedings Volumes Appearing Before the Conference

1. LNCS 14684, Human-Computer Interaction: Part I, edited by Masaaki Kurosu and Ayako Hashizume
2. LNCS 14685, Human-Computer Interaction: Part II, edited by Masaaki Kurosu and Ayako Hashizume
3. LNCS 14686, Human-Computer Interaction: Part III, edited by Masaaki Kurosu and Ayako Hashizume
4. LNCS 14687, Human-Computer Interaction: Part IV, edited by Masaaki Kurosu and Ayako Hashizume
5. LNCS 14688, Human-Computer Interaction: Part V, edited by Masaaki Kurosu and Ayako Hashizume
6. LNCS 14689, Human Interface and the Management of Information: Part I, edited by Hirohiko Mori and Yumi Asahi
7. LNCS 14690, Human Interface and the Management of Information: Part II, edited by Hirohiko Mori and Yumi Asahi
8. LNCS 14691, Human Interface and the Management of Information: Part III, edited by Hirohiko Mori and Yumi Asahi
9. LNAI 14692, Engineering Psychology and Cognitive Ergonomics: Part I, edited by Don Harris and Wen-Chin Li
10. LNAI 14693, Engineering Psychology and Cognitive Ergonomics: Part II, edited by Don Harris and Wen-Chin Li
11. LNAI 14694, Augmented Cognition, Part I, edited by Dylan D. Schmorrow and Cali M. Fidopiastis
12. LNAI 14695, Augmented Cognition, Part II, edited by Dylan D. Schmorrow and Cali M. Fidopiastis
13. LNCS 14696, Universal Access in Human-Computer Interaction: Part I, edited by Margherita Antona and Constantine Stephanidis
14. LNCS 14697, Universal Access in Human-Computer Interaction: Part II, edited by Margherita Antona and Constantine Stephanidis
15. LNCS 14698, Universal Access in Human-Computer Interaction: Part III, edited by Margherita Antona and Constantine Stephanidis
16. LNCS 14699, Cross-Cultural Design: Part I, edited by Pei-Luen Patrick Rau
17. LNCS 14700, Cross-Cultural Design: Part II, edited by Pei-Luen Patrick Rau
18. LNCS 14701, Cross-Cultural Design: Part III, edited by Pei-Luen Patrick Rau
19. LNCS 14702, Cross-Cultural Design: Part IV, edited by Pei-Luen Patrick Rau
20. LNCS 14703, Social Computing and Social Media: Part I, edited by Adela Coman and Simona Vasilache
21. LNCS 14704, Social Computing and Social Media: Part II, edited by Adela Coman and Simona Vasilache
22. LNCS 14705, Social Computing and Social Media: Part III, edited by Adela Coman and Simona Vasilache

23. LNCS 14706, Virtual, Augmented and Mixed Reality: Part I, edited by Jessie Y. C. Chen and Gino Fragomeni

24. LNCS 14707, Virtual, Augmented and Mixed Reality: Part II, edited by Jessie Y. C. Chen and Gino Fragomeni

25. LNCS 14708, Virtual, Augmented and Mixed Reality: Part III, edited by Jessie Y. C. Chen and Gino Fragomeni

26. LNCS 14709, Digital Human Modeling and Applications in Health, Safety, Ergonomics and Risk Management: Part I, edited by Vincent G. Duffy

27. LNCS 14710, Digital Human Modeling and Applications in Health, Safety, Ergonomics and Risk Management: Part II, edited by Vincent G. Duffy

28. LNCS 14711, Digital Human Modeling and Applications in Health, Safety, Ergonomics and Risk Management: Part III, edited by Vincent G. Duffy

29. LNCS 14712, Design, User Experience, and Usability: Part I, edited by Aaron Marcus, Elizabeth Rosenzweig and Marcelo M. Soares

30. LNCS 14713, Design, User Experience, and Usability: Part II, edited by Aaron Marcus, Elizabeth Rosenzweig and Marcelo M. Soares

31. LNCS 14714, Design, User Experience, and Usability: Part III, edited by Aaron Marcus, Elizabeth Rosenzweig and Marcelo M. Soares

32. LNCS 14715, Design, User Experience, and Usability: Part IV, edited by Aaron Marcus, Elizabeth Rosenzweig and Marcelo M. Soares

33. LNCS 14716, Design, User Experience, and Usability: Part V, edited by Aaron Marcus, Elizabeth Rosenzweig and Marcelo M. Soares

34. LNCS 14717, Culture and Computing, edited by Matthias Rauterberg

35. LNCS 14718, Distributed, Ambient and Pervasive Interactions: Part I, edited by Norbert A. Streitz and Shin'ichi Konomi

36. LNCS 14719, Distributed, Ambient and Pervasive Interactions: Part II, edited by Norbert A. Streitz and Shin'ichi Konomi

37. LNCS 14720, HCI in Business, Government and Organizations: Part I, edited by Fiona Fui-Hoon Nah and Keng Leng Siau

38. LNCS 14721, HCI in Business, Government and Organizations: Part II, edited by Fiona Fui-Hoon Nah and Keng Leng Siau

39. LNCS 14722, Learning and Collaboration Technologies: Part I, edited by Panayiotis Zaphiris and Andri Ioannou

40. LNCS 14723, Learning and Collaboration Technologies: Part II, edited by Panayiotis Zaphiris and Andri Ioannou

41. LNCS 14724, Learning and Collaboration Technologies: Part III, edited by Panayiotis Zaphiris and Andri Ioannou

42. LNCS 14725, Human Aspects of IT for the Aged Population: Part I, edited by Qin Gao and Jia Zhou

43. LNCS 14726, Human Aspects of IT for the Aged Population: Part II, edited by Qin Gao and Jia Zhou

44. LNCS 14727, Adaptive Instructional System, edited by Robert A. Sottilare and Jessica Schwarz

45. LNCS 14728, HCI for Cybersecurity, Privacy and Trust: Part I, edited by Abbas Moallem

46. LNCS 14729, HCI for Cybersecurity, Privacy and Trust: Part II, edited by Abbas Moallem

47. LNCS 14730, HCI in Games: Part I, edited by Xiaowen Fang
48. LNCS 14731, HCI in Games: Part II, edited by Xiaowen Fang
49. LNCS 14732, HCI in Mobility, Transport and Automotive Systems: Part I, edited by Heidi Krömker
50. LNCS 14733, HCI in Mobility, Transport and Automotive Systems: Part II, edited by Heidi Krömker
51. LNAI 14734, Artificial Intelligence in HCI: Part I, edited by Helmut Degen and Stavroula Ntoa
52. LNAI 14735, Artificial Intelligence in HCI: Part II, edited by Helmut Degen and Stavroula Ntoa
53. LNAI 14736, Artificial Intelligence in HCI: Part III, edited by Helmut Degen and Stavroula Ntoa
54. LNCS 14737, Design, Operation and Evaluation of Mobile Communications: Part I, edited by June Wei and George Margetis
55. LNCS 14738, Design, Operation and Evaluation of Mobile Communications: Part II, edited by June Wei and George Margetis
56. CCIS 2114, HCI International 2024 Posters - Part I, edited by Constantine Stephanidis, Margherita Antona, Stavroula Ntoa and Gavriel Salvendy
57. CCIS 2115, HCI International 2024 Posters - Part II, edited by Constantine Stephanidis, Margherita Antona, Stavroula Ntoa and Gavriel Salvendy
58. CCIS 2116, HCI International 2024 Posters - Part III, edited by Constantine Stephanidis, Margherita Antona, Stavroula Ntoa and Gavriel Salvendy
59. CCIS 2117, HCI International 2024 Posters - Part IV, edited by Constantine Stephanidis, Margherita Antona, Stavroula Ntoa and Gavriel Salvendy
60. CCIS 2118, HCI International 2024 Posters - Part V, edited by Constantine Stephanidis, Margherita Antona, Stavroula Ntoa and Gavriel Salvendy
61. CCIS 2119, HCI International 2024 Posters - Part VI, edited by Constantine Stephanidis, Margherita Antona, Stavroula Ntoa and Gavriel Salvendy
62. CCIS 2120, HCI International 2024 Posters - Part VII, edited by Constantine Stephanidis, Margherita Antona, Stavroula Ntoa and Gavriel Salvendy

https://2024.hci.international/proceedings

Preface

The use and role of technology in the business and organizational context have always been at the heart of human-computer interaction (HCI) since the start of management information systems. In general, HCI research in such a context is concerned with the ways humans interact with information, technologies, and tasks in the business, managerial, and organizational contexts. Hence, the focus lies in understanding the relationships and interactions between people (e.g., management, users, implementers, designers, developers, senior executives, and vendors), tasks, contexts, information, and technology. Today, with the explosion of the metaverse, social media, big data, and the Internet of Things, new pathways are opening in this direction, which need to be investigated and exploited.

The 11th International Conference on HCI in Business, Government and Organizations (HCIBGO 2024), an affiliated conference of the HCI International (HCII) conference, promoted and supported multidisciplinary dialogue, cross-fertilization of ideas, and greater synergies between research, academia, and stakeholders in the business, managerial, and organizational domain.

HCI in business, government, and organizations ranges across a broad spectrum of topics from digital transformation to customer engagement. The HCIBGO conference facilitates the advancement of HCI research and practice for individuals, groups, enterprises, and society at large. A considerable number of papers accepted to HCIBGO 2024 encompass recent developments in the area of digital commerce and marketing, expanding on virtual influencer strategies and impact, e-commerce trends, consumer attitudes, gamification, and dark patterns' effects. Moreover, motivated by the digital transformation and reshaping of business operations, submissions have explored productivity, gamification strategies in work environments, and professional digital well-being, as well as teleworking and virtual collaboration. Particular emphasis was given by conference authors to the impact of Artificial Intelligence (AI) in business, addressing generative AI, automated content creation, predictive analysis, fraud detection, as well as transparency and explainability in public services. Finally, a considerable number of submissions were devoted to user experience and service efficiency, focusing on service quality assessment and improvement, consumer requirements, and user experience studies in businesses and e-government services.

Two volumes of the HCII 2024 proceedings are dedicated to this year's edition of the HCIBGO conference. The first covers topics related to Digital Commerce and Marketing, Artificial Intelligence in Business, and Workplace, Well-being and Productivity. The second focuses on topics related to Teleworking and Virtual Collaboration, and Improving User Experience and Service Efficiency.

The papers of these volumes were accepted for publication after a minimum of two single-blind reviews from the members of the HCIBGO Program Board or, in some cases, from members of the Program Boards of other affiliated conferences. We would like to thank all of them for their invaluable contribution, support, and efforts.

July 2024   Fiona Fui-Hoon Nah
                                                           Keng Leng Siau

11th International Conference on HCI in Business, Government and Organizations (HCIBGO 2024)

The full list with the Program Board Chairs and the members of the Program Boards of all thematic areas and affiliated conferences of HCII 2024 is available online at:

http://www.hci.international/board-members-2024.php

HCI International 2025 Conference

The 27th International Conference on Human-Computer Interaction, HCI International 2025, will be held jointly with the affiliated conferences at the Swedish Exhibition & Congress Centre and Gothia Towers Hotel, Gothenburg, Sweden, June 22–27, 2025. It will cover a broad spectrum of themes related to Human-Computer Interaction, including theoretical issues, methods, tools, processes, and case studies in HCI design, as well as novel interaction techniques, interfaces, and applications. The proceedings will be published by Springer. More information will become available on the conference website: https://2025.hci.international/.

General Chair
Prof. Constantine Stephanidis
University of Crete and ICS-FORTH
Heraklion, Crete, Greece
Email: general_chair@2025.hci.international

https://2025.hci.international/

Contents – Part I

Digital Commerce and Marketing

The Impact of Virtual Shopping Presentation Modes on Consumer
Satisfaction and Purchase Intention 3
  Tzuhsuan Chen and Yinghsiu Huang

Investigating Consumer Attitudes Toward Recessive Advertising
in Short-Form Videos 23
  *Xiaoliang Fu, Lili Liu, Siyi Liang, Zipei Ling, Xiarui Qian,
  and Zhewei Mao*

Digital Trends Changing Solution Selling: An Overview of Use Cases 33
  Sabine Gerster

Influence of Streamer Characteristics on Trust and Purchase Intention
in Live Stream Shopping 46
  Franziska Grassauer and Andreas Auinger

Are Dark Patterns Self-destructive for Service Providers?: Revealing Their
Impacts on Usability and User Satisfaction 66
  *Toi Kojima, Tomoya Aiba, Soshi Maeda, Hiromi Arai,
  Masakatsu Nishigaki, and Tetsushi Ohki*

Designing Inspiration: A Study of the Impact of Gamification in Virtual
Try-On Technology 79
  *Sebastian Weber, Bastian Kordyaka, Marc Wyszynski,
  and Bjoern Niehaves*

Virtual Influencers' Lifecycle: An Exploratory Study Utilizing a 4-Stage
Framework of Planning, Production, Debut, and Retirement 91
  Joosun Yum, Youjin Sung, Yurhee Jin, and Kwang-Yun Wohn

Exploring the Impact of Virtual Influencers on Social Media User's
Purchase Intention in Germany: An Empirical Study 108
  Silvia Zaharia and Jasmin Asici

Assessing the Influence Mechanism of Media Richness on Customer
Experience, Trust and Swift Guanxi in Social Commerce 127
  Kaiyan Zhu, Caroline Swee Lin Tan, and Tarun Panwar

Artificial Intelligence in Business

ChatGPT and the Medical Industry: A Topic Modeling of Online
Discussions by Medical Professionals 145
  Langtao Chen, Brenda Eschenbrenner, and Youhong Hu

Exploring Segmentation in eTourism: Clustering User Characteristics
in Hotel Booking Situations Using k-Means 157
  Stefan Eibl, Robert A. Fina, and Andreas Auinger

Keywords Effectiveness in Textile Product Sales Performance: A Case
Study of the Shopee Website 176
  Pei-Hsuan Hsieh and Ambrose Phong

Predictive Analysis for Personal Loans by Using Machine Learning 187
  Hui-I. Huang, Chou-Wen Wang, and Chin-Wen Wu

UX-Optimized Lottery Customer Acquisition Processes Through
Automated Content Creation: Framework of an Industry-University
Cooperation 200
  Diana Kolbe, Andrea Müller, Annebeth Demaeght, and Barbara Woerz

Application of Machine Learning in Credit Card Fraud Detection: A Case
Study of F Bank 210
  Yuan-Fa Lin, Chou-Wen Wang, and Chin-Wen Wu

Explainable AI in Machine Learning Regression: Creating Transparency
of a Regression Model 223
  Robbie T. Nakatsu

Designing for AI Transparency in Public Services: A User-Centred Study
of Citizens' Preferences 237
  *Stefan Schmager, Samrat Gupta, Ilias Pappas,
  and Polyxeni Vassilakopoulou*

Sustainable E-commerce Marketplace: Reshaping Consumer Purchasing
Behavior Through Generative AI (Artificial Intelligence) 254
  Jung Joo Sohn, Nickolas Guo, and Youri Chung

Equipping Participation Formats with Generative AI: A Case Study
Predicting the Future of a Metropolitan City in the Year 2040 270
  *Constantin von Brackel-Schmidt, Emir Kučević, Stephan Leible,
  Dejan Simic, Gian-Luca Gücük, and Felix N. Schmidt*

Workplace, Wellbeing and Productivity

Requirements of People with Disabilities and Caregivers for Robotics:
A Case Study 289
Anke Fischer-Janzen, Markus Gapp, Marcus Götten,
Katrin-Misel Ponomarjova, Jennifer J. Blöchle, Thomas M. Wendt,
Kristof Van Laerhoven, and Thomas Bartscherer

Professional Digital Well-Being: An In Situ Investigation into the Impact
of Using Screen Time Regulation at Work 302
Katrin Gratzer, Stephan Schlögl, and Aleksander Groth

Developing Gamification Strategies to Reduce Handling Damages in High
Precision Production Environments 314
David Kessing and Manuel Löwer

Research on Solving the Problem of Children's Doctor-Patient
Relationship from the Perspective of System Theory - The Example
of Children's Medical IP "Courageous Planet" 334
Cheng Peng and Yajie Wang

LingglePolish: Elevating Writing Proficiency Through Comprehensive
Grammar and Lexical Refinement 349
Kai-Wen Tuan, Alison Chi, Hai-Lun Tu, Zi-Han Liao,
and Jason S. Chang

Author Index 365

Contents – Part II

Teleworking and Virtual Collaboration

A Bibliometric Approach to Existing Literature on Teleworking   3
   Jorge Cruz-Cárdenas, Carlos Ramos-Galarza, Ekaterina Zabelina,
   Olga Deyneka, and Andrés Palacio-Fierro

Configurational Perspectives in Social Media Research: A Systematic
Literature Review 13
   Kailing Deng and Langtao Chen

IT Project Management Complexity Framework: Managing
and Understanding Complexity in IT Projects in a Remote Working
Environment 27
   Megan Rebecca Evans and Tevin Moodley

Virtual Reality for Home-Based Citizen Participation in Urban
Planning – An Exploratory User Study 38
   Martin Guler, Valmir Bekiri, Matthias Baldauf,
   and Hans-Dieter Zimmermann

From Use to Value: Monitoring the User Adoption Journey 50
   Marvin Heuer and Philip Ostermann

Virtual Reality: Curse or Blessing for Cultural Organizations and Its
Consequences on Individuals' Intentions to Attend 63
   Kai Israel and Christopher Zerres

Maturity Measurement Framework for Evaluating BIM-Based AR/VR
Systems Adapted from ISO/IEC 15939 Standard 79
   Ziad Monla, Ahlem Assila, Djaoued Beladjine, and Mourad Zghal

Evaluating the Usability of Online Tools During Participatory Enterprise
Modelling, Using the Business Model Canvas 96
   Anthea Venter and Marné de Vries

Research Status and Trends of Virtual Simulation Technology in Clothing
Design 115
   Zichan Wang

Improving User Experience and Service Efficiency

An Investigation of Readability, User Engagement, and Popularity
of E-Government Websites in Saudi Arabia 133
  Obead Alhadreti

The Implementation of Advanced AIS and the Accounting Data Quality:
The Case of Jordanian SMEs 149
  Esraa Esam Alharasis and Abeer F. Alkhwaldi

A Study on Speech Emotion Recognition in the Context of Voice User
Experience 174
  Annebeth Demaeght, Josef Nerb, and Andrea Müller

China's Evidence for the Determinants of Green Business Environment
from a Dynamic fsQCA 189
  Hang Jiang, Yongle Wang, Jiangqiu Wu, and Beini Zhuang

A Multimodal Analysis of Streaming Subscription 200
  Yi-Cheng Lee, Yu-chen Yang, Yen-Hsien Lee, and Tsai-Hsin Chu

Digital Transformation in Banking–How Do Customers Assess the Quality
of Digital Banking Services? 209
  *Andrea Müller, Annebeth Demaeght, Larissa Greschuchna,
  and Joachim Reiter*

The Innovation Generated by Blockchain in Accounting Systems:
A Bibliometric Study 221
  *Javier Alfonso Ramírez, Evaristo Navarro, Joaquín Sierra,
  Johny García-Tirado, Rosmery Suarez Ramirez, and Carlos Barros*

EV Explorer 2.0: An Online Vehicle Cost Calculator for Gig Drivers
Considering Going Electric 233
  *Angela Sanguinetti, Kate Hirschfelt, Debapriya Chakraborty,
  Matthew Favetti, Nathaniel Kong, Eli Alston-Stepnitz,
  and Angelika Cimene*

Blockchain for Food Traceability - Consumer Requirements in Austria 253
  Robert Zimmermann, Magdalena Richter, and Patrick Brandtner

Author Index 277

Digital Commerce and Marketing

The Impact of Virtual Shopping Presentation Modes on Consumer Satisfaction and Purchase Intention

Tzuhsuan Chen(✉) and Yinghsiu Huang

Department of Industrial Design, National Kaohsiung Normal University, Taiwan, No. 62, Shenjhong Road, Yanchao District, Kaohsiung City 82446, Taiwan (R.O.C.)
{611272003,yinghsiu}@mail.nknu.edu.tw

Abstract. Amidst the swift evolution of technology, notably the ascendancy of virtual reality (VR) and augmented reality (AR), consumers' shopping paradigms are gradually veering from traditional to virtual realms. This study endeavors to meticulously juxtapose the perceptual disparities of products of distinct scales under diverse presentation modes, while delving into their ramifications on consumer satisfaction and purchase intent. This research has curated three distinct products—wristwatches, hairdryers, and chairs—of varying scales, presented through physical, AR, and VR modalities. Through surveys, participants' sentiments and cognitions regarding disparate presentation methods are meticulously collected. The findings unveil that VR exerts a pronounced influence on satisfaction and purchase intent, especially concerning substantial-sized products, whereas AR necessitates attention to the interactivity of such sizable items. Caution is advised when employing AR as a presentation mode for diminutive products to mitigate adverse effects on purchase intent. Future research could delve deeper into the impact of emotional factors in virtual shopping environments on consumers' perceptions of product information. Additionally, understanding consumers' expectations and demands for products of diverse sizes could further enhance the quality of the virtual shopping experience.

Keywords: Virtual shopping · Purchase intention · Consumer Satisfaction

1 Introduction

In the wake of rapid technological advancements, consumer consumption patterns are progressively diversifying, transitioning from traditional physical purchasing to remote virtual modalities. The immersive nature and the advantages of realism unbounded by scale, time, and space in virtual reality (VR) and augmented reality (AR) have propelled virtual shopping into a new trend. According to the Global Fashion Industry: Trends, Consumer Shifts, and Outlook survey released by the research institution GoodFirms in 2022, over sixty percent of respondents believe that virtual shopping is becoming the norm. Additionally, a consumer survey conducted by BigCommerce in September 2022 revealed that nearly half (46%) of global consumers are willing to engage in virtual shopping. While remote virtual shopping is a burgeoning phenomenon facilitated by technology, current research predominantly focuses on factors influencing consumers in physical or web-based shopping scenarios, with limited understanding and impact assessments of AR and VR shopping.

Purchase intent refers to consumers' inclination or intention to purchase specific products or services. Marketing strategies often adjust based on an understanding of consumer purchase intent to prompt consumers to make purchase decisions. Purchase intent is influenced by various factors, including personal needs and preferences, pricing, brand reputation, and product characteristics. Previous research has found that consumer satisfaction with the shopping experience significantly impacts purchase intent (Mittal and Kamakura, 2001). The distinct characteristics of different extended reality (XR) environments, such as virtual reality's heightened immersion and augmented reality's emphasis on the interaction and experience of virtual objects within the real environment, contribute to variances in consumer perceptions and conveyed product information.

Therefore, this study aims to explore the differences in consumer experience satisfaction between virtual and traditional physical shopping. Some studies suggest that users perceive differences in the size perception of virtual objects in AR and VR, with users being more sensitive to size changes in VR compared to AR (Wang, 2023). Through a comparative analysis of products of different scales presented in augmented reality, virtual reality, and physical environments, this research seeks to understand the differential consumer perceptions and experiences in various contexts. Furthermore, it aims to analyze the impact of customer satisfaction and purchase intent under different product presentation modes.

This study will focus on three different-sized products: watches, hair dryers, and chairs. Participants will view these products through physical, AR, and VR presentation methods, and their perceptions and experiences will be assessed using a questionnaire. Statistical analyses, including t-tests, two-way ANOVA, and regression analysis using SPSS, will be conducted to delve into the effects of different product presentation modes on consumer satisfaction and purchase intent. The research aims to draw conclusions and provide recommendations for suitable presentation modes for different-sized products, contributing to future research in this field. The ultimate goal is to offer designers and businesses guidance on appropriate product display modes based on the advantages and disadvantages of AR and VR, thereby enhancing consumer satisfaction and purchase intent in virtual shopping experiences.

2   Related Works

2.1   Virtual Shopping

Virtual shopping refers to the simulation of real shopping interactions in a virtual environment through digital technology and online platforms. It encompasses interactive methods such as virtual reality (VR) or augmented reality (AR), allowing consumers to browse products, engage in virtual try-ons or trials, and complete shopping transactions within a virtual space. Virtual shopping is often integrated into online shopping platforms, enabling consumers to browse, select, and make online transactions on computers or mobile devices. Through virtual shopping, consumers can enjoy a more convenient and personalized shopping experience, with the opportunity to participate in socially engaging shopping environments and interactions.

The development of virtual reality (VR) and augmented reality (AR) has brought diverse possibilities to virtual shopping, and numerous studies have explored their applications in e-commerce. In a 2005 study by Suh and Lee, VR interactive interfaces were found to provide consumers with a higher sense of telepresence in online shopping compared to two-dimensional web-based interfaces. This heightened telepresence had a significant positive impact on consumer behavior. The study further indicated that VR shopping websites improved consumers' brand recognition and perception of online retailers (Suh and Lee, 2005).

A study by Schlosser in 2006 highlighted the potential benefits of 3D virtual shopping. Results showed that compared to traditional online shopping, 3D environments of virtual stores could enhance consumers' pleasure and evoke positive emotions towards products. Customized content and effective guidance in operation were identified as crucial factors determining user satisfaction (Schlosser, 2006). VR shopping presents products more vividly, providing users with a shopping experience distinct from traditional physical stores. Despite some practical challenges and technological difficulties such as operational complexities and limited device ubiquity, the immersive and interactive nature of VR still has a positive impact on the shopping experience.

In addition to virtual reality, augmented reality also contributes to a unique consumer experience in virtual shopping. According to a 2018 study by Hilken et al., augmented reality allows consumers to interact with furniture products or try on clothing virtually, enhancing the online shopping experience with real-world interactions. The study showed that AR shopping increased the likelihood of product purchase and purchase intent. Other research also found that AR shopping enhances entertainment and usability (Rese et al., 2017). While Javornik's (2016) study pointed out technical difficulties and privacy issues in virtual shopping, advancements in hardware technology and software development have gradually addressed and improved these challenges.

2.2 Purchase Intent

Purchase intention is widely applied in consumer behavior and marketing research. Purchase intention refers to the subjective expectation of a consumer to buy a particular product or service (Dodds et al., 1991). In other words, it represents the consumer's willingness to purchase a product if the opportunity and capability allow, serving as a crucial precursor to the actual purchase behavior (Schlosser et al., 2006). Compared to simple product or brand attitudes, purchase intention is considered more effective in predicting actual purchase behavior (Jamieson and Bass, 1989), especially in high-involvement purchase decisions (Morwitz and Schmittlein, 1992).

Consumer purchase intention is influenced by various factors, such as product characteristics, brand effects, perceived risks, price sensitivity, and consumer characteristics (Schlosser et al., 2006). The higher the purchase intention, the greater the likelihood that consumers will actually purchase the product or service. Enhancing consumer purchase intention is deemed crucial in driving actual product sales (Morwitz and Schmittlein, 1992). In the realm of e-commerce, purchase intention is extensively applied to predict the purchasing behavior of online consumers.

3 Methods

3.1 Hypothesis

To investigate how different product presentation modes affect consumer satisfaction with the shopping experience, this study's model postulates three key factors: Interactivity, Immersion, and Informativeness. These factors are hypothesized to determine the impact on customer satisfaction and purchase intent in various virtual shopping environments. In the following sections, the study will delve into an in-depth exploration of each hypothesis (Fig. 1).

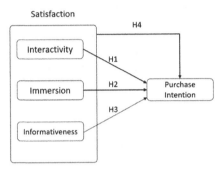

Fig. 1. Research model of our study.

1. Hypothesis 1 (H1): The interactivity of virtual shopping significantly influences purchase intent.

    Consumer willingness to purchase a product is closely related to the interactivity of their consumption process (McMahanetal., 2013). In terms of user agency, interactivity provides a heightened sense of control over the simulated environment. Therefore, increased interactivity fosters a better experience and enhances users' willingness to make a purchase (Shin Det al., 2018).

2. Hypothesis 2 (H2): The immersion of virtual shopping significantly influences purchase intent.

    Consumer willingness to purchase a product is highly correlated with the immersion experienced during the process. Immersive technologies extend people's imagination, allowing them to develop and experience a consumption process distinct from reality and become immersed in it (Baceviciute et al., 2021). Therefore, virtual shopping satisfies consumers' desires and imagination distinct from reality, resulting in an enhanced consumer experience and shopping desire.

3. Hypothesis 3 (H3): The amount of information conveyed in virtual shopping significantly influences the willingness to purchase a product.

    Gaining more information during the virtual shopping process brings additional benefits to the consumer experience, such as saving time and fostering more positive product considerations. Therefore, enhancing information accessibility can significantly improve consumer satisfaction with the consumption process and purchase intent (Kranzbühler et al., 2018).

4. Hypothesis 4 (H4): Satisfaction during the virtual shopping process significantly influences purchase intent.

Numerous past studies uniformly acknowledge the correlation between satisfaction during the consumption and service process and consumers' willingness to repurchase under various circumstances (Yi, 1990). When businesses enhance customer satisfaction, the resulting added value of the product significantly influences purchase intent. In the presence of high satisfaction, the willingness to repurchase also proportionally increases (Mittal and Kamakura, 2001).

3.2 Methodology

This study aims to explore the differences in consumer experiential satisfaction between virtual shopping and traditional physical shopping. By comparing the presentation of products at different scales in augmented reality, virtual reality, and physical settings, the research seeks to understand the divergences in consumer perceptions and experiences with products across various contexts. The study will select three different-scale products: a wristwatch, a hair dryer, and a chair. Participants will observe these products through physical, AR, and VR display methods for cross-comparison (see Table 1). Utilizing the results from questionnaire scales, the study will conduct t-tests, ANOVA and regression analysis to comprehend participants' satisfaction with the consumption process. Finally, through semi-structured interviews, the research aims to gain insights into consumers' cognitive experiences during the virtual shopping process.

Table 1. Cross Analysis Table for Experimental Comparison.

test stimulus (Scale)	Watch (Small)	Hair Dryer (Middle)	Chair (Large)
VR vs. Physical Shopping	A1	B1	C1
AR vs. Physical Shopping	A2	B2	C2

3.3 A. Sample and Data Collection

This study selects a sample of 91 participants aged between 20 and 50 with purchasing power as the primary consumer group. Participants will be categorized into three groups based on different-scale products (wristwatch, hair dryer, and chair). Each group will fill out a questionnaire after observing the respective product in different presentation modes (physical, AR, and VR).

For the VR virtual shopping experience, 3D models will be integrated into Sketchfab, allowing participants to use Oculus Quest 2 for immersive VR shopping. The AR shopping component will be created using Spark AR for an interactive platform. The questionnaire for this study (see Fig. 2) employs a Likert 7-point scale, consisting of 5 major constructs with 2 to 4 assessment items each, totaling 16 questions (see Table 2).

Fig. 2. The example of the questionnaire.

Table 2. Questionnaire aspects and items.

VR	Indicators	AR	Indicators
Interactivity	1.VR shopping is an interactive experience 2. Interacting with VR shopping is more interesting than real world shopping 3. Engaging in shopping through VR enhances my impression 4.Interacting with VR shopping is quick	Interactivity	1.AR shopping is an interactive experience 2. Interacting with AR shopping is more interesting than real world shopping 3. Engaging in shopping through AR enhances my impression 4.Interacting with AR shopping is quick
Immersion	5.While using VR shopping time seemed to go very quickly 6. I was completely immersed in the content while experiencing VR shopping 7.Interacting with VR shopping keeps the attention	Immersion	5.While using AR shopping time seemed to go very quickly 6. I was completely immersed in the content while experiencing AR shopping 7.Interacting with AR shopping keeps the attention
Purchase intention	8. Using VR shopping gives you more options to view or try products in the real world 9. Compared to physical shopping, does Using VR for shopping make you more inclined to procure this product?	Purchase intention	8. Using AR shopping gives you more options to view or try products in the real world 9. Compared to physical shopping, does Using AR for shopping make you more inclined to procure this product?
Informativeness	10.The information provided by VR shopping is comprehensive 11. VR shopping offers accurate information on the product 12.Relative to physical shopping, VR shopping allows me to acquire a more extensive array of product information 13.Compared with physical shopping, VR shopping provides more information about products	Informativeness	10.The information provided by AR shopping is comprehensive 11. AR shopping offers accurate information on the product 12.Relative to physical shopping, AR shopping allows me to acquire a more extensive array of product information 13.Compared with physical shopping, AR shopping provides more information about products
Satisfaction	14. I am satisfied with this VR Shopping 15. VR shopping has met my expectation 16. I will continue to use VR shopping in the future	Satisfaction	14. I am satisfied with this AR Shopping 15. AR shopping has met my expectation 16. I will continue to use AR shopping in the future

3.4 Measures

This study conducted an exploratory review of relevant literature to identify indicators for measuring each construct. The five constructs include Interactivity, Immersion, Informativeness, Purchase Intent, and Satisfaction. The reference scales for each construct are presented in the table (Table 3).

4 Data Analysis and Results

The analysis of the experimental results will be divided into three main parts. In the first part, a One-sample t-test will be employed to examine the differences between various experimental groups in five dimensions (interactivity, immersion, informativeness, satisfaction, and purchase intention) concerning physical shopping. The second part will utilize Two-Way ANOVA to investigate whether there are differences in the effects of three different product scales, presented in two display modes (VR, AR), across the five dimensions. Post hoc tests using LSD method will be conducted for further analysis. In the third part, explanatory regression analysis will be applied to verify the experimental hypotheses for different experimental groups (A1, A2, B1, B2, C1, C2). This analysis aims to understand the impact of interactivity, immersion, informativeness, and

Table 3. The reference scales for each construct.

Factors	Reference
Interactivity	Deepak Arumugam Ravindran *et al.*, (2023) Leonnard et al. , (2019) Flavian, (2019)
Immersion	Deepak Arumugam Ravindran *et al.*, (2023)
Purchase intention	Leonnard *et al.*, (2019) Flavian, (2019)
Informativeness	Deepak Arumugam Ravindran *et al.*, (2023) Leonnard *et al.*, (2019)
Satisfaction	Deepak Arumugam Ravindran *et al.*, (2023)

satisfaction on consumer purchase intention across different product scales and display modes.

4.1 One-Sample T-Test

In the one-sample t-test for the interactivity dimension, all groups achieved statistical significance ($p < 0.05$, as shown in Table 4), with t-values greater than zero. This indicates that, compared to the physical display mode, both VR and AR virtual display modes exhibit significantly higher interactivity for products of any scale. Across different product scales, the mean scores obtained with VR as the display mode were consistently higher than those with AR, with t-values of 11.205, 16.393, and 21.181, respectively. This suggests that, regardless of product size, using VR as the display mode provides participants with higher interactivity than using AR.

Table 4. Table of one-sample t-Test results for the interactivity dimension.

Interactivity	Test value = 4					
	t	DF	Sig. (2-tailed)	Std. Error Difference	95% confidence interval of the Difference	
					Lower	Upper
Small-VR (A1)	11.205	30	< .001**	1.354	1.107	1.6018
Small-AR (A2)	4.059	30	< .001**	0.379	0.188	0.5697
Middle-VR (B1)	16.393	29	< .001**	1.458	1.276	1.6403
Middle-AR (B2)	7.326	29	< .001**	0.958	0.69	1.2259
Large-VR (C1)	21.181	29	< .001**	2.083	1.882	2.2845
Large-AR (C2)	11.705	29	< .001**	1.425	1.176	1.674

In the one-sample t-test for the immersion dimension, all groups, except for group A2, achieved statistical significance (p < 0.05, as shown in Table 5), with t-values greater than zero. This indicates that, in virtual shopping, if small-scale products use AR as the display mode, their immersion is not significantly different from physical shopping. However, the immersion for the other five virtual display modes is superior to physical display. Across different product scales, using VR as the display mode consistently yielded higher mean scores for immersion than AR, with t-values of 10.603, 7.805, and 18.687, respectively. This suggests that, regardless of product size, using VR as the display mode provides superior immersion compared to AR.

Table 5. Table of one-sample t-Test results for the immersion dimension.

Immersion	Test value = 4					
	t	DF	Sig. (2-tailed)	Std. Error Difference	95% confidence interval of the Difference	
					Lower	Upper
Small-VR (A1)	10.603	30	< .001**	1.118	0.902	1.333
Small-AR (A2)	-0.577	30	0.284	0.086	0.39	0.218
Middle-VR (B1)	7.805	29	< .001**	1.111	0.819	1.402
Middle-AR (B2)	3.128	29	0.002**	0.433	0.149	0.716
Large-VR (C1)	18.687	29	< .001**	2.033	1.81	2.255
Large-AR (C2)	8.394	29	< .001**	0.955	0.722	1.188

In the one-sample t-test for the informativeness dimension, all groups achieved statistical significance (p < 0.05, as shown in Table 6), indicating that in terms of informativeness, using virtual display modes for products of any scale yields significantly different effects compared to physical display. However, in group A2, the t-value is less than zero (t = −3.436), suggesting that if small-scale products use AR as the display mode, the informativeness provided to participants is lower than that of physical display.

Regardless of product scale, if VR is used as the presentation mode, the mean scores and t-values for informativeness are consistently higher than those with AR. This indicates that, in virtual shopping, using VR for products of any scale provides better informativeness compared to AR. In group C1, with a mean score of 5.9417 and a t-value of 23.966, it shows that, for participants, using VR as the presentation mode for large-scale products in virtual shopping can offer optimal informativeness.

In the one-sample t-test for the satisfaction dimension, all groups, except for group A2, achieved statistical significance (p < 0.05, as shown in Table 7), with t-values greater than zero. This indicates that, in virtual shopping, if small-scale products use AR as the display mode, their satisfaction is not significantly different from physical shopping. However, the satisfaction for the other five virtual display modes is superior to physical display.

Table 6. Table of one-sample t-Test results for the informativeness dimension.

Informativeness	Test value = 4					
	t	DF	Sig. (2-tailed)	Std. Error Difference	95% confidence interval of the Difference	
					Lower	Upper
Small-VR (A1)	8.582	30	< .001**	0.919	0.7	1.138
Small-AR (A2)	-3.436	30	.002**	-0.491	-0.784	-0.199
Middle-VR (B1)	9.421	29	< .001**	1.133	0.887	1.379
Middle-AR (B2)	6.157	29	< .001**	0.791	0.528	1.054
Large-VR (C1)	23.966	29	< .001**	1.941	1.776	2.107
Large-AR (C2)	9.214	29	< .001**	1.125	0.875	1.374

Across different product scales, if VR is used as the display mode, the mean scores for satisfaction are consistently higher than those with AR, with t-values of 13.522, 12.544, and 20.743, respectively. This suggests that, regardless of product size, using VR as the display mode provides higher satisfaction to participants compared to AR.

Table 7. Table of one-sample t-Test results for the satisfaction dimension.

Satisfaction	Test value = 4					
	t	DF	Sig. (2-tailed)	Std. Error Difference	95% confidence interval of the Difference	
					Lower	Upper
Small-VR (A1)	13.522	30	< .001**	1.505	1.278	1.732
Small-AR (A2)	1.847	30	0.075	0.258	0.027	0.543
Middle-VR (B1)	12.544	29	< .001**	1.577	1.32	1.835
Middle-AR (B2)	7.507	29	< .001**	1.122	0.816	1.427
Large-VR (C1)	20.743	29	< .001**	2.211	1.993	2.429
Large-AR (C2)	11.181	29	< .001**	1.444	1.18	1.708

In the one-sample t-test for the purchase intention dimension, all groups, except for group A2, achieved statistical significance ($p < 0.05$, as shown in Table 8), with t-values greater than zero. This indicates that, in virtual shopping, if small-scale products use AR as the display mode, their purchase intention is not significantly different from physical shopping. However, the purchase intention for the other five virtual display modes is higher than physical display.

Across different product scales, if VR is used as the display mode, the mean scores for purchase intention are consistently higher than those with AR, with t-values of 11.869, 3.971, and 13.725, respectively. This suggests that, regardless of product size, using VR as the display mode elicits higher purchase intention from participants compared to AR.

Table 8. Table of one-sample t-Test results for the purchase intention dimension.

Purchase Intention	Test value = 4					
	t	DF	Sig. (2-tailed)	Std. Error Difference	95% confidence interval of the Difference	
					Lower	Upper
Small	11.869	30	< .001**	1.225	1.015	1.437
Small-AR (A2)	1.975	30	0.057	0.274	-0.009	0.558
Middle-VR (B1)	3.971	29	< .001**	0.716	0.348	1.086
Middle-AR (B2)	3.485	29	0.002**	0.55	0.227	0.873
Large-VR (C1)	13.725	29	< .001**	1.816	1.546	2.087
Middle	7.341	29	< .001**	1.066	0.77	1.364

In summary, across the five dimensions, groups A1, B1, B2, C1, and C2 all achieved statistical significance, with t-values greater than zero. This indicates that these five virtual shopping display modes differ significantly from physical display and outperform it across all five dimensions. Group A2 achieved statistical significance with a positive t-value in the interactivity dimension, suggesting that if small-scale products use AR as the display mode, their interactivity is superior to physical display. However, in the informativeness dimension, group A2 reached statistical significance but with a negative t-value. Therefore, for participants, if small-scale products use AR as the display mode, the informativeness is lower than physical display. Group A2 did not reach statistical significance in the dimensions of immersion, satisfaction, and purchase intention, indicating that using AR to display small-scale products does not differ from physical display in terms of immersion, satisfaction, and purchase intention.

From the above results, it can be observed that regardless of the product scale, if VR is used as the display mode, the t-values across all five dimensions are higher than when using AR. To further verify whether there is a difference in the display effects between VR and AR, this study conducted independent sample t-tests for the five dimensions for both virtual product presentation modes (VR and AR).

In the independent sample t-tests, both small-scale and large-scale products achieved statistical significance across all five dimensions in both VR and AR, with t-values greater than 0 (as shown in the Table 9). This indicates that for both small-scale and large-scale products, using VR as the display mode is superior to using AR across all five dimensions. However, for medium-scale products, statistical significance was only reached in the dimensions of interactivity, immersion, and satisfaction. This suggests that if medium-scale products use VR as the display mode, their performance in interactivity, immersion, and satisfaction will be superior to using AR, while there is no significant difference in informativeness and purchase intention dimensions.

Table 9. Independent sample t-test table for VR and AR.

		Test value = 4				
		t	DF	Sig.(2-tailed)	95% confidence interval of the Difference	
					Lower	Upper
Small	Interactivity	6.387	60	< .001**	0.67	1.281
	Immersion	6.596	60	< .001**	0.839	1.57
	Informativeness	7.893	60	< .001**	1.054	1.769
	Satisfaction	6.982	60	< .001**	0.89	1.605
	Purchase Intention	5.5	60	< .001**	0.606	1.298
Middle	Interactivity	3.161	58	0.003**	0.183	0.817
	Immersion	3.412	58	0.001**	0.28	1.075
	Informativeness	1.94	58	0.057	-0.011	0.694
	Satisfaction	2.332	58	0.023*	0.065	0.847
	Purchase Intention	0.695	58	0.49	-0.313	0.647
Large	Interactivity	4.206	58	< .001**	0.345	0.972
	Immersion	6.844	58	< .001**	0.763	1.393
	Informativeness	5.573	58	< .001**	0.523	1.11
	Satisfaction	4.578	58	< .001**	0.431	1.102
	Purchase Intention	3.816	58	< .001**	0.357	1.143

4.2 Two-Way ANOVA

In the Two-Way ANOVA for interactivity, both factors, virtual display mode (AR, VR), and product scale (Small, Middle, Large), reached significance ($p < 0.05$), and they don't exhibit interaction. Therefore, it can be inferred that different product scales and display modes have a significant impact on interactivity. In the post-hoc multiple comparison table (as shown in Table 10), it is evident that when VR is used as the display mode, the interactivity of large-scale products is greater than that of small-scale and medium-scale products, and there is no significant difference in interactivity between medium-scale and small-scale products in the VR display. When AR is used as the display mode, the interactivity ranks from high to low as large-scale, medium-scale, and small-scale.

Table 10. Interactivity Post-hoc Multiple Comparison Table

Interactivity LSD			average deviation	Std. Error Difference	p	95% confidence interval of the Difference	
						Lower	Upper
VR	Small	Middle	−.100	0.146	0.483	−0.395	0.1882
		Large	−.728	0.146	<.001**	−1.020	−0.4368
	Middle	Small	.103	0.146	0.483	−0.188	0.3952
		Large	−.625*	0.147	<.001**	−0.919	−0.331
	Large	Small	.728	0.146	<.001**	0.436	1.0202
		Middle	.625	0.147	<.001**	0.331	0.919
AR	Small	Middle	−.579	0.163	<.001**	−0.904	−0.2543
		Large	−1.045	0.163	<.001**	−1.371	−0.7209
	Middle	Small	.579	0.163	<.001**	0.254	0.9043
		Large	−.466	0.164	.006**	−0.794	−0.139
	Large	Small	1.045	0.163	<.001**	0.721	1.371
		Middle	.466	0.164	.006**	0.139	0.7943

In the immersion Two-way ANOVA analysis, both factors, virtual presentation mode, and product scale, reached significance ($p < 0.05$), and there was no interaction effect. Thus, it can be concluded that different product scales and presentation modes have a significant impact on immersion. The post-hoc multiple comparison table (as shown in Table 11) reveals that when VR is used as the presentation mode, the immersion of large-scale products is greater than that of small and medium-scale products, while the immersion of medium and small-scale products under VR presentation does not differ significantly. When AR is used as the presentation mode, the level of immersion decreases from large to medium and then to small scale.

In the informativeness Two-way ANOVA analysis, both factors, virtual presentation mode, and product scale, reached significance ($p < 0.05$), and they have an interaction effect. Thus, it can be concluded that different product scales and presentation modes have a significant impact on informativeness. The post-hoc multiple comparison table (as shown in Table 12) reveals that when VR is used as the presentation mode, the informativeness of large-scale products is greater than that of small and medium-scale products, while the informativeness of medium and small-scale products under VR presentation does not differ significantly. When AR is used as the presentation mode, the informativeness of small-scale products is less than that of large and medium-scale products, while the informativeness of large and medium-scale products under AR presentation does not differ significantly.

Table 11. Immersion Post-hoc Multiple Comparison Table

Immersion LSD			average deviation	Std. Error Difference	p	95% confidence interval of the Difference	
						Lower	Upper
VR	Small	Middle	0.007	0.169	1	−0.328	0.343
		Large	−0.915	0.169	<.001**	−1.251	−0.579
	Middle	Small	−0.007	0.169	1	−0.343	0.328
		Large	−0.922	0.170	<.001**	−1.261	−0.584
	Large	Small	0.915	0.169	<.001**	0.579	1.251
		Middle	0.922	0.170	<.001**	0.584	1.261
AR	Small	Middle	−0.519	0.190	<.001**	−0.898	−0.141
		Large	−1.042	0.190	<.001**	−1.42	−0.663
	Middle	Small	0.519	0.190	<.001**	0.141	0.898
		Large	−0.522	0.192	<.001**	−0.904	−0.141
	Large	Small	1.042	0.190	<.001**	0.663	1.42
		Middle	0.522	0.192	<.001**	0.141	0.904

Table 12. Informativeness Post-hoc Multiple Comparison Table

Informativeness LSD			average deviation	Std. Error Difference	p	95% confidence interval of the Difference	
						Lower	Upper
VR	Small	Middle	−0.214	0.146	0.149	−0.506	0.078
		Large	−1.022	0.146	<.001**	−1.314	−0.730
	Middle	Small	0.214	0.146	0.149	−0.078	0.506
		Large	−0.808	0.148	<.001**	−1.103	−0.514
	Large	Small	1.022	0.146	<.001**	0.73	1.314
		Middle	0.808	0.148	<.001**	0.514	1.103
AR	Small	Middle	−1.284	0.186	<.001**	−1.653	−0.914
		Large	−1.617	0.186	<.001**	−1.987	−1.247
	Middle	Small	1.284	0.186	<.001**	0.914	1.653
		Large	−0.333	0.187	0.079	−0.706	0.039
	Large	Small	1.617	0.186	<.001**	1.247	1.987
		Middle	0.333	0.187	0.079	−0.039	0.706

In the satisfaction Two-way ANOVA analysis, both factors, virtual presentation mode, and product scale, reached significance ($p < 0.05$), and they do not have an interaction effect. Thus, it can be concluded that different product scales and presentation modes have a significant impact on satisfaction. The post-hoc multiple comparison table (as shown in Table 13) reveals that when VR is used as the presentation mode, the satisfaction of large-scale products is greater than that of small and medium-scale products, while the satisfaction of medium and small-scale products under VR presentation does not differ significantly. When AR is used as the presentation mode, satisfaction from high to low is in the order of large scale, medium scale, and small scale.

Table 13. Satisfaction Post-hoc Multiple Comparison Table

Satisfaction LSD			average deviation	Std. Error Difference	p	95% confidence interval of the Difference	
						Lower	Upper
VR	Small	Middle	−0.072	0.162	1	−0.394	0.249
		Large	−0.706	0.162	<.001**	−1.028	−0.384
	Middle	Small	0.072	0.162	1	−0.249	0.394
		Large	−0.633	0.163	<.001**	−0.958	−0.309
	Large	Small	0.706	0.162	<.001**	0.384	1.028
		Middle	0.633	0.163	<.001**	0.309	0.958
AR	Small	Middle	−0.864	0.197	<.001**	−1.256	−0.472
		Large	−1.186	0.197	<.001**	−1.578	−0.795
	Middle	Small	0.864	0.197	<.001**	0.472	1.256
		Large	−0.322	0.199	<.001**	−0.717	0.073
	Large	Small	1.186	0.197	<.001**	0.795	1.578
		Middle	0.322	0.199	<.001**	−0.073	0.717

In the purchase intention Two-way ANOVA analysis, both factors, virtual presentation mode, and product scale, reached significance ($p < 0.05$), and they do not have an interaction effect. Thus, it can be concluded that different product scales and presentation modes have a significant impact on purchase intention. The post-hoc multiple comparison table (as shown in Table 14) reveals that when VR is used as the presentation mode, the purchase intention from high to low is in the order of large scale, small scale, and finally medium scale. When AR is used as the presentation mode, the purchase intention for large-scale products is greater than that for small and medium-scale products, while the purchase intention for medium and small-scale products under AR presentation does not differ significantly.

Table 14. Purchase intention Post-hoc Multiple Comparison Table

Purchase Intention LSD			average deviation	Std. Error Difference	p	95% confidence interval of the Difference	
						Lower	Upper
VR	Small	Middle	0.509	0.2	.013*	0.112	0.906
		Large	−0.591	0.2	.004**	−0.988	−0.194
	Middle	Small	−0.509	0.2	.013*	−0.906	−0.112
		Large	−1.1	0.201	<.001**	−1.5	−0.7
	Large	Small	0.591	0.2	.004**	0.194	0.988
		Middle	1.1	0.201	<.001**	0.7	1.5
AR	Small	Middle	−0.276	0.208	0.188	−0.689	0.137
		Large	−0.792	0.208	<.001**	−1.206	−0.379
	Middle	Small	0.276	0.208	0.188	−0.137	0.689
		Large	−0.517	0.21	.016*	−0.933	−0.1
	Large	Small	0.792	0.208	<.001**	0.379	1.206
		Middle	0.517	0.21	.016*	0.1	0.933

From the above, it is evident that when using VR as the product display mode, on the dimensions of interaction, immersion, informativeness, and satisfaction, large-scale products surpass medium and small-scale ones. However, the relationship between medium and small-scale products on these four dimensions does not exhibit significant differences. In VR presentations, the intensity of purchase intention is in the order of large scale, small scale, and finally medium scale.

When AR is utilized as the product display mode, on the dimensions of interaction, immersion, and satisfaction, the intensity follows the order of large scale, medium scale, and finally small scale. When AR is used as the product display mode, the informativeness of small-scale products is less than that of large and medium-scale products, while the informativeness of large and medium-scale products does not differ significantly. On the dimension of purchase intention, the purchase intention of large-scale products is greater than that of small and medium-scale products, while the purchase intention of medium and small-scale products under AR display does not exhibit significant differences.

4.3 Regression Analysis

In the regression analysis for group A1 (as shown in Table 15), interactivity ($\beta = 0.463$) and immersion ($\beta = 0.679$) on both dimensions reached significance ($p < 0.05$). Therefore, it can be inferred that if small-sized products utilize VR as their presentation mode, both interactivity and immersion will significantly influence purchase intention. Furthermore, immersion has a more substantial impact on participants' purchase intention compared to interactivity.

Table 15. Regression Analysis for Group A1

A1(VR-Small)	Dependent Variable: Purchase Intention					
	B	Std. Error	β	T	p	
(Constant)	1.116	0.891		1.253	0.221	
Interactivity	0.396	0.161	0.463	2.454	.021*	
Immersion	0.665	0.188	0.679	3.527	.002**	
Informativeness	-0.089	0.223	-0.093	-0.401	0.692	
Satisfaction	-0.177	0.248	-0.19	-0.713	0.482	

In the regression analysis for group A2 (as shown in Table 16), the facet of immersion ($\beta = 0.601$) attains significance ($p < 0.05$). This indicates that if small-scale products opt for AR as their display mode, immersion will exert a significant impact on purchase intention. The impact levels of the remaining four facets on purchase intention are deemed not significant.

Table 16. Regression Analysis for Group A2

A2(AR-Small)	Dependent Variable: Purchase Intention					
	B	Std. Error	β	T	p	
(Constant)	2.094	0.925		2.263	0.032	
Interactivity	-0.228	0.284	-0.153	-0.803	0.429	
Immersion	0.560	0.170	0.601	3.286	0.003**	
Informativeness	0.156	0.200	0.161	0.783	0.441	
Satisfaction	0.103	0.243	0.104	0.424	0.675	

In the regression analysis for group B1 (as shown in Table 17), the facet of satisfaction ($\beta = 0.552$) achieves significance ($p < 0.05$). This indicates that if medium-scale products opt for VR as their display mode, satisfaction will exert a significant impact on purchase intention. The direct impact of the remaining three facets on purchase intention is considered not significant.

Table 17. Regression Analysis for Group B1

B1(VR-Middle)	Dependent Variable: Purchase Intention					
	B	Std. Error	β	T	p	
(Constant)	-3.145	1.377		-2.284	0.031	
Interactivity	0.273	0.296	0.135	0.922	0.365	
Immersion	0.256	0.173	0.202	1.484	0.15	
Informativeness	0.125	0.268	0.083	0.465	0.646	
Satisfaction	0.792	0.263	0.552	3.016	0.006**	

In the regression analysis for group B2 (as shown in Table 18), the facets of immersion ($\beta = 0.352$) and satisfaction ($\beta = 0.472$) both achieve significance ($p < 0.05$). Hence, it can be inferred that if small-scale products opt for VR as their display mode, both interactivity and immersion will exert a significant impact on purchase intention.

Table 18. Regression Analysis for Group B2

B2(AR-Middle)	Dependent Variable: Purchase Intention					
	B	Std. Error	β	T	p	
(Constant)	-0.325	0.743		-0.437	0.666	
Interactivity	0.352	0.186	0.291	1.89	0.07	
Immersion	0.401	0.172	0.352	2.332	0.028*	
Informativeness	-0.25	0.191	-0.204	-1.31	0.202	
Satisfaction	0.498	0.182	0.472	2.735	0.011*	

In the regression analysis for group C1 (as shown in Table 19), none of the four facets reached significance ($p < 0.05$). Thus, if large-scale products adopt VR as their display mode, interactivity, immersion, informativeness, and satisfaction will not significantly influence purchase intention.

Table 19. Regression Analysis for Group C1

C1(VR-Large)	Dependent Variable: Purchase Intention					
	B	Std. Error	β	T	p	
(Constant)	0.883	1.752		0.504	0.619	
Interactivity	0.022	0.416	0.016	0.053	0.958	
Immersion	0.174	0.326	0.143	0.535	0.598	
Informativeness	0.569	0.569	0.348	1	0.327	
Satisfaction	0.059	0.390	0.048	0.151	0.881	

In the regression analysis for group C2 (as shown in Table 20), the facet of interactivity ($\beta = 0.432$) reached significance ($p < 0.05$). Thus, if large-scale products utilize AR as their display mode, interactivity will significantly influence purchase intention, while the impact of the other four facets on purchase intention is not significant.

Table 20. Regression Analysis for Group C2

C2(AR-Large)	Dependent Variable: Purchase Intention					
	B	Std. Error	β	T	p	
(Constant)	-1.216	0.731		-1.664	0.109	
Interactivity	0.515	0.192	0.432	2.689	0.013*	
Immersion	0.324	0.214	0.254	1.517	0.142	
Informativeness	0.269	0.164	0.226	1.642	0.113	
Satisfaction	0.092	0.176	0.082	0.523	0.605	

5 Conclusion and Suggestions

This study delves into the impact of products of three different scales under various virtual display modes on satisfaction and purchase intention using three statistical methods. The results of the single-sample t-tests indicate that groups A1, B1, B2, C1, and C2 all achieved significance on the five facets, suggesting a significant influence on consumer satisfaction and purchase intention when using virtual shopping, aligning with the experimental hypothesis. However, for group A2, the t-tests on the facets of immersion, satisfaction, and purchase intention did not reach significance, and a negative significant difference was found in the information facet. This indicates that if small-scale products use AR as a display mode, there is no significant difference in consumer satisfaction and purchase intention compared to traditional in-person shopping, but there might be a negative impact on information. The reason may be related to AR's inability to convey detailed information about small-sized products, reducing consumers' perception of product information.

Additionally, t-test results show that, in virtual shopping, regardless of the product scale, using VR results in higher consumer satisfaction and purchase intention compared to AR. This may be attributed to the stronger interactivity and immersion of VR compared to AR, positively influencing consumers' purchase intention and instilling confidence and motivation in shopping behavior. Novelty could also contribute to higher satisfaction and purchase intention in VR. As VR is currently a higher-priced and less common device, consumers using VR for virtual shopping may experience a unique and distinct shopping experience from traditional methods, enhancing their satisfaction and purchase intention.

The results of the two-way ANOVA reveal that when using VR as the product display mode, large-sized products outperform medium-sized and small-sized products on all five facets. This may be because VR provides a more expansive and immersive virtual space and perspective, allowing large-sized products to appear more realistic and immersive, leading to higher satisfaction and purchase intention. However, when medium-sized and small-sized products use VR as the display mode, there is no significant difference in interactivity, immersion, information, and satisfaction, while the purchase intention of small-sized products is higher than that of medium-sized products. The reason may be related to inherent differences in product nature, warranting further research and understanding.

   Regression analysis results indicate that immersion significantly influences purchase intention in groups A1, A2, and B2. The heightened involvement of consumers in the virtual shopping environment induced by immersion deepens their understanding and emotional connection with the product, further influencing their purchase intention. However, in group C1, none of the four facets significantly affects purchase intention, possibly due to the large size of products, making the perceived sensations in VR inconsistent with the actual spatial experience. In group C2, when large-sized products use AR as the display mode, interactivity becomes a key factor influencing purchase intention, while information and immersion do not significantly impact purchase intention. This could be attributed to the operational needs related to the large size of products, where consumers demand more practical interactions such as rotation, zooming, sliding to view details, and the desire for a more realistic perception in physical space, making interactivity a direct and crucial factor influencing purchase intention.

   Future research could delve into the impact of emotional factors in virtual shopping environments on consumers' perception of product information. Additionally, understanding consumers' expectations and demands for products of different sizes can contribute to enhancing the quality of the virtual shopping experience.

References

1. Beatty, S.E., Smith, S.M.: External search effort: an investigation across several product categories. J. Consum. Res. **14**, 83–95 (1987)
2. Ravindran, D.A.: Customer experience through virtual reality online shopping. Int. J. Res. Appl. Sci. Eng. Technol. (IJRASET) (2023)
3. Echo, H.: Online experiences and virtual goods purchase intention (2012)
4. Flavian, C.: The impact of virtual, augmented and mixed reality technologies on the customer experience. J. Bus. Res. **100**, 547–560 (2019)
5. Holloway, B.B.: Consumer satisfaction and post-purchase cognitive dissonance. J. Mark. Res. **4**(3), 294–299 (1967)
6. Leonnard, L.: The effect of augmented reality shopping applications on purchase intention (2019)
7. Mittelstaedt, R.A.: The effect of cognitive dissonance on consumer dissatisfaction. J. Consum. Res. **21**(2), 235–243 (1969)
8. Zhang, T., Yu, W., Wang, C., Cao, L., Wang, Y.: The role of virtual try-on technology in online purchase decision from consumers' aspect (2019)
9. Kaur, P., Dhir, A., Chen, S., Malibari, A., Almtairi, M.: Why do people purchase virtual goods? A uses and gratification (U&G) theory perspective (2020)
10. Hilken, et al.: Augmented reality to enhance the shopping experience: an exploration of perceived value. In: Proceedings of the 51st Hawaii International Conference on System Sciences (2018)
11. Dodds, W.B., Monroe, K.B., Grewal, D.: Effects of price, brand, and store information on buyers' product evaluations. J. Mark. Res. **28**(3), 307–319 (1991)
12. Morwitz, V.G., Schmittlein, D.: Using segmentation to improve sales forecasts based on purchase intent: which "intenders" actually buy? J. Mark. Res. **29**(4), 391–405 (1992)
13. Mittal, V., Kamakura, W.A.: Satisfaction, repurchase intent, and repurchase behavior: investigating the moderating effect of customer characteristics. J. Mark. Res. **38**(1), 131–142 (2001)

14. Wang, L.: Perceptual thresholds of visual size discrimination in augmented and virtual reality. Comput. Graph. **117**, 105–113 (2023)
15. Javornik, A.: Augmented reality: research agenda for studying the impact of its media characteristics on consumer behaviour. J. Retail. Consum. Serv. **30**, 252–261 (2016)
16. Schlosser, A.E.: Converting web site visitors into buyers: how web site investment increases consumer trusting beliefs and online purchase intentions. J. Mark. Am. Mark. Assoc. **70**(2), 133–148 (2006)

Investigating Consumer Attitudes Toward Recessive Advertising in Short-Form Videos

Xiaoliang Fu[1], Lili Liu[1(✉)], Siyi Liang[1], Zipei Ling[1], Xiarui Qian[1], and Zhewei Mao[2]

[1] College of Economics and Management, Nanjing University of Aeronautics and Astronautics, Nanjing, China
llili85@nuaa.edu.cn
[2] Chengdu Jiaxiang Foreign Language Senior High School, Chengdu, China

Abstract. In we-media era, recessive advertising frequently appears in the user-generated contents, especially in the short-form videos. Yet we know little about whether consumers are able to recognize recessive ads and their responses to recessive ads. In this study, we conduct data mining to explore how consumers recognize recessive ads and their attitudes toward these ads. We use Python to crawl 178,000 structured bullet comments from three short videos in a representative Chinese short-form video app Bilibili.com. We first extract the common features of recessive ads (suddenness, persistence, familiarity) by visualizing the bullet comments. Thereafter, E-DIAF model is developed to explore how the common features affect consumers' identification of recessive ads. Additionally, we establish a Multidimensional Emotion Computing Model (MDE-CM) to conduct sentiment analysis, which uncovers consumers' significant emotional shifts, especially the negative emotions such as "badness" and "fright" during ad segments. Findings indicate that seamless integration of ads and video content could minimize consumers' negative emotional responses toward ads, meanwhile enhance brand awareness and affinity. Theoretical and practical contributions are discussed.

Keywords: Recessive Advertising · Consumer Attitudes · Short-form Videos · Suddenness · Persistence · Familiarity

1 Introduction

Recessive advertising refers to the advertisements that been skillfully integrated into content (e.g., films, TV series, and online short-form videos), which appears like use experience sharing rather than promoting products [1]. Compared with traditional advertising, recessive ads are convert and difficult to be identified, thus the products could be subtly accepted and absorbed by the audiences, achieving more satisfactory results [2]. Recessive advertising has been favored by advertisers [3–7]. In we-media era, recessive advertising frequently appears in the user-generated contents, especially in the short-form videos (e.g., TikTok videos). However, as audiences become more familiar with recessive ads, some of them are able to recognize whether a vlogger is sharing his/her use experience or promoting a product based on certain clues. When audiences identify

F. F.-H. Nah and K. L. Siau (Eds.): HCII 2024, LNCS 14720, pp. 23–32, 2024.
https://doi.org/10.1007/978-3-031-61315-9_2

a recessive ad, the ad may become less effective. Audiences are unlikely to buy the product in the recessive ad, even worse, they may generate negative attitude towards the product [8]. Limited research has explored how consumers identify recessive ads and their attitudes toward recessive ads. In this study, we crawl and analyze bullet comments from three short-form videos that have inserted recessive ads, in order to: (1) extract the features based on which users identify recessive ads, and (2) explore users' attitudes toward recessive ads via sentiment analysis (Fig. 1).

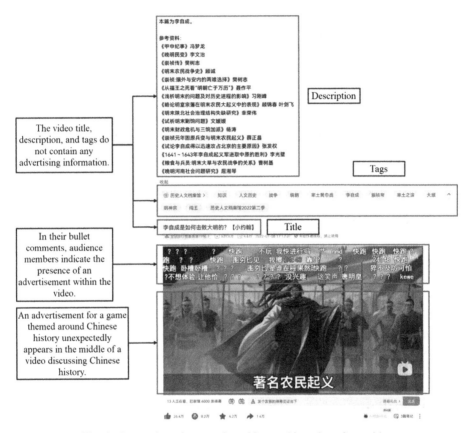

Fig. 1. Screenshot of a recessive ad inserted in a short-form video

2 Data Collection

We conducted data mining to explore the features of recessive ads inserted in short-form videos, as well as user attitudes toward the ads. First, three short-form videos were selected from Bilibili.com, a top-ranking Chinese video platform. Details of the three videos and recessive ads were summarized in Table 1. On one hand, the three videos shared several common features. For instance, one recessive ad was implanted in each video. Furthermore, the total video duration, recessive ad duration, and view counts were comparable. On the other hand, the three videos differed in themes, advertised products, product-video integration method, and ad placement. It is worth mentioning that, in the third video, the ad was seamlessly integrated with the entire video, since the product (car) is highly correlated with the video theme (travel).

Table 1. Details of the three videos and recessive ads

Video	Video Theme	Advertised Product	Product-Video Theme Integration Method	Total duration (minutes)	Advertising Duration (seconds)	Advertising Placement	View Count
1	History	A history themed mobile game	Thematic Relevance	29	161	End of the video	6,648,000
2	Sports	A drink	Causal Association	24	145	At the end	4,422,000
3	Travel	An automobile product	Causal Association	26	202 (Main Advertising period)	Throughout the entire video	1,479,000

We employed Python scripting to parse the 'seg.so' files associated with the three selected videos, which enabled us to extract a comprehensive dataset comprising more than 178,000 structured bullet comments (fields include the dates bullet comments were posted, the timestamps of bullet comments within the video, and the text content of the bullet comments). In Fig. 2, the visualization of the number of bullet comments when a recessive ad appears in video 1 was displayed.

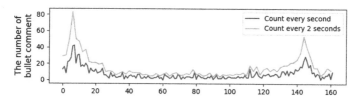

Fig. 2. Visualization of the number of bullet comments when a recessive ad appears in video 1

3 Identification of Recessive Advertising

3.1 Recessive Ad Feature Extraction

Given that the text content of bullet comments were dispersed and did not follow a structured question and answer thread, visual analysis approach based on word frequency statistics is more suitable for the identification of representative common features of recessive ad, compared with theme-word analysis method. First, in the data processing before analysis, we removed non-Chinese characters from the bullet comments and imported customized dictionaries and stop-word lists into the jieba module for Python, to ensure the accuracy of segmenting the text [9]. Second, we utilized Python scripts to conduct word frequency analysis for the bullet comments that posted within the advertising segments of these three videos (see Fig. 3).

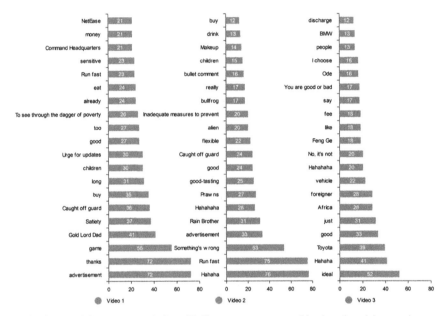

Fig. 3. Word frequency statistics of bullet comments posted in the advertising sections

High-frequency words such as "Caught off guard", "Something's wrong", "long", "game", "NetEase" and "alien" indicated that some of the consumers had recognized the recessive ads inserted in the short videos. To summarize, recessive ads exposed due to three frequently mentioned common features: suddenness, persistence, and familiarity.

Suddenness. At a particular time, where the short video suddenly changed in its content, speech rate, or background music, might make the consumers aware that a recessive ad appeared. For instance, "Caught off guard" were frequently posted during the recessive advertising segments in Video 1 and Video 2, indicating that consumers perceived a sudden shift in the video content delivery manner. If the vlogger could minimize the suddenness of presenting recessive ad in the short video, the recessive ad would be more inconspicuous. For instance, in Video 3, the recessive ad (car) was seamlessly integrated with the entire video them and content, we found that fewer consumers had detected the ad.

Persistence. Prolonged exposure to a product within the video might remind consumers the existence of recessive ads. For instance, "long" appeared 31 times in the advertising segment of Video 1, implying that the longer time the product was displayed, the easier the consumers recognized the recessive ad. If vloggers could effectively promote a product with less time, it might reduce the possibility of consumers' awareness and identification of recessive ads.

Familiarity. In Video 1, "Game" and "NetEase" frequently appeared in the advertising segment, indicating that consumers had recognized a recessive ad which intended to promote a game of NetEase. Consumers might have been exposed to some game advertisements elsewhere (e.g., in other short-form videos or social media websites), thus were familiar with this marketing approach. Although different from existing game advertising mode, the vlogger implicitly inserted a recessive game ad in the video, consumers could still easily recognize the ad. If users had never encountered similar game ads elsewhere, they were less likely to identify the recessive ad. Similarly, in Videos 2 and 3, frequently appeared words such as "alien" "drink" "ideal" and "car" indicated that consumers were familiar with the products in the recessive ad.

3.2 Developing a Model for Consumers' Identification of Recessive Advertising

Considering suddenness, persistence, and familiarity, we proposed a preliminary quantitative analysis model, the E-DIAF model, which was designed to calculate the ease η and probability P of a consumer's awareness of a recessive ad. The formula for the model was displayed below:

$$\eta = \gamma \cdot \mu \cdot f(N) \qquad (1)$$

In the formula, γ represented the suddenness, characterized by the abrupt appearance of a recessive ad, a factor that was challenging to be quantified; μ denoted persistence, defined by the ad's duration relative to the entire video; and N reflected the frequency with which the audience had encountered the same product elsewhere. The term $f(N)$ represented familiarity, which was simplified to a logarithmic function of N. The following discussion aimed to quantitatively break down the expressions for μ and $f(N)$:

$$\mu = \frac{length_{ad}}{length_{whole}} \qquad (2)$$

where, the variable μ was defined as the ratio of the recessive advertisement's duration to the total duration of the video.

$$f(N) = \log_2(x + b) + c \qquad (3)$$

The function $f(N)$ was modeled as a logarithmic function, reflecting the diminishing incremental increase in a consumer's familiarity with recessive ads for similar products as they encountered more advertisements. To ensure $f(N)$ remains positive, appropriate values for the constants b and c were selected within the ranges $b \in (0, +\infty)$ and $c \in (-\log_2 b, +\infty)$, respectively.

Ultimately, the degree of ease was mapped from the interval $(0, +\infty)$ to the interval $(0,1)$ using function (4), which quantified the probability P of an audience detecting a recessive advertisement. The formula was presented below:

$$P = 1 - e^{(-x^2)} \qquad (4)$$

4 Consumers' Attitudes Toward Recessive Advertising

In this study, we developed a Multidimensional Emotion Computing Model (MDE-CM) leveraging the emotion vocabulary ontology from Dalian University of Technology [10], along with a degree adverb table and a negative word list [11]. This model was designed to generate the emotional profile of consumers exposed to videos containing recessive advertisements and to categorize users' attitudes toward these ads as positive, negative, or neutral.

4.1 Developing an Sentiment Analysis Model

Following the segmentation of a bullet comment into n words, the corresponding seven-dimensional emotional vector was calculated as follows:

$$\vec{E} = \sum_{i=1}^{n} \vec{e}_i \tag{5}$$

In the following formula, $\vec{e}_i$ denoted the emotional vector associated with the i-th word in the bullet comment:

$$\vec{e}_i = \begin{cases} (a_{i1}, 0, 0, 0, 0, 0, 0) & c_i = \text{joy} \\ (0, a_{i2}, 0, 0, 0, 0, 0) & c_i = \text{good} \\ (0, 0, a_{i3}, 0, 0, 0, 0) & c_i = \text{anger} \\ (0, 0, 0, a_{i4}, 0, 0, 0) & c_i = \text{sorrow} \\ (0, 0, 0, 0, a_{i5}, 0, 0) & c_i = \text{fear} \\ (0, 0, 0, 0, 0, a_{i6}, 0) & c_i = \text{badness} \\ (0, 0, 0, 0, 0, 0, a_{i7}) & c_i = \text{fright} \end{cases} \tag{6}$$

where c_i denoted the emotion category to which the i-th word belongs, $c_i \in \{\text{joy, good, anger, sorrow, fear, badness, fright}\}$. A_{ij} represented the type j emotional intensity of the i-th word.

Assuming that there were p consecutive negative words preceding the i-th word and q successive degree adverbs (considering sequences with interleaved negative words and degree adverbs as continuous) in the text, the calculation formula for a_{ij} was depicted below:

$$a_{ij} = (-1)^p \prod_{k=1}^{q} \lambda_{ik} \cdot v_i \tag{7}$$

In this formula, $(-1)^p$ indicated that if a word was modified by an odd number of negative elements, the intensity of its emotional category reversed. Conversely, an even number of negative modifications left the emotional intensity unchanged. Furthermore, λ_{ik} represented the intensity of the k-th degree adverb for the i-th word, $\lambda_{ik} \in \{1, 1.5, 2\}$. Additionally, v_i denoted the base intensity of the emotional category for the i-th word.

The formula to calculate the emotional tendency value (M) of a bullet comment was presented as follows:

$$M = \sum_{i=1}^{n} \beta_i \tag{8}$$

β_i represented the intensity of the emotional tendency for the i-th word within a bullet comment, and its detailed expansion was outlined as follows:

$$\beta_i = \begin{cases} (-1)^{p+1} \prod_{k=1}^{q} \lambda_{ik} \cdot v_i & p_i = \text{negative} \\ (-1)^p \prod_{k=1}^{q} \lambda_{ik} \cdot v_i & p_i = \text{positive} \\ 0 & p_i = \text{neutral} \end{cases} \tag{9}$$

Analogous to formula (7), this expanded expression also accounted for the impact of negative words and degree adverbs on the foundational intensity of emotional tendency. Here, p_i denoted the polarity of the i-th word, with possible values being {positive, negative, neutral}.

4.2   Results of Sentiment Analysis

Inputting three selected videos into the MDE-CM model, we obtained a graph illustrating the changes in user emotions (see Fig. 4, Fig. 5 and Fig. 6).

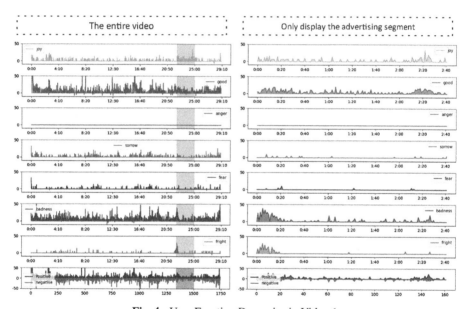

Fig. 4. User Emotion Dynamics in Video 1

Figure 4, Fig. 5, and Fig. 6 illustrated the emotional fluctuations experienced by consumers of the three videos, in which the horizontal axis indicating the video timeline and the vertical axis showing the intensity of consumers' emotions. The shaded sections represented the advertisement segments within the videos (notably, the shaded portion in Video 3 corresponds to the main advertisement segment). In general, distinct shifts in consumer emotions were observed from the advertising segments in Videos 1 and 2. Specifically, in Video 1's advertisement section, the emotion of "joy" escalated sharply

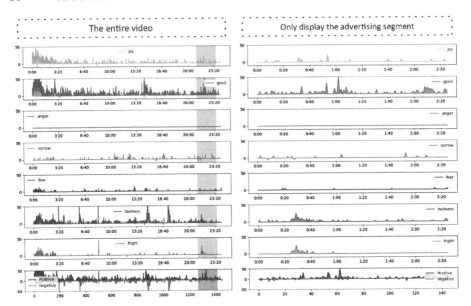

Fig. 5. User Emotion Dynamics in Video 2

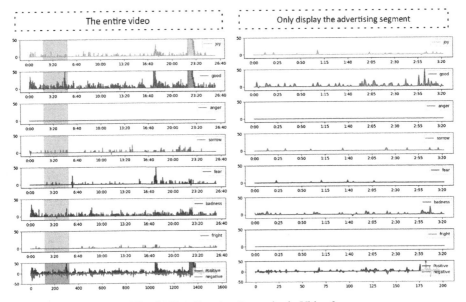

Fig. 6. User Emotion Dynamics in Video 3

and sustained a heightened state, while "badness" and "fright" experienced a substantial initial surge yet rapidly subsided thereafter. Conversely, Video 2's advertisement phase featured a modest uptick in "badness" and "fright," which swiftly tapered off. The advertisement portion of Video 3 did not register a notable emotional response shift (the abrupt emotional fluctuation towards the ad's end resulted from the video creator emphasizing national pride). We drew several inferences from above observations:

The suddenness of advertisements was positively correlated with the degree of consumers' emotional changes. Our findings indicated that the greater the suddenness of the advertisements, the higher the degree of emotional changes. The suddenness of advertisements was ranked as Video 1 > Video 2 > Video 3, and the degree of emotional changes observed in the advertisement segments was consistent with the suddenness ranking (see Figs. 4, 5, and 6).

The suddenness of advertisements had a more pronounced impact on the degree of change in certain negative emotions. For instance, the variations in "badness" and "fright" in the advertisement segments of Video 1 and Video 2 were considerably greater than those in other emotional dimensions (see Figs. 4 and 5).

The suddenness of advertisements had no influence on the persistence of the advertisement, yet positively associated with consumers' perception of this persistence. We found that the word "long" only appeared in the recessive advertising segment in Video 1 (the video with the greatest suddenness). Therefore, the E-DIAF model could be further optimized by replacing the persistence μ with the perceived persistence $g(\gamma,\mu)$:

$$\eta = \gamma \cdot g(\gamma, \mu) \cdot f(N) \tag{10}$$

The function $g(\gamma,\mu)$ was monotonically increasing with respect to both γ and μ, and the specific form of this function was not further decomposed.

An interesting observation was the substantial increase in the "joy" emotion in Video 1. During the advertising segment in Video 1, instead of paying attention to the ad, lots of consumers entertained themselves by interacting and joking via bullet comments. For instance, the "joy" emotion was aligned with self-amusement, while "good" emotion was highly correlated with external approval [5].

5 Conclusion

This research contributes to both theory and practice. Firstly, it offers a guiding framework for the identification of features upon which users recognize recessive advertising—suddenness, persistence, and familiarity—and introduces the E-DIAF model, which lays the groundwork for future studies to conduct qualitative analysis through function fitting. Secondly, the proposed MDE-CM model can reflect, in a timely and accurate manner, users' attitudes towards recessive advertising post-video broadcast, thereby aiding advertisers in the design of advertising strategies. Lastly, this study offers several recommendations for advertisers and vloggers who pursue greater ad effectiveness. For instance, while inserting recessive ads into short-form videos, vloggers should put great effort to seamlessly integrating the advertisement with the video content, in order to minimize its suddenness and consumers' negative attitudes toward the advertised product. Besides, vloggers are suggested to sufficiently manage the ad disclosure time, in order to promote the brand recognition, meanwhile minimize the ad intrusiveness.

References

1. Fan, C.G., Li, Z.C., Zhang, N.: Analysis of the phenomenon of recessive advertising. Contemp. Commun. **5**, 53–55 (2005)
2. Wang, F.: Short video implantable advertising analysis. Chin. Cities Dissolut. (10), 48–50 (2022). https://doi.org/10.16763/j.cnki.1007-4643.2022.10.013
3. Wen, Y.: Discussion on the design of film and television recessive advertising. In: Proceedings of the 3rd International Conference on Art Studies: Science, Experience, Education (ICASSEE 2019). Atlantis Press (2019)
4. Cain, R.M.: Embedded advertising on television: disclosure, deception, and free speech rights. J. Publ. Policy Mark. **30**(2), 226–238 (2011)
5. Zhang, Y., Liu, L., Bi, C.Q.: Influencer marketing and commercialization of social media platforms: how do consumers react to blurred-boundary advertisements? In: PACIS 2022 Proceedings, p. 188 (2022)
6. Sun, Y., Ly, T.P.: The influence of word-of-web on customers' purchasing process: the case of Xiaohongshu. J. China Tour. Res. 1–24 (2022)
7. Yihan, M.: Active user age distribution of Xiaohongshu in 2020 (2021)
8. Wang, Q., Tian, Y.L.: Web celebrity recessive advertisement transparent regulation: necessity, regulation mode and standard. J. Publish. Sci. **28**(02), 74–81 (2020). https://doi.org/10.13363/j.publishingjournal.2020.02.011
9. Is xiao li ah: Chinese NLP stop list (with Harbin institute of stop list code). CSDN, 18 May 2023. https://blog.csdn.net/qq_44543774/article/details/120742737
10. Zhang, X.K.: Emotional vocabulary ontology - dictionary. Dalian University of Technology Information Retrieval Laboratory, 3 August 2021. http://ir.dlut.edu.cn/info/1013/1142.htm
11. C01acat: Dalian university of technology based emotional vocabulary of Chinese sentiment analysis. CSDN, 8 May 2021. https://blog.csdn.net/qq_43342294/article/details/116545928

Digital Trends Changing Solution Selling: An Overview of Use Cases

Sabine Gerster[✉]

Hochschule Offenburg, Klosterstraße 14, 77723 Gengenbach, Germany
sabine.gerster@hs-offenburg.de

Abstract. Digital technologies have the potential to change well-established practices in business-to-business (B2B) sales. This paper gives an overview to understand how digital trends (e.g., emerging technologies) will impact the future of solution selling. For this purpose, solution business is seen from a process-oriented point of view. This article presents insights into digital use case scenarios changing the solution selling process.

Keywords: Solution Business · Solution Selling · Digital Technologies · Digital Trends · B2B Sales · Artificial Intelligence · Mixed Reality · Augmented reality · Virtual reality Information platforms · Digitized products · Intelligent Products · Digital Products

1 Introduction

Lots of research has focused on how digitalization affects marketing and sales. But most of this research looks at how digitalization impacts selling products directly to consumers (B2C). Scholars have focused on topics like mobile retailing [16], online advertising [20], online communities [21] and online product reviews [28]. However, not as much research has focused on how digitalization influences selling between businesses (B2B) [15].

The B2B sales environment is not immune to the fundamental changes brought about by digitalization. Today, many technological innovations have shaped B2B sales, most notably the profound advances in information and communications technology [22]. Customer interactions increasingly take place through the use of digital communication tools. Gupta predicts that by 2025, approximately 80% of B2B engagements will be digital [14]. In addition, emerging technologies like AI (artificial intelligence) and XR (e.g. augmented reality) are transforming business-to-business relationships [10].

In the B2B context, European and US companies often respond to competitive pressures, particularly from lower-priced emerging markets, by combining their products with services [7, 36]. Following IBM's successful implementation of this strategy, these combined offerings, termed "hybrid offerings" [33], are often marketed as "solutions" -"unique combinations of products and services tailored to address specific business challenges faced by customers" [5]. While digital technologies and trends have made positive contributions in the B2B context, their application in the B2B solutions business

F. F.-H. Nah and K. L. Siau (Eds.): HCII 2024, LNCS 14720, pp. 33–45, 2024.
https://doi.org/10.1007/978-3-031-61315-9_3

has not been well researched. This article addresses this research gap by answering the question of how the use of emerging digital technologies and tools can positively impact the customer interactions in solution business. This is relevant because selling from a solution marketing point of view does not mean simply exchanging products. Rather, it points to the need for intensive interactions between the supplier and the customer to cocreate an integrated offering [34].

Conceptually, this paper is based on a four-step solution process that has been used in existing research and is applicable to most solution selling situations [32]. In this paper, this process represents the customer interaction and is called "solution selling process". According to the "one face to the customer" principle, the people who work at this customer interface are the salespeople. They are organizationally part of the sales and/or service department. This paper describes the key sales tasks of the solution selling process and explain how digital trends add value to the customer-supplier interaction to accomplish these tasks. At this point, it is worth mentioning that the tasks are idealized tasks. Depending on the customer problem and framework conditions, the influence of the digital trends can be stronger/weaker, or cross process phase. Specifically, this article focuses on four digital trends. The theoretical framework of this paper is shown in Fig. 1.

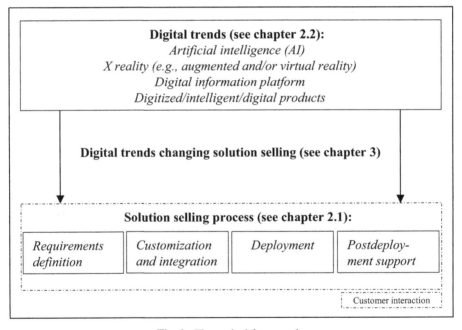

Fig. 1. Theoretical framework.

2 Theoretical Foundations

2.1 Solution Selling Process

Previous research on solution marketing points to the importance of understanding the solution business as a long-term, relational process between customer and provider. This perspective is largely supported by Tuli/Kohli/Bharadwaj (2007). The authors consider the concept of a solution from both the customer's and the supplier's perspective. They find significant differences between the two perspectives. Suppliers define a solution primarily in terms of its level of customization and integration. Demanding customers, on the other hand, see a series of interaction processes with the solution provider as the core of a solution. These exchange processes between market participants can be divided into four phases. The essence of the solution business is embedded in the notion that customers and suppliers work together to create superior value that cannot be delivered by solution providers alone. Based on their research, the authors advocate a shift from a product-oriented to a process-oriented perspective regarding the concept of a solution [32]. Figure 2 shows the solution process. Furthermore, the four process phases are briefly discussed below [1, 32].

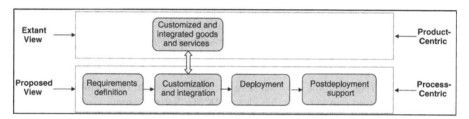

Fig. 2. Definition of a solution as a process (Source: Based on [32]).

Requirements Definition. The starting point for any customer solution is a clear understanding of the customer's needs. Therefore, at the beginning of the solution selling process, the provider should attempt to clearly define the customer's problem. A thorough understanding of the customer's business processes and models will facilitate this task. In addition, future customer needs should be considered at this stage so that they can be incorporated into the solution design.

Customization and Integration. In the second phase, the results from the first process step feed into the provider-side solution creation. Here, the solution provider selects compatible products and services and/or modifies them. Integration involves bringing together and aligning the individual components.

Deployment. The subsequent implementation phase includes delivering the solution and commissioning it at the customer's end. Additional services such as training might be essential measures to enhance the customer-side solution benefits.

Postdeployment Support. The final step in the process is customer follow-up. This takes place after the solution has been embedded into the customer's value system. It is at this stage that the configuration of the solution in terms of service and support friendliness becomes apparent. The long-term goal of a solution provider is to build a lasting relationship with the customer, for example, to secure future business from the customer. Within the aftercare phase, there are several options for stabilizing a customer relationship in the long term. These include traditional maintenance and repair as well as extended warranty services.

2.2   Digital Trends

This article describes digital trends having a strong potential to influence the solution selling process. To better understand the meaning of the considered trends, the digital trends are shortly introduced in this section.

Artificial Intelligence (AI).   By interpreting data, learning from them, and making flexible assumptions AI applications perform a well-defined set of tasks with little or no human intervention [8, 10]. The most common use cases are applications dealing with classification, estimation and clustering tasks [6]. The area where AI could have the biggest impact is in customer interaction [23].

XR.   There are two technologies that enhance or substitute reality with virtual components: Augmented reality (AR) and virtual reality (VR). AR integrates digital information and data into the real-world setting. For example, digital objects can be placed between real physical objects [9, 11]. VR, on the other hand, immerses users in virtual, computer-generated 3-D virtual representations of the real world [9]. The term XR encompasses both phenomena (AR and VR) by spanning the spectrum from the entirely real world to the entirely virtual one [11, 27]. The letter "X"-in XR-represents a placeholder for any form of new reality [25]. XR plays a central role in B2B business. For example, the lion's share of spending on AR glasses is currently made in the B2B sector [3].

Digital Information Platform.   A service information platform is a system that provides information and tools for managing and delivering services and information to customers. This can include information about available products and services, pricing and billing, support resources, manuals and maintenance information. A typical digital information platform is the seller's website.

Digitized/Intelligent/Digital Products.   Today, products that were once just physical components have evolved: They can collect environmental and operational data using GPS, for example. Technologies such as RFID (Radio Frequency Identification), NFC (Near-Field Communication), Bluetooth, ZigBee (standard for wireless radio networks) or WLAN (Wireless Local Area Network) are used for wireless data exchange with IT systems, machines, employees or customers [2]. In addition, hardware products are complemented by software to configure and parameterize them for different customer problems. In this paper, this new generation of products is referred to as digitized products. Digitized products that independently evaluate the collected data and react to the analysis results are called intelligent products [2]. Before this development, caused by new technical possibilities, the complexity was in the products themselves. Now, many products that once consisted of mechanical and electrical components have evolved into complex systems [12]. In addition to developments towards digitized/intelligent products, digital products are increasingly emerging in the digital age. Digital products differ from a digitized/intelligent product in that the product either provides digital data on a data carrier (e.g. e-book, CD, DVD) or does not exist physically, but only immaterially in the form of digital data [2].

3 Digital Trends Changing Solutions Selling

This chapter presents the changes of digital trends on the solution process. It discusses the extent to which the digital trends add value to the processing of tasks during the solution process steps. Table 1 summarizes the results.

3.1 Artificial Intelligence

Requirements Definition. AI can automate the solution selling process in different ways. For example, it can automate interaction with customers in requirements definition phase. While salespeople today still interact with prospects in person to advise them or acquire them as customers, AI could take over in the future. An avatar or a bot could imitate the behavior of salespeople and have real-time conversations with customers. By interpreting customers' emotions, AI can advise customers individually by finding the right solution [23]. For example, replika (replika.ai) gives a practical impression of what such AI might look like [8]. The AI software is intended to serve as an AI companion, programmed to ask "meaningful" questions about the customer's business and problem, and to provide emotional support [26].

Customization and Integration. In the customization and integration phase, salespeople demonstrate the problem-solving features of the offering to the customer, which often includes a presentation of the solution to be delivered. AI systems also improve the task of presenting a solution. Slide bots transform text content into visually appealing and engaging slide decks. The text content can come from existing information (e.g., bids, technical specifications, timetables) that have already been exchanged between salespeople and customer. Thus, these bots help salespeople deliver more engaging presentation materials. In addition, AI can be used to detect verbal and nonverbal cues in the communication. For example, AI applications can conduct sentiment analysis on a presenter's words, voice, and tone. AI also can analyze audience emotions to gauge sentiment or identify themes that offer valuable insights into potential customer concerns or objections [23]. An existing AI presentation tool is beautiful.ai. This AI tool analyzes entered texts and automatically generates suitable presentations including layouts. The tool is also cloud-based to eliminate the need to passing around revised files and to facilitate collaboration with colleagues. This saves time and effort when designing the presentation and allows sales staff to concentrate on the customer-specific content to enhance the customer experience [4].

Postdeployment Support. Some existing predictive maintenance systems have analyzed time-series data from embedded sensors, such as those that monitor temperature or vibration, to detect anomalies or make predictions about the remaining useful life of components [2]. Deep learning's ability to analyze very large amounts of high-dimensional

Table 1. Digital trends for value creation in solution selling process.

Solution selling process phase	Traditional tasks	The changes brought about by the digital trend that add value to tasks
Requirements definition	Finding potential customers, making contacts and acquiring more information about customers need	• "To be found by customer" via information platforms (e.g., avoiding contact formulas) • Making contact via chatbots and emotion AI (e.g., a bot or an avatar) for emotion interpretation and effective response
Customization and integration	Communicating the problem-solving characteristics of the offering	• AI- and XR-enabled presentations (e.g., slidebots, emotion AI, virtual customer solution) • Customization of hardware via software (digital/intelligent products) • Expanded service portfolio (digital products)
Deployment	Install and make running the solution at customer site	• XR-enabled training conditions (e.g., in silent environment)
Postdeployment support	Follow up beyond the current order	• Expanded service portfolio (e.g., software updates, access to support portal) • AI-automated workflow (e.g., predictive maintenance) • XR-enabled maintenance (e.g. AR glasses)

data can take this to a new level by being able to overlay additional data such as audio and video from digitized/intelligent products or third-party sensors. The capability AI to forecast malfunctions and facilitate scheduled intervention can be harnessed to minimize downtime and operational expenses [6].

3.2 XR

Customization and Integration. XR can support solution selling because it may extend the sales environment and lead to a better customer experience. In B2C environment, XR-supported tools are well-known. For example, they form the basis of discussions with customers when selling kitchens or houses. In these cases, customers expect a detailed and realistic presentation of their solution before placing an order [27]. The German sensor manufacturer SICK provides a concrete example of this type of benefit: The company is working on a virtual customer center. This will allow customers and sales stuff to meet virtually. Among other things, SICK offers individual customer solutions for logistic applications (e.g., a logistics solution in which objects move on a conveyor belt). In the virtual customer center, the customer can be shown where SICK components are placed on the conveyor belt, what their dimensions are and how many are needed for this particular application. This gives the customer a realistic idea of what the solution to their problem will look like. But more than just presenting customer solutions, solution providers can use the XR technology to create an enhanced customer experience by presenting the company as a whole [10].

Deployment and Postdeployment Support. This use case opens the opportunity to train customers on how to use the customized solution before or after the installation. The supplier can provide detailed instructions to the customer, supporting the learning process. XR improves customer interactions by facilitating the transfer of complex content through visualization [10]. However, XR technology is not only used to train customers on how to use the solution. It also makes a valuable contribution to training the sales department of the solution provider to ensure a smooth commissioning [27]. Moreover, the XR application supports the sales team by resolving issues on the solution already running at the customer site. A practical example is given by China Southern Airlines. The company is a leader in digitization, having implemented AR smart glasses in its maintenance system to replace paper job cards in aircraft maintenance. AR glasses are used in maintenance scenarios to provide a step-by-step, near-eye display of key information. This approach provides guidance and shows the current technical status of the aircraft component being serviced. Workers can see scenarios unfold as they work, and have access to tools such as assisted photography and automated measurement. The AR-smart glasses also facilitate access to manuals, experience documents, and back-office experts can be called at any time for support (Fig. 3) [17].

3.3 Digital Information Platform

Requirements Definition. A customer needs to identify a potential solution provider to communicate their requirements. The increased accessibility of information on the Internet is changing the way B2B customers obtain information about potential suppliers [31]. The increased accessibility of information on the Internet enables customers to obtain information more independently. As a result, the salesperson's job becomes more passive. It is no longer about actively generating leads through cold calling or trade shows. It is much more about being found by the customer. [37]. In this context, McKinsey & Company speaks of the "consumerization of business buying" [18]. This term is used to describe the fact that the boundary between B2B and B2C buyers is slowly becoming blurred, and that organizational buyers are converging in their information gathering habits and requirements, and in their expectations regarding the availability of information to B2C customers. A study by Google concludes that 71% of B2B buyers start their purchases with a personal online search, and 57% of the buying process is

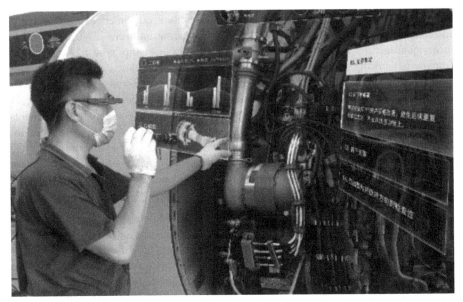

Fig. 3. Wearing AR glasses, the aircraft maintenance personnel can receive real-time guidance and diagnoses from experts (Source: Based on [32]).

complete by the time these buyers contact the provider [30]. For a solution provider this means focusing on well-structured information platforms provided with up-to-date information.

In a qualitative study of B2B buyers, Lüders/Klarmann 2019 provide insight into what a "well structured" information platform means to buyers in a B2B context. The researchers found that many supplier websites lack direct information on how to contact sales representatives and technical consultants. Based on the study findings, the researchers recommend avoiding contact forms or general email addresses (e.g. contact@XY.com) or phone numbers (e.g. ending with a "0") [19]. The following image shows a screenshot from the website of the German company Trumpf. The firm sells machine tools and lasers for industrial manufacturing. The screenshot serves as negative example as it shows both a contact form and a general phone number as (Fig. 4) contact option.

As noted on Sect. 2.2, digitized/intelligent/digital products allow solution providers to create a variety of customer solutions by flexibly combining, configuring, and parameterizing the solution components. This multitude of options and the associated complex technical specifications make it difficult for customers to find a suitable solution provider. Customers feel lost in this complexity. Filling information platforms with technical specifications is therefore pointless [19]. Solution providers need to segment their customers. Identified target segments can then be addressed effectively by providing the segment-specific information. Thus, the content of information platforms should focus on clustered customer problems and how the company's products and services can solve them [12].

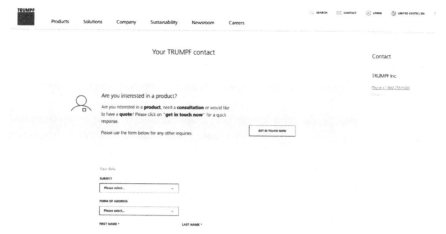

Fig. 4. Negative example of how to motivate potential customers to contact a supplier (Source: Based on [32]).

3.4 Digitized/Intelligent/Digital Products

Customization and Integration. The move to digitized/intelligent products means that salespeople need to be very knowledgeable about the technical capabilities of these new types of products. Customers are often overwhelmed by the level of complexity that comes with the various ways to parameterize and configure digitized/intelligent/digital products. As a result, customers need more support and explanation. Therefore, the role of salespeople is to help customers to effectively make the complex decisions associated with digitized/intelligent/digital products. In addition, the salespeople also need to be familiar with the customer processes. Knowledge of the customer's processes is essential for the solution provider's internal processes. By understanding them in detail, the solution supplier is able to pre-program the solution components in the factory according to the customer's individual requirements [12]. Digitization is a driver for digital products [35]. Formerly separate, physical products are complemented by digital products [12]. Both the technical possibilities of digitized/intelligent products and the expansion of the portfolio with digital products facilitate the development of customized solutions. On the other hand, the essence of solution selling is consequently embedded in the notion that salespeople and customers work together to create superior value that neither party can create alone.

Postdeployment Support. In the post-deployment support phase, digital products offer new ways to maintain customer relationships through the expansion of the service portfolio. Software updates, support portals, and remote maintenance are three typical examples of such digital products: When product features are implemented through software rather than electrical or mechanical components, they can be optimized over time through software updates - both in terms of the quality of existing features and the creation of new functions. Specialists (customer and supplier) around the world are digitally connected to the solution provider's support portal. Customers can easily create support tickets and send messages to the solution provider's specialists. The support portal also provides the latest information on product releases and offers a user forum for sharing experiences [29]. Remote maintenance can be offered to the customer as a complementary service to a customer solution. The operating data of the solution (e.g. via the interfaces of the digitized products) is analyzed in the IT system of a service provider or the solution provider so that maintenance or repair activities can be triggered if threshold values are exceeded [23].

By offering digital products, solution companies are able to move away from traditional product-centric business models and create greater value. This trend also presents a number of challenges for customer interaction in solution selling. First, once digital services are created, the cost of producing new units is very low. This is where digital products differ from traditional products. Consequently, sellers must adopt new pricing approaches [24]. Digital products can also affect the customer's perception of the value proposition [15]. As a result, solution selling companies need to pay close attention to these challenges to ensure they can truly benefit from digital products.

4 Discussion and Conclusion

This study offers both theoretical as well as managerial contributions. Digitalization is pervasive in today's business markets. New insights in this area, especially in B2B business markets, are of particular interest [15]. By focusing on customer interaction influenced by digital trends in the context of solution selling, this study provides a much-needed addition to the literature on solution selling, as well as to the burgeoning field related to digitalization in B2B marketing in general.

From an additional theoretical point of view, this article considers a process-oriented perspective of the solution business. Thus, this study contributes to the understanding of solutions as processes by exploring how digital trends influence the solution business not only at the level of individual applications, but also from a process-oriented and process-phase-oriented perspective. As a result, this study provides an expanded perspective on digital trends that have valuable potential to change the customer interaction at each stage of the solution selling process. At the same time, the paper also highlights the potential challenges and benefits of the digital use cases.

Greater cohesion and transparency throughout the sales process in business markets leads to more predictable and smoother customer journeys, which are key to creating valuable customer experiences [31]. This paper also contributes from a management perspective, as it provides insights into where the customer journey in the solutions business can be extended, enhanced or supported by digital trends (e.g., AI or digital information platform). The implementation of digital trends in the solution business is expected to be a long-term process. With the overview provided in this paper, managers should start the implementation process by considering which digital use case is most valuable to their customers. Simultaneously, this article offers digital use cases that managers can implement with little effort and resources. For example, it shows how to easily connect with potential customers. On the other hand, this article also offers use cases that provide a more visionary perspective by showing how new technologies can virtually shape customer interactions.

In addition, this paper takes an indirect look at the role of the sales in the solution selling process and how it is being challenged by the digital use cases. Two examples are used to illustrate these challenges. First, a salesperson selling digitized/intelligent/digital products should be able to develop customized bundles of products and convince customers of their superior performance and contribution to customers' profit [24]. Since this task is complex and uncertain, selling digitized/intelligent/digital products requires different capabilities than traditional product selling. Selling digitized/intelligent/digital products requires any salesperson to have a strong task-specific motivation to develop the abilities to engage in extensive coordination with multiple functions internally and customer networking externally [13]. Thus, the main challenge is to change the mindset of persons interacting with customers in a solution context. These people need other skills, e.g. technical know-how, knowledge of internal processes, methodological competence in project management, knowledge of value-based selling approach [12]. Second, although some of the introduced digital trends (e.g., AI tools, XR applications) can be used to effectively perform many tasks within the solution selling process, some tasks remain exclusively with sales professionals: Salespeople must continue to build relationships with prospective and current customers [22]. As in all stages of the solution selling process, a personal touch is important to build trust and credibility, and sales professionals are needed to address any potential ambiguity [32]. The challenge is to take advantage of the digital trends to improve the customer experience and, at the same time, not to lose sight of the personal interactions.

In conclusion, the digital trends highlighted in this paper simplify customer interactions in the solution selling process. However, the complexity associated with solution selling due to the individuality of customer problems makes it difficult to implement digital trends in the solution selling process (e.g., implementing XR-enabled showrooms that must cover all types of customer solutions). Consequently, a "one size fits all" approach will not work. Hopefully, this article will further advance the discussion and research agenda in the area of digital customer interaction in the solutions business. This will enable solution providers to translate their digitization strategy into effective digital customer interactions in the context of solution selling.

Disclosure of Interests.   The author has no competing interests to declare that are relevant to the content of this article.

References

1. Ahlert, D., Kawohl, J.: Best Practice des Solution Sellings: Projektbericht Nr. 1. Münster (2008)
2. Appelfeller, W., Feldmann, C.: Die digitale Transformation des Unternehmens. Springer Gabler, Berlin (2018). https://doi.org/10.1007/978-3-662-54061-9
3. AR Insider: What's the Outlook for AR Glasses? https://arinsider.co/2021/09/03/xr-talks-whats-the-outlook-for-ar-glasses/. Accessed 24 Jan 2024
4. Beautiful Slides: The presentation software for everyone. https://www.beautiful.ai/presentation-software. Accessed 24 Jan 2024
5. Brady, T., Davies, A., Gann, D.: Creating value by delivering integrated solutions. Int. J. Project Manag. **23**(5), 360–365 (2005)

6. Chui, M., et al.: Notes from the AI Frontier - Insights from Hundreds of Use Cases. McKinsey Global Institute, Washington, DC (2018)
7. Colm, L., Ordanini, A., Bornemann, T.: Dynamic governance matching in solution development. J. Mark. **84**(1), 105–124 (2020)
8. Davenport, T., Guha, A., Grewal, D., Bressgott, T.: How artificial intelligence will change the future of marketing. J. Acad. Mark. Sci. **48**(1), 24–42 (2020)
9. Farshid, M., Paschen, J., Eriksson, T., Kietzmann, J.: Go boldly! Explore augmented reality (AR), virtual reality (VR), and mixed reality (MR) for business. Bus. Horiz. **61**(5), 657–663 (2018)
10. Fischer, H., Seidenstricker, S., Berger, T., Holopainen, T.: Digital sales in B2B: status and application. In: Markopoulos, E., Goonetilleke, R., Ho, A., Luximon, Y. (eds.) AHFE 2021. LNNS, vol. 276, pp. 369–375. Springer, Cham (2021). https://doi.org/10.1007/978-3-030-80094-9_44
11. Fischer, H., Seidenstricker, S., Poeppelbuss, J.: Extended reality in business-to-business sales - an exploration of adoption factors. In: Ahram T., Taiar R., Groff F. (eds.): IHIET-AI 2021. AISC, vol. 1378, pp. 123–130. Springer, Cham (2021). https://doi.org/10.1007/978-3-030-74009-2_16
12. Gerster, S.: Funktionsübergreifende Zusammenarbeit im Lösungsgeschäft - Eine Empirische Betrachtung aus Anbietersicht. Springer, Cham (2018). https://doi.org/10.1007/978-3-658-23345-7
13. Guenzi, P., Nijssen, E.: The relationship between digital solution selling and value-based selling - a motivation-opportunity-ability perspective. Eur. J. Mark. **57**(3), 745–770 (2023)
14. Gupta, S.: Top 5 Emerging B2B Software Marketing Trends for 2022. https://www.gartner.com/en/digital-markets/insights/b2b-marketing-trends-2022. Accessed 24 Jan 2024
15. Hofacker, C., Golgeci, I., Pillai, K., Gligor, D.: Digital marketing and business-to-business relationships: a close look at the interface and a roadmap for the future. Eur. J. Mark. **54**(6), 1161–1179 (2020)
16. Hofacker, C., Ruyter, K., de Lurie, N., Manchanda, P., Donaldson, J.: Gamification Mob. Mark. Effect. J. Interact. Mark. **34**(1), 25–36 (2016)
17. Li, T., Li, C.: Flying into the Metaverse - China Southern Airlines' Digital Transformation Journey, RSM Case Development Centre, Rotterdam (2023)
18. Lingqvist, O., Plotkin, C., Stanley, J.: Do you really understand how your business customers buy? McKinsey Quarterly **1**, 74–85 (2015)
19. Lüders, M., Klarmann, M.: A contingency perspective on B2B purchasing information behavior in the digital age. Unpublished manuscript, Karlsruher Institut für Technologie (KIT), Karlsruhe (2019)
20. Miclau, C., Peuker, V., Gailer, C., Panitz, A., Müller, A.: Increasing customer interaction of an online magazine for beauty and fashion articles within a media and tech company. In: Nah, F., Siau, K. (eds.): HCII 2023. LNCS, vol. 14038, pp. 401–420, Springer, Cham (2023). https://doi.org/10.1007/978-3-031-35969-9_27
21. Park, E., Rishika, R., Janakiraman, R., Houston, M., Yoo, B.: Social dollars in online communities - the effect of product, user, and network characteristics. J. Mark. **82**(1), 93–114 (2018)
22. Paschen, J., Kietzmann, J., Kietzmann, T.: Artificial intelligence (AI) and its implications for market knowledge in B2B marketing. J. Bus. Ind. Mark. **34**(7), 1410–1419 (2019)
23. Paschen, J., Wilson, M., Ferreira, J.: Collaborative intelligence - how human and artificial intelligence create value along the B2B sales funnel. Bus. Horiz. **63**(3), 403–414 (2020)
24. Raja, J., Frandsen, T., Kowalkowski, C., Jarmatz, M.: Learning to discover value - value-based pricing and selling capabilities for services and solutions. J. Bus. Res. **114**, 142–159 (2020)
25. Rauschnabel, P., Felix, R., Hinsch, C., Shahab, H., Alt, F.: What is XR? Towards a framework for augmented and virtual reality. Comput. Hum. Behav. **133**, 1–18 (2022)

26. Replika: The AI companion who cares. https://replika.com/. Accessed 24 Jan 2024
27. Rietz, S., Steinhoff, F.: Die Schaffung und Nutzung von Virtualität - Jederzeit den Blick auf das Wesentliche fokussieren. In: Graumann, M., Burkhardt, A., Wenger, T. (eds.) Aspekte des Managements der Digitalisierung, pp. 115–139. Springer, Wiesbaden (2022). https://doi.org/10.1007/978-3-658-36889-0_6
28. Roelen-Blasberg, T., Habel, J., Klarmann, M.: Automated inference of product attributes and their importance from user-generated content: can we replace traditional market research? Int. J. Res. Mark. **40**(1), 164–188 (2023)
29. SICK AG: SICK Support Portal. https://www.sick.com/de/de/catalog/service/produkt-und-systemsupport/sick-support-portal/c/g356652. Accessed 24 Jan 2024
30. Snyder, K., Pashmeena, H.: The Changing Face of B2B Marketing. https://www.thinkwithgoogle.com/consumer-insights/consumer-trends/the-changing-face-b2b-marketing/. Accessed 24 Jan 2024
31. Terho, H., Mero, J., Siutla, L., Jaakkola, E.: Digital content marketing in business markets - activities, consequences, and contingencies along the customer journey. Ind. Mark. Manag. **105**, 294–310 (2022)
32. Tuli, K., Kohli, A., Bharadwaj, S.: Rethinking customer solutions: from product bundles to relational processes. J. Mark. **71**(3), 1–17 (2007)
33. Ulaga, W., Reinartz, W.: Hybrid offerings - how manufacturing firms combine goods and services successfully. J. Mark. **75**(6), 5–23 (2011)
34. Vargo, S., Lusch, R.: Institutions and axioms - an extension and update of service-dominant logic. J. Acad. Mark. Sci. **44**(1), 5–23 (2016)
35. Vendrell-Herrero, F., Bustinza, O., Parry, G., Georgantzis, N.: Servitization, digitization and supply chain interdependency. Ind. Mark. Manag. **60**, 69–81 (2017)
36. Worm, S., Bharadwaj, S., Ulaga, W., Reinartz, W.: When and why do customer solutions pay off in business markets? J. Acad. Mark. Sci. **45**(4), 490–512 (2017)
37. Zoltners, A., Sinha, P., Lorimer, S.: How more accessible information is forcing B2B sales to adapt. https://hbr.org/2016/01/how-more-accessible-information-is-forcing-b2b-sales-to-adapt. Accessed 24 Jan 2024

Influence of Streamer Characteristics on Trust and Purchase Intention in Live Stream Shopping

Franziska Grassauer[1] and Andreas Auinger[2(✉)]

[1] Pulpmedia GmbH, Linzer Straße 1, 4040 Linz, Austria
[2] FH Upper Austria, Digital Business Institute, Wehrgrabengasse 1-3, 4400 Steyr, Austria
Andreas.auinger@fh-steyr.at

Abstract. Live stream shopping is gaining global popularity, notably in China and Austria, with major brands like BIPA, L'Oréal, and IKEA adopting this approach. This new format, featuring interactive product showcases by streamers, addresses traditional online shopping limitations such as insufficient product information and lack of retailer trust. This study investigates how streamer characteristics affect trust and purchase intentions in Austrian live stream shoppers, focusing on two streamer roles: experts and influencers. Based on the SOR model and source models, the research involved a quantitative experiment with 188 participants viewing live stream scenarios. Results indicate streamer attractiveness and expertise significantly boost trust and purchase intent. Attractiveness has a stronger trust correlation with influencers, while expertise is more influential for experts. Trust positively impacts purchase intentions for both roles. However, 63% of Austrians aged 18–59 show reluctance towards buying via live streams. The study also finds experts rate higher in attractiveness and expertise than influencers.

Keywords: Live stream shopping · electronic commerce · streamer

1 Introduction

Live stream shopping is becoming increasingly prominent on a global scale. Live stream shopping is the sale of products in real time via a live stream on a website. In this live stream, streamers such as experts or influencers present certain products in an interactive way, provide additional information, and demonstrate their functionality [1]. At the same time, viewers can interact with the streamers through written comments or the 'Like' button [2]. All features and functionalities of products can be shown to minimize viewer uncertainty [3].

In China, live stream shopping is already an established part of e-commerce. Its growth was particularly rapid during the Covid 19 pandemic [4]. According to a study by Academy of China Council for the Promotion of International Trade [5], the number of users of live stream shopping platform users in China reached 469 million by June 2022 (204 million in March 2020), accounting for 44.6% of total Internet users. to the report, China's largest short video platform Douyin (comparable to TikTok) hosted more than 9 million live streams every month between May 2021 and April 2022, and sold

F. F.-H. Nah and K. L. Siau (Eds.): HCII 2024, LNCS 14720, pp. 46–65, 2024.
https://doi.org/10.1007/978-3-031-61315-9_4

at least 10 billion goods [5]. Due to the growing popularity of live stream shopping in China, the number of academic studies on this issue has increased significantly in recent years. Most of the existing studies focus on streamers [6, 7]. In addition, the motivation of Chinese people to participate in live-stream shopping and the purchasing behavior of live-stream shopping have also been studied [8]. Most of the existing studies focus primarily on the Asian region, while there are almost no academic studies on the topic in Europe.

Meanwhile, European companies are also using live stream shopping to sell their products. An Austrian study found that 25% of the respondents had already purchased products presented in live stream shopping events. In addition, another 15% could imagine expanding their shopping experience by participating in such events in future [9]. With the growing importance of online retail, the drawbacks of traditional, digital forms of sales are becoming more apparent. First and foremost, contacting online retailers to obtain additional product information is often very time-consuming and interrupts the buying process. In addition, the product information that users receive before making a purchase is usually limited to product photos and text descriptions. Another problem is the uncertainty about the authenticity of online retailers. Consumers perceive an increased risk when buying online if there is no direct contact with the seller [10]. This is mainly because traditional online stores offer limited opportunities to convey emotions, facial expressions, or gestures. As a result, trust in the retailer is reduced because a personal connection cannot be established [11]. Purchasing decisions are often made based on product photos, which are often edited after the fact, and the appearance of the products shown does not always reflect reality. During a live stream shopping event, consumers' concerns can be reduced, and products can be presented truthfully. Heinemann [12] also believes that live video will continue to grow in importance, and that live videos are more authentic, convey more closeness and make companies seem more tangible. In addition, they can build trust and a more effective retain customers.

Previous research has focused primarily on the characteristics of streamers in live stream shopping and little on the roles that streamers play. There are numerous studies dealing with the attractiveness, expertise and trust of streamers [6, 13–16]. Although the effects of attractiveness and expertise have been confirmed several times in published studies, the results vary depending on the study, the mediators used, and the dependent variables. In general, most studies on live stream shopping have been conducted in Asia, which is a significant difference because the society and culture in Asia is not comparable to that in Europe. Despite extensive research on live stream shopping, the literature review did not identify any study in Austria or German-speaking countries that specifically addresses the effects of streamer characteristics and roles on trust and ultimately purchase intention. Therefore, this study aims to fill this research gap. The study will provide information on which streamer characteristics, according to the Austrian population, can have an effect on mediator trust and ultimately lead to purchase intention. The role of the streamers is also considered.

Subsequently, this paper investigates whether Austrians would build up trust depending on streamer characteristics and roles and whether they would ultimately purchase products in live stream shopping or not. To investigate purchase intention, independent variables such as streamer attractiveness and credibility are linked with the construct of trust as a mediator. From these problem areas, the authors derive the following research questions:

RQ1: What are the effects of streamer characteristics in live stream shopping on the trust and purchase intention of Austrians?

RQ2: What specific differences in terms of attractiveness and expertise can be identified among streamers depending on their role (expert or influencer)?

RQ3: What is the effect of streamer characteristics in relation to the respective role (expert or influencer) on viewer trust and purchase intention in live stream shopping?

The paper is organized in the following manner: Following this introduction, the theoretical background delves into the subject matter, while the method section synthesizes both our literature review and the survey conducted. Afterwards, the presented and discussed results from the literature review and survey are provided. The paper concludes by summarizing the outcomes and highlighting the implications for the field of HCI, along with identifying potential avenues for future research.

2 Theoretical Background

The theoretical background is derived from recent practical studies and an extensive literature review. It includes an overview of live stream shopping in Europe and explores the roles of streamers.

2.1 Live Stream Shopping in Europe

Live stream shopping has expanded from China to the European market in recent years. A 2022 German study revealed, that 56% of respondents aged 18–34 were aware of live stream shopping. Similarly, nearly half of those surveyed in the 35+ age group has heard of live stream shopping [17]. Despite this, live stream shopping events are still infrequently used in Germany. To date, only 10% of individuals aged between 18 and 34 have utilized live stream shopping as an alternative to traditional shopping. In comparison, less than 10% of purchasers aged 35 or older have made live stream purchases [17]. The results of the study further demonstrate that 74% of German participants hold a positive attitude towards the ability to ask questions directly during live stream shopping. Additionally, 61% of respondents found a well-detailed product explanation to be appealing. Participants described live stream shopping as an engaging format [9]. A 2021 study conducted by the Austrian Retail Association revealed that 25% of surveyed participants had already made purchases of products showcased during live stream shopping events. In addition, a further 15% can imagine expanding their shopping experience in the future by participating in such events [18]. More than one-third of the top beauty and fashion brands and retailers in Germany offer live stream shopping. Live stream shopping is gaining popularity not only in the three main e-commerce markets in Europe - Germany, the UK, and France - but also in the Nordic countries. Numerous Swedish and Danish brands have already gained expertise in this field. A sector comparison demonstrates that beauty brands frequently hold live stream shopping events, while fashion brands only host them occasionally. The study shows that 70% of European brands favor integrating their own webshops and promoting events via social media. Companies benefit from using their own channels, as they provide comprehensive control over the live stream shopping environment and stream quality. However, one disadvantage is that existing customers are reached primarily. In order to make the live stream shopping event known to a wider public, additional marketing on other platforms is required, which in turn is associated with costs [19].

2.2 Role of Streamers in Live Stream Shopping

The people who present the products during the live stream are referred to as 'streamers' or 'hosts' [20, 21]. They include, for example, influencers, well-known personalities, independent salespeople, employees and ordinary people without a specific role assignment. During the live stream, viewers can interact with the streamers and other viewers by sending messages [22]. In Europe, streamers are classified as influencers or product experts. The streamers of a live stream shopping event play a vital role in achieving the goal of increasing brand awareness, launching a new collection, or boosting sales to the audience who already have an interest in your brand and/or webshop. The concept of live stream shopping presents a chance to take advantage of the growing popularity of influencer marketing among select customer segments. It is logical to assume that the incorporation of celebrities or subject matter experts from social media can increase the audience for live stream shopping events. Influencers play a crucial role in the sphere of social media among young customers, while social media is progressively becoming a critical influencer for consumer choices. According to a 2023 survey, 45% of German Generation Z participants stated that influencers should take on the role of streamers at live stream shopping events. In contrast, only ten percent of baby boomers shared this opinion. However, 70% of baby boomers stated that experts for the product being presented should be used as streamers for live stream shopping events. Only 21% of Generation Z agreed with this view [22]. As consumers age, practicality and efficiency become increasingly important factors in their product preferences, overtaking the influence of salespeople or other presenters. Forrester recently conducted a survey on consumer sentiment towards live stream shopping in Europe, specifically asking which type of streamer the participants preferred. Across all age groups, local product experts such as in-store salespeople were favored. Product experts are more popular than social media influencers, even among adults aged 18 to 34. Generally, European brands and retailers maintain a balanced distribution of influencers and/or internal product experts. Beauty companies differ from fashion brands by tending to integrate both influencers and experts during live shopping events [19]. The results of this survey are especially relevant for this paper, because beauty and cosmetic products were also utilized for the research context of the conducted study.

2.3 Literature Review

2.4 The Author of This Paper has Developed the Procedure for the Literature Research and Analysis Based on [23–26]

The objective of this study is to accurately document and present the current findings from existing literature. To achieve this, Cooper's taxonomy is utilized to determine the scope of the literature search. The initial phase of preparation involves identifying relevant databases and defining search terms, while also considering synonyms and additional relevant terms. The databases of Web of Science, Scopus, IEEE and ACM were utilized, while search terms like [live shopping, live stream shopping, live commerce, live stream commerce, e-commerce livestream, livestream retail, live video shopping and taobao live] were used in different combinations and synonymous versions (Fig. 1).

Afterwards, inclusion and exclusion criteria are established in order to effectively filter the located literature sources. The content of the sources was evaluated based on how relevant they are to the research questions. The initial phase involves conducting a literature search wherein search criteria are established and search terms are executed in corresponding databases. The content of the sources was evaluated based on how relevant they are to the research questions. After applying the search strategy to titles, abstracts, and keywords, a total of 405 search results were yielded. The content of the sources was evaluated based on how relevant they are to the research questions. The content of the sources was evaluated based on how relevant they are to the research questions. This assessment reduced the number of sources to a total of 55, and duplicates were also eliminated during this phase. The subsequent evaluation step involved thoroughly reviewing the complete texts of each publication, as well as conducting a backward and forward search [26].

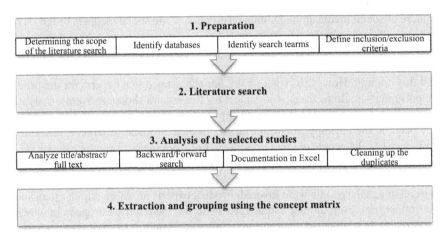

Fig. 1. Literature review procedure

After completing the analysis, we identified 39 relevant publications. Following [1] recommended conceptual approach, we formed an organizational framework for analyzing the factors that influence live stream shopping. In this situation, various concepts shape these factors. After analyzing the selected sources, a total of eight identified influencing factors for the use of live stream shopping were recorded in the concept matrix created for this thesis. These influencing factors were formed based on very frequently examined variables from previous research on live stream shopping in general as well as regarding trust and purchase intention. Upon analyzing the frequencies of the individual factors, it is evident that the two primary factors that influenced the scientific papers were 'host/streamer' and 'social aspects' which include the streamer character-istics of attractiveness, expertise and trust. The remaining factors include 'information', 'entertainment', 'interaction', 'product and price', 'technical aspects', and other aspects (Table 1).

The authors mainly focused on the factors of attractiveness and expertise as well as trust in streamers [13, 15, 6–10]. The existing literature confirms the positive influence of streamer expertise and the physical attractiveness of streamers. Streamer expertise is often considered more important than attractiveness when researching viewer trust and the dependent variable is purchase intent [27]. In contrast, physical attractiveness repeatedly has positive effects when hedonistic factors or user engagement are exam-ined [27]. The familiarity of the streamers was also examined in the studies, and it was found that the attitude towards live stream shopping is more positive and the interac-tion higher when it is a well-known person, for example influencers [29]. In addition, the identification of viewers with the streamers was also extensively examined in the studies. Furthermore, many relevant studies focused on social aspects and interpersonal interaction factors. Social presence has been analyzed in detail in numerous studies and its influence has been repeatedly confirmed. However, several studies have not been able to prove a direct and significant influence of interactivity between streamers and viewers on the purchase intention. Instead, the effect is usually indirect via mediators such as trust, immersion, or social presence [10, 31].

3   Research Model

The forthcoming section will explicate the specific variables. Attractiveness and exper-tise, as independent variables, are analyzed in the context of streamers. Purchase intention for live stream shopping serves as the dependent variable. Trustworthiness, serving as the mediator, connects the independent and dependent variables.

3.1 Attractiveness

Attractiveness is defined as the evaluation of the physical aspects, temperament and similarity of a person by another person [16]. Attractiveness does not just mean physical attractiveness but includes a range of positive characteristics that consumers may per-ceive in a celebrity, for example intellectual ability, personality traits, lifestyle or athletic ability [56]. Physically attractive streamers can provide a sense of security and foster trust [45]. As demonstrated by [30] the physical attractiveness of streamers has a positive impact on viewer engagement during live stream shopping. Consumers tend to place a significant emphasis on physical appearance and consequently direct all their attention and interest towards attractive streamers. Eye-catching streamers capture attention and can induce behavioral responses.

3.2 Expertise

Perceived expertise is the extent to which communicators are viewed as a reliable source of information. The term encompasses the knowledge, experience, and skills that com-municators possess. It is not mandatory that they are experts but rather, it is important how

Table 1. Identified influencing factors in live stream shopping

Study	Identified influencing factors							
	Hosts/Streamers	Social aspects	information	entertainment	interaction	Product and price	Technical aspects	other aspects
[28] Cai et al. (2018)	x					x		
[32] Chen/Lin (2018)	x	x		x	x			
[10] Sun et al. (2019)		x		x				
[33] Hu/Chaudhry (2020)			x			x		
[14] Heo/Kim/Yan (2020)	x	x						
[35] Ko/Chen (2020)	x	x			x			
[27] Park/Lin (2020)	x							
[2] Wongkitrungrueng/Assarut (2020)			x					
[13] Chen et al. (2021)	x							
[35] Li X. et al. (2021)		x						
[36] Li/Li/Cai (2021)		x					x	
[7] Li/Peng (2021)	x							
[37] Liu/Kim (2021)			x	x				
[38] Ming et al. (2021)		x						
[39] Zhou et al. (2021)				x				
[40] Zhu et al. (2021)	x							
[41] Ma (2021a)		x						
[42] Ma (2021b)							x	
[43] Bao/Zhu (2022)					x			

(continued)

Table 1. (*continued*)

Study	Identified influencing factors							
	Hosts/Streamers	Social aspects	information	entertainment	interaction	Product and price	Technical aspects	other aspects
[21] Chen et al. (2022)	x			x				
[6] Guo et al. (2022)	x							
[29] Li et al. (2022)	x				x			
[44] Lin et al.. (2022)				x				
[45] Liu et al. (2022)	x							
[16] Rungruangji (2022)	x							
[31] Sawmong (2022)	x		x	x				
[15] Wang et al. (2022)	x							
[46] Zeng et al. (2022)								x
[31]Dang-Van et al. (2023)	x							
[47] Dong/Liu/Xiao (2023)					x		x	
[48] Gao et al. (2023)	x	x						
[49] Hwang/Youn (2023)			x			x		
[50] Joo/Yang (2023)				x	x			
[51] Liao et al. (2023)	x	x			x			
[52] Tian et al. (2023)	x				x			
[53] Wu et al. (2023)						x		
[54] Yin et al. (2023)		x						
[55] Zhang/Chen/Zamil (2023)	x						x	
[52] Zheng et al. (2023)		x						
Sum	**20**	**12**	**5**	**8**	**7**	**4**	**3**	**1**

their audience perceives them [56]. Expertise is a crucial factor in enhancing the credibility of communicators [57]. [15] Proving expertise has a positive impact on consumer trust.

3.3 Trustworthiness

Trustworthiness pertain to the truthfulness, honorability, and credibility of communicators. As per [27], trustworthiness refers to "the listener's level of confidence in and acceptance of the speaker and the message". The perception of communicator's trustworthiness varies among the target audience. Studies have indicated that acclaimed advertisers with high trustworthiness can significantly sway consumers' attitudes and intentions to purchase [58]. In their study, [59] confirms the influence of streamers' trustworthiness on purchase intentions.

3.4 Purchase Intention

Purchase intention is described by the willingness of consumers to buy a certain product or service based on their subjective assessment and overall evaluation [60]. As articulated by [61], purchase intention represents an individual's perception of the likelihood of purchasing a product in a particular time frame. Consumer analyses for tangible goods frequently take into account purchase intentions. However, the reliability of purchase intentions is limited since not all stated intentions are always put into practice. Prior research has explored the impact of celebrities on purchase intentions. Regarding live stream shopping, [27] confirmed that celebrities do influence the purchase intentions of Chinese consumers (Fig. 2).

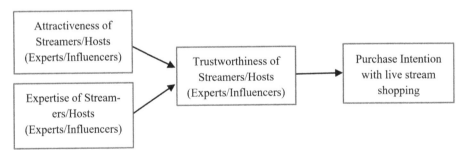

Fig. 2. Research model

4 Methodology

Based on the findings of the literature review, an online survey was carried out as an empirical investigation. The outcomes of this study shed light on the three proposed research questions.

4.1 Online Survey

An online survey was conducted in Austria from July 17 to July 31, 2023, to determine the impact of streamer characteristics on trust and purchase intention in live stream shopping among Austrian consumers. Prior to this survey, a pretest was conducted with participants from various fields to ensure its comprehensiveness and clarity. The language used in the survey was kept unambiguous to avoid any confusion. Furthermore, the roles of streamers, including experts and influencers, were taken into account. The online survey was conducted through the use of the Lime Survey tool.

The survey follows a standardized structure with the demographic inquiry appearing at the end. It includes a total of 38 questions or statements, along with the demographic query. Attractiveness, expertise, and trustworthiness are the selected influencing factors of streamers, adapted and queried using the question items of Gupta, Kishor, and Verma [16] and Ohanian [58], as in the study by Rungruangjit [62].

A 5-point Likert scale is utilized for this purpose, with the dependent variable of purchase intention being tested through the question items from Chen et al. [13], also using a 5-point Likert scale. The following table displays the variables alongside the question items, which renders the theoretical construct measurable (Table 2).

Table 2. Measured Constructs and Research Questions

Construct	Question Items
Attractiveness [16]	ATT1: The streamer is attractive ATT2: The streamer has a beautiful face ATT3: The streamer has a very convincing voice ATT4: The streamer has a very professional appearance ATT5: The streamer attracts my attention
Expertise [16, 58]	EXP1: The streamer has expertise in his/her field EXP2: The streamer has expertise in his/her field EXP3: The streamer has extensive product knowledge EXP4: The streamer is highly qualified in sales
Trustworthiness [16, 58]	TRU1: The streamer is a sincere person TRU2: The streamer is an honest person TRU3: The streamer is trustworthy TRU4: The streamer is a reliable source of information
Purchase Intention [13]	PUI1: I would like to buy products through live stream shopping PUI2: I would recommend my friends and families to buy products through live stream shopping PUI3: If I want to buy a product, I would like to buy it from live stream shopping

The paper's authors chose their scale based on previous studies examining similar constructs. The items are presented according to variables, with the order of the items being randomly rotated for participants. The online survey begins with a brief general introduction and definition of live stream shopping. General questions about online shopping behavior are also asked at the beginning. The participants then watched an excerpt from a live stream shopping event featuring both experts and influencers, whereby cosmetic products were presented. The participants then evaluate the two streamer roles they saw in the previous video. At the end, the participants are asked about their socio-demographic data, including age, gender, country and level of education. The statements to be evaluated were randomly organized to prevent any primacy or recency effects, ensuring that each respondent receives a uniquely arranged list of items [63]. The findings of this empirical study, conducted via LimeSurvey, are analyzed with the aid of SPSS.

4.2 Survey Sample

In this study, a non-probabilistic sampling method was used, in which the respondents were selected according to their availability or other predefined criteria without a specific system. It is therefore an random selection method [63]. The online survey received 334 responses between July 17 and July 31, 2023, with 202 complete questionnaires. All responses from individuals who did not reside in Austria at the time of survey completion and those over 59 years old who couldn't answer the control question correctly were excluded during data cleaning. After removing invalid cases, 188 valid responses were available for further analysis. The average response time was 6 min and 17 s. Of the 188 participants, 132 identified as female and 56 identified as male, resulting in over 70% of participants being female and close to 30% being male. The average age of the sample was 29 years old, with ages ranging from 18 to 59. A substantial proportion of participants, namely 71.2%, fell between the ages of 18 and 29 years old. 17.6% of participants are aged between 30 and 39. Of all the participants, 39.4% hold a bachelor's degree, while 20.7% have a master's degree and 19.1% possess a baccalaureate. The remaining 20.8% are distributed among other categories, with apprenticeships accounting for the largest portion.

The survey revealed that 42.6% of respondents engage in online shopping two to three times a month, followed by 20.2% who shop online once a month and 9% who shop every two to three months. In addition, 19.1% shop online at least once a week. Following the online shopping survey, respondents were polled about their familiarity and prior experience with live streaming shopping. The survey results indicate that 56.9% of the participants were aware of live stream shopping, 3.2% were unsure, and 39.9% were unfamiliar. Of those surveyed, 9.6% had participated in live stream shopping while 1.6% were unsure, and 88.8% had never watched a live stream shopping event before this survey.

Reliability Analysis. The reliability of a test refers to the reliability with which a measurement instrument provides consistent measurement results for a group of items in relation to a specific characteristic. Various methods are available in SPSS, whereby Cronbach's alpha is frequently used as a measure of reliability. For a sufficiently large number of similar indicators, the degree of internal consistency of a scale is measured using Cronbach's alpha. This measure is based on the average correlation between the indicators [64]. In this consistency analysis, the calculated value of Cronbach's alpha ranges from 0 to 1. A value closer to 1 indicates higher reliability and it can be assumed that the items consistently measure the same characteristic [65]. Recommendations tend towards minimum values of 0.7 or 0.8 [3.]. The results show that all items of the individual variables display a high value of Cronbach's alpha, they are therefore clearly reliable and the items correlate (Table 3).

Table 3. Reliability Analysis

Item	Cronbach's Alpha	Number of Items
ATTEX	0,769	5
EXPEX	0,795	4
TRUEX	0,852	4
ATTIN	0,884	5
EXPIN	0,886	4
TRUIN	0,909	4
ATT	0,867	10
EXP	0,873	8
TRU	0,898	8
PUI	0,898	3

5   Results

The section below outlines the results obtained from a survey of 188 participants in Austria. First, we present the descriptive findings, followed by the statistical analysis that addresses the research questions.

5.1   Descriptive Statistics

The survey findings revealed that 42.6% of participants shop online two to three times monthly, while 20.2% make online purchases once a month, and only 9% shop online bi-monthly. However, 19.1% of those surveyed shop online on a weekly basis. Following the online shopping survey, respondents were queried about their prior familiarity and engagement with live stream shopping. The survey found that 56.9% of respondents were aware of live stream shopping, while 3.2% were not sure and 39.9% were unfamiliar with the concept. The language used in the survey was precise and unambiguous while adhering to proper grammar, metrics, and units. Among the respondents, 9.6% reported participation in live stream shopping, 1.6% were unsure, and 88.8% had never watched a live stream shopping event.

5.2   Statistical Tests to Verify the Research Questions

Before exploring any correlation or difference, it is crucial to conduct a normal distribution test. The mathematical test for normal distribution can be performed using tests like the Kolmogorov-Smirnov or Shapiro-Wilk tests. The variables tested, namely EXP (expertise of both roles), EXPEX (expertise of the experts), EXPIN (expertise of the influencers), ATTEX (Attractiveness of the experts), ATTIN (Attractiveness of the influencers), TRU (Trust of both roles), TRUEX (Trust of the experts), TRUIN (Trust of the influencers), and PUI (Purchase intention during live stream shopping) show significant results ($p < 0.05$) in both Kolmogorov-Smirnov and Shapiro-Wilk tests [65]. These results clearly show that the data are not normally distributed. In contrast, the ATT (attractiveness of both roles) variable clearly shows that it is not significant and is therefore normally distributed (Table 4).

Table 4. Descriptive Statistics

Item	MV	SD	K-S/p[a]	S-W/p
ATT	3,63	0,631	0,066	0,055
ATTEX	3,71	0,657	<0,001	0,001
ATTIN	3,54	0,793	<0,001	0,003
EXP	3,21	0,659	<0,001	0,004
EXPEX	3,43	0,673	0,001	<0,001
EXPIN	2,99	0,821	<0,001	0,004
TRU	3,20	0,659	<0,001	0,002
TRUEX	3,15	0,686	<0,001	<0,001
TRUIN	3,25	0,796	<0,001	<0,001
PUI	2,19	0,907	<0,001	<0,001

(Note: Mean Value (MV), Standard Deviation (SD), Kolmogorov-Smirnov test (K-S), Shapiro-Wilk (S-W) test).

RQ1. To answer RQ1, we will examine the correlation. The Pearson correlation ($r = 0.619$) and the Spearman correlation ($r = 0.580$) both demonstrate a significant, positive, and medium correlation between attractiveness and trust. We will proceed to test the correlation between expertise and trust. The Spearman correlation results showcase a significant outcome ($p = < 0.001$), with a correlation coefficient of 0.551 ($r = 0.551$). This indicates that there is a substantial, favorable, and average correlation between expertise and confidence level. Additionally, we analyze the correlation between trust and purchase intention in live stream shopping. The Spearman-Rho results demonstrate a significant correlation between trust and purchase intention, as their significance is nearly zero ($p < 0.001$). The correlation coefficient is 0.449 ($r = 0.476$). There is a significant, positive but low correlation between trust and purchase intention (Table 5).

RQ2. The following test aims to determine if stream-ers' characteristics differ based on whether experts or influencers moderate live stream shopping events. To assess differences between two dependent groups, utilize either the Wilcoxon or T-test for dependent samples, depending on the normal distribution and scale level. The Wilcoxon test allows for the immediate comparison of two measured values for the same individuals. The

Table 5. Correlations ATT, EXP, TRU and PUI

Item	r_p	Sig. (2-tailed)	r_s	Sig. (2-tailed)
ATT - TRU	0,619	<0,001	0,580	<0,001
EXP - TRU	-	-	0,551	<0,001
TRU - PUI	-	-	0,449	<0,001

Note: Pearson correlation (rp), Spearman correlation (rs), Significance (Sig.)

Wilcoxon test, a parameter-free method, is used to answer the second research question (RQ2), as the variables are not normally distributed. The mean difference between ATTEX and ATTIN is deemed highly significant ($Z(n = 188) = -3.248$, p = 0.001). The test statistic for expertise is $Z = -7.416$ and the associated significance value is p = < 0.001, indicating a significant difference. The central tendencies of the two variables differ significantly ($Z(n = 188) = -7.416$, p < 0.001). Additionally, the linked samples demonstrated a difference in attractiveness and expertise. The respondents rated the expert as more attractive and knowledgeable than the influencer (Table 6).

Table 6. Differences ATTEX-ATTIN and EXPEX-EXPIN

Item	MV	SD	Z	Asymp. Sig. (2-tailed)
ATTEX	3,71	0,657	−3,248	0,001
ATTIN	3,54	0,793		
EXPEX	3,43	0,673	**−7,416**	<0,001
EXPIN	2,99	0,821		

Note: Mean Value (MV), Standard Deviation (SD), Asymthotic Significance (Asymp. Sig.),

RQ3. To address the third research question (RQ3), we examined the correlation between variables using Spearman's correlation test due to the non-normal distribution and ordinal data level. First, we present all variables related to the expert. The correlation test reveals a significant, positive, and low correlation (r = 0.476) between the expert's attractiveness (ATTEX) and trustworthiness (TRUEX). A significant medium, positive correlation exists between expertise (EXPEX) and trust (TRUEX) variables, as demonstrated by the Spearman correlation analysis. To test the correlation between a veteran's trust (TRUEX) and purchase intention (PUI) during live stream shopping, we conducted another analysis that showed a significant result (p < 0.001) with a correlation coefficient of 0.409 (r = 0.409). Therefore, there is a significant, positive, but low correlation between trust in the expert (TRUEX) and purchase intention (PUI). The following section outlines the variables related to the influencer and the results of the Spearman correlation analysis. The analysis reveals a significant result (p < 0.001) with a correlation coefficient of 0.522 (r = 0.522), indicating a significant, positive, medium correlation between the influencer's attractiveness (ATTIN) and trust (TRUIN). The Spearman-Rho analysis indicates a substantial association between influencer expertise and trust. The correlation coefficient is 0.476 (r = 0.476), indicating a positive yet weak connection between EXPIN and TRUIN. It can be presumed that TRUIN and PUI in live stream shopping also display a significant, positive but low correlation (Table 7).

Table 7. Correlations among variables

Item	r_s	Sig. (2-tailed)
ATTEX - TRUEX	0,476	<0,001
ATTIN- TRUIN	0,522	<0,001
EXPEX - TRUEX	0,522	<0,001
EXPIN - TRUIN	0,476	<0,001
TRUEX - PUI	0,409	<0,001
TRUIN - PUI	0,390	<0,001

Note: Spearman correlation (rs), Significance (Sig.)

In summary, there is a significant, positive correlation between the expertise of experts and the level of trust they inspire, as well as between the attractiveness of influencers and the level of trust they inspire. Additionally, there is a significant, positive but low correlation between the attractiveness of experts and the expertise and trust of influencers, as well as the trust levels towards both experts and influencers, and the purchase intention in live stream shopping. The table above displays the correlation between the variables that were tested.

6 Discussion

The statistical tests revealed significant correlations and variations. The overall findings indicate that streamer attractiveness and expertise have a positive and moderate effect on trust. Furthermore, there is a significant yet small positive correlation between trust and purchase intention. A higher level of trust in streamers is positively associated with a greater intention to purchase during live stream shopping. Differences in the appeal and competence of streamers were observed, contingent upon their position as experts or influencers. This study aimed to analyze the influence of specific streamer traits on trust and purchase inclination during live stream shopping. Additionally, it scrutinized the function of streamers, encompassing both experts and influencers. Through our research, we were able to examine the issue previously raised regarding streamers in live stream shopping in Austria. In this section, we answer the individual research questions and provide an outlook.

The first research question (RQ1) could be answered based on empirical research. Consequently, the attractiveness and expertise of the streamers positively influence trust. The more attractive and expert the streamers are, the higher the trust towards the streamers by Austrians aged 18 to 59. Increased trust in the streamers is positively correlated with an increased purchase intention for live stream shopping. These results are also consistent with previous studies, which were presented in the research statements. Nevertheless, the descriptive results show that Austrians between the ages of 18 and 59 are less likely to buy products from live stream shopping.

The study's RQ2 utilized statistical methods to assess differences in the attraction and competency of streamers, based on whether they were experts or influencers. The findings indicate that experts were rated significantly higher than influencers in both aspects.

The third research question (RQ3) was also answered based on empirical results. All correlations could be confirmed significantly. Results indicate that attractiveness and trust have a significantly stronger relationship for influencers than experts. Conversely, expertise and trust tend to have a higher correlation for experts than influencers. Regarding the relationship between trust and intent to purchase, there is a significant and similar correlation for both factors.

Important Role of Experts in Livestream Shopping. The results suggest that the expert received higher ratings than the influencer in both attractiveness and expertise. Among experts, there were significant, positive correlations between trust and both attractiveness and expertise, with a stronger correlation between trust and expertise. Influencers showed a similar pattern, with a moderate correlation between trust and attractiveness, and a weaker correlation between trust and expertise. The results substantiate previous research, such as the study conducted by [15], which proved that expertise positively impacts trust. The correlation between trust and purchase intention was significant, albeit small and positive. In conclusion, higher trust in streamers is closely linked to increased intention to make a purchase during live stream shopping. This result is in line with the study results of [31] and [52], which confirmed a significant, positive influence of trustworthiness on the intention to buy.

Regardless of the statistical tests, the results of the descriptive analysis indicate that the participants' intention to buy products during live stream shopping is rather low. Regarding purchase intention, it is unclear if test participants were able to answer realistically due to the fact that a significant number of the respondents had never before participated in a live stream shopping event.

7 Conclusion

Live stream shopping is highly prevalent and thoroughly researched in Asia, while only a few studies have been conducted in Europe due to its nascent stage of development. Despite this, the format shows great potential and provides numerous opportunities for businesses to promote their products and services. The live stream shopping format stands out for its engaging video content, use of streamers, and high level of interaction. Studies show that in Austria, experts are perceived as more attractive and knowledgeable than influencers in live stream shopping. As electronic and social commerce continue to evolve, along with ongoing digitalization, new opportunities for research are opening up. This creates new avenues for further investigation in this field.

References

1. Guo, L., Hu, X., Lu, J., Ma, L.: Effects of customer trust on engagement in live streaming commerce: mediating role of swift guanxi. INTR **31**, 1718–1744 (2021). https://doi.org/10.1108/INTR-02-2020-0078
2. Wongkitrungrueng, A., Assarut, N.: The role of live streaming in building consumer trust and engagement with social commerce sellers. J. Bus. Res. **117**, 543–556 (2020). https://doi.org/10.1016/j.jbusres.2018.08.032
3. Kang, K., Lu, J., Guo, L., Li, W.: The dynamic effect of interactivity on customer engagement behavior through tie strength: Evidence from live streaming commerce platforms. Int. J. Inf. Manag. **56**, 102251 (2022). https://doi.org/10.1016/j.ijinfomgt.2020.102251
4. Zhang, L.: The Development of Livestream Commerce in China (2022)
5. China Daily Information Co: Users of live streaming e-commerce increase. China Daily Information Co, vol. 2022 (2022)
6. Guo, Y., Zhang, K., Wang, C.: Way to success: understanding top streamer's popularity and influence from the perspective of source characteristics. J. Retail. Consum. Serv. **64**, 102786 (2022). https://doi.org/10.1016/j.jretconser.2021.102786
7. Li, Y., Peng, Y.: What drives gift-giving intention in live streaming? The perspectives of emotional attachment and flow experience. Int. J. Hum.-Comput. Interact. **37**, 1317–1329 (2021). https://doi.org/10.1080/10447318.2021.1885224
8. Xu, X., Huang, D., Shang, X.: Social presence or physical presence? Determinants of purchasing behaviour in tourism live-streamed shopping. Tourism Manag. Perspect. **40**, 100917 (2021). https://doi.org/10.1016/j.tmp.2021.100917
9. Handelsverband Österreich, Mindtake: Consumer-Check zum Thema Social Media Nutzung in den Generationen X, Y & Z (2021)
10. Sun, Y., Shao, X., Li, X., Guo, Y., Nie, K.: How live streaming influences purchase intentions in social commerce: an IT affordance perspective. Electron. Commer. Res. Appl. **37**, 100886 (2019). https://doi.org/10.1016/j.elerap.2019.100886
11. Addo, C., Fang, J., Asare, A.O., Kulbo, N.B.: Customer engagement and purchase intention in live-streaming digital marketing platforms. Serv. Ind. J. **41**, 767–786 (2021). https://doi.org/10.1080/02642069.2021.1905798
12. Heinemann, G.: Der neue Online-Handel. Geschäftsmodelle, Geschäftssysteme und Benchmarks im E-Commerce. Springer, Wiesbaden (2022). https://doi.org/10.1007/978-3-658-20354-2
13. Chen, H., Zhang, S., Shao, B., Gao, W., Xu, Y.: How do interpersonal interaction factors affect buyers' purchase intention in live stream shopping? The mediating effects of swift guanxi. INTR **32**, 335–361 (2021). https://doi.org/10.1108/INTR-05-2020-0252
14. Heo, J., Kim, Y., Yan, J.: Sustainability of live video streamer's strategies: live streaming video platform and audience's social capital in South Korea. Sustainability **12**, 1969 (2020). https://doi.org/10.3390/su12051969
15. Wang, X., Aisihaer, N., Aihemaiti, A.: Research on the impact of live streaming marketing by online influencers on consumer purchasing intentions. Front. Psychol. **13**, 1021256 (2022). https://doi.org/10.3389/fpsyg.2022.1021256
16. Rungruangjit, W.: What drives Taobao live streaming commerce? The role of parasocial relationships, congruence and source credibility in Chinese consumers' purchase intentions. Heliyon **8**, e09676 (2022). https://doi.org/10.1016/j.heliyon.2022.e09676
17. Lohmeier, L.: Hast du bereits über Live Shopping eingekauft? (2022). https://de-1statista-1com-1007e9aly0e5b.han.ubl.jku.at/statistik/daten/studie/1377751/umfrage/nutzung-liveshopping-altersgruppen/

18. Lohmeier, L.: Was findest du besonders attraktiv an Live Shopping im Internet? (2022). https://de-1statista-1com-1007e9aly0e5b.han.ubl.jku.at/statistik/daten/studie/1377762/umfrage/gruende-attraktivitaet-liveshopping/
19. Arvato Supply Chain Solutions, Tolouee, A., Dittmer, R.: Live-Shopping in Europa. Das neue Must-have im Fashion-und Beauty-E-Commerce? (2021)
20. Richter, C.: E-Commerce Trends in China. Springer, Wiesbaden (2021). https://doi.org/10.1007/978-3-658-33345-4
21. Chen, C.-D., Zhao, Q., Wang, J.-L.: How livestreaming increases product sales: role of trust transfer and elaboration likelihood model. Behav. Inf. Technol. **41**, 558–573 (2020). https://doi.org/10.1080/0144929X.2020.1827457
22. Lohmeier, L.: Wer sollte Liveshopping-Events am besten durchführen: Influencer oder Produktexperten? (2023). https://de-1statista-1com-1007e9aly0e5b.han.ubl.jku.at/statistik/daten/studie/1387073/umfrage/umfrage-liveshopping-events-hosts-influencer-produktexperten-generation/
23. Webster, J., Watson, R.T.: Analyzing the past to prepare for the future: writing a literature review. MIS Q. **26** (2002)
24. Cooper, H.M.: Organizing knowledge syntheses: a taxonomy of literature reviews. Knowl. Soc. **1**, 100–123 (1988)
25. Kitchenham, B.: Guidelines for performing Systematic Literature Reviews in Software Engineering (2007)
26. Vom Brocke, J., Simons, A., Niehaves, B., Plattfaut, R., Cleven, A.: Reconstructing the giant: on the importance of rigour in documenting the literature search process (2009)
27. Park, H.J., Lin, L.M.: The effects of match-ups on the consumer attitudes toward internet celebrities and their live streaming contents in the context of product endorsement. J. Retail. Consum. Serv. **52**, 101934 (2020). https://doi.org/10.1016/j.jretconser.2019.101934
28. Cai, J., Wohn, D.Y., Mittal, A., Sureshbabu, D.: Utilitarian and hedonic motivations for live streaming shopping. In: Ryu, H., Kim, J., Chambel, T., Bartindale, T., Vinayagamoorthy, V., Tsang Ooi, W. (eds.) Proceedings of the 2018 ACM International Conference on Interactive Experiences for TV and Online Video, pp. 81–88. ACM, New York, NY, USA (2018). https://doi.org/10.1145/3210825.3210837
29. Li, L., Kang, K., Zhao, A., Feng, Y.: The impact of social presence and facilitation factors on online consumers' impulse buying in live shopping – celebrity endorsement as a moderating factor. ITP (2022). https://doi.org/10.1108/ITP-03-2021-0203
30. Dang-Van, T., Vo-Thanh, T., Vu, T.T., Wang, J., Nguyen, N.: Do consumers stick with good-looking broadcasters? The mediating and moderating mechanisms of motivation and emotion. J. Bus. Res. **156**, 113483 (2023). https://doi.org/10.1016/j.jbusres.2022.113483
31. Sawmong, S.: Examining the key factors that drives live stream shopping behavior. Emerg. Sci. J. **6**, 1394–1408 (2022). https://doi.org/10.28991/ESJ-2022-06-06-011
32. Chen, C.-C., Lin, Y.-C.: What drives live-stream usage intention? The perspectives of flow, entertainment, social interaction, and endorsement. Telematics Inform. **35**, 293–303 (2018). https://doi.org/10.1016/j.tele.2017.12.003
33. Hu, M., Chaudhry, S.S.: Enhancing consumer engagement in e-commerce live streaming via relational bonds. INTR **30**, 1019–1041 (2020). https://doi.org/10.1108/INTR-03-2019-0082
34. Ko, H.-C., Chen, Z.-Y.: Exploring the factors driving live streaming shopping intention, pp. 36–40. https://doi.org/10.1145/3409891.3409901
35. Li, X., Li, Y., Cai, J., Cao, Y., Li, L.: Understanding the psychological mechanisms of impulse buying in live streaming: a shopping motivations perspective, 811–820 (2021)
36. Li, Y., Li, X., Cai, J.: How attachment affects user stickiness on live streaming platforms: a socio-technical approach perspective. J. Retail. Consum. Serv. **60**, 1024781 (2021). https://doi.org/10.1016/j.jretconser.2021.102478

37. Liu, X., Kim, S.H.: Beyond shopping: the motivations and experience of live stream shopping viewers, vol. 187–192 (2021). https://doi.org/10.1109/QoMEX51781.2021.9465387
38. Ming, J., Jianqiu, Z., Bilal, M., Akram, U., Fan, M.: How social presence influences impulse buying behavior in live streaming commerce? The role of S-O-R theory. IJWIS 17, 300–320 (2021). https://doi.org/10.1108/IJWIS-02-2021-0012
39. Zhou, M., Huang, J., Wu, K., Huang, X., Kong, N., Campy, K.S.: Characterizing Chinese consumers' intention to use live e-commerce shopping. Technol. Soc. 67, 101767 (2021). https://doi.org/10.1016/j.techsoc.2021.101767
40. Zhu, L., Li, H., Nie, K., Gu, C.: How do anchors' characteristics influence consumers' behavioural intention in livestream shopping? A moderated chain-mediation explanatory model. Front. Psychol. 12, 730636 (2021). https://doi.org/10.3389/fpsyg.2021.730636
41. Ma, Y.: To shop or not: Understanding Chinese consumers' live-stream shopping intentions from the perspectives of uses and gratifications, perceived network size, perceptions of digital celebrities, and shopping orientations. Telematics Inform. 59, 101562 (2021). https://doi.org/10.1016/j.tele.2021.101562
42. Ma, Y.: Elucidating determinants of customer satisfaction with live-stream shopping: an extension of the information systems success model. Telematics Inform. 65, 101707 (2021). https://doi.org/10.1016/j.tele.2021.101707
43. Bao, Z., Zhu, Y.: Understanding customers' stickiness of live streaming commerce platforms: an empirical study based on modified e-commerce system success model. APJML 35, 775–793 (2023). https://doi.org/10.1108/APJML-09-2021-0707
44. Lin, S.-C., Tseng, H.-T., Shirazi, F., Hajli, N., Tsai, P.-T.: Exploring factors influencing impulse buying in live streaming shopping: a stimulus-organism-response (SOR) perspective. APJML 35, 1383–1403 (2022). https://doi.org/10.1108/APJML-12-2021-0903
45. Liu, X., Wang, D., Gu, M., Yang, J.: Research on the influence mechanism of anchors' professionalism on consumers' impulse buying intention in the livestream shopping scenario. Enterp. Inf. Syst. 17 (2022). https://doi.org/10.1080/17517575.2022.2065457
46. Zeng, Q., Guo, Q., Zhuang, W., Zhang, Y., Fan, W.: Do Real-time reviews m? Examining how bullet screen influences consumers' purchase intention in live streaming commerce. Inf. Syst. Front. (2022). https://doi.org/10.1007/s10796-022-10356-4
47. Dong, X., Liu, X., Xiao, X.: Understanding the influencing mechanism of users' participation in live streaming shopping: a socio-technical perspective. Front. Psychol. 13, 1082981 (2023). https://doi.org/10.3389/fpsyg.2022.1082981
48. Gao, W., Jiang, N., Guo, Q.: How do virtual streamers affect purchase intention in the live streaming context? A presence perspective. J. Retail. Consum. Serv. 73, 103356 (2023). https://doi.org/10.1016/j.jretconser.2023.103356
49. Hwang, J., Youn, S.: From brick-and-mortar to livestream shopping: product information acquisition from the uncertainty reduction perspective. Fash Text 10 (2023). https://doi.org/10.1186/s40691-022-00327-3
50. Joo, E., Yang, J.: How perceived interactivity affects consumers' shopping intentions in live stream commerce: roles of immersion, user gratification and product involvement. JRIM (2023). https://doi.org/10.1108/JRIM-02-2022-0037
51. Liao, J., Chen, K., Qi, J., Li, J., Yu, I.Y.: Creating immersive and parasocial live shopping experience for viewers: the role of streamers' interactional communication style. JRIM 17, 140–155 (2023). https://doi.org/10.1108/JRIM-04-2021-0114
52. Tian, B., Chen, J., Zhang, J., Wang, W., Zhang, L.: Antecedents and consequences of streamer trust in livestreaming commerce. Behav. Sci. (Basel, Switzerland) 13 (2023). https://doi.org/10.3390/bs13040308
53. Wu, D., Wang, X., Ye, H.J.: Transparentizing the "Black Box" of live streaming: impacts of live interactivity on viewers' experience and purchase. IEEE Trans. Eng. Manag. 1–12 (2023). https://doi.org/10.1109/TEM.2023.3237852

54. Yin, J., Huang, Y., Ma, Z.: Explore the feeling of presence and purchase intention in livestream shopping: a flow-based model. JTAER **18**, 237–256 (2023). https://doi.org/10.3390/jtaer1801 0013

55. Zhang, L., Chen, M., Zamil, A.M.A.: Live stream marketing and consumers' purchase intention: an IT affordance perspective using the S-O-R paradigm. Front. Psychol. **14**, 1069050 (2023). https://doi.org/10.3389/fpsyg.2023.1069050

56. Erdogan, B.Z.: Celebrity endorsement: a literature review. J. Mark. Manag. **15**, 291–314 (2010). https://doi.org/10.1362/026725799784870379

57. Riley, M.W., Hovland, C.I., Janis, I.L., Kelley, H.H.: Communication and persuasion: psychological studies of opinion change. Am. Sociol. Rev. **19**, 355 (1954). https://doi.org/10.2307/2087772

58. Ohanian, R.: Construction and Validation of a Scale to Measure Celebrity Endorsers' Perceived Expertise, Trustworthiness, and Attractiveness. J. Advert. **19**, 39–52 (1990). https://doi.org/10.1080/00913367.1990.10673191

59. Mat, W.R.W., Kim, H.J., Manaf, A.A.A., Ing, G.P., Adis, A.-A.A.: Young Malaysian consumers' attitude and intention to imitate Korean celebrity endorsements. AJBR **9** (2020). https://doi.org/10.14707/ajbr.190065

60. Dodds, W.B., Monroe, K.B., Grewal, D.: Effects of Price, Brand, and Store Information on Buyers' Product Evaluations. J. Mark. Res. **28**, 307 (1991). https://doi.org/10.2307/3172866

61. Schweiger, G., Schrattenecker, G.: Werbung. Einführung in die Markt- und Markenkommunikation: mit Expertenbeiträgen von Andreas Strebinger sowie Barbara Khayat und Stefan Schiel. UTB; UVK Brtlsh, Konstanz, München (2021)

62. Gupta, R., Kishor, N., Verma, D.: Construction-and-Validation-of-a-Five-Dimensional-Celebrity-Endorsement-Scale-Introducing-the-Pater-Model (2017)

63. Brosius, H.-B., Haas, A., Unkel, J.: Methoden der empirischen Kommunikationsforschung. Eine Einführung. Springer, Wiesbaden (2022). https://doi.org/10.1007/978-3-531-94214-8

64. Eckstein, P.P.: Angewandte Statistik mit SPSS. Springer, Wiesbaden (2016). https://doi.org/10.1007/978-3-658-10918-9

65. Braunecker, C.: How to do Empirie, how to do SPSS. Eine Gebrauchsanleitung. UTB GmbH; facultas, Stuttgart, Wien (2016)

66. Janssen, J.: Statistische Datenanalyse mit SPSS. Eine anwendungsorientierte Einführung in das Basissystem und das Modul Exakte Tests. Springer, Heidelberg (2017). https://doi.org/10.1007/978-3-662-53477-9

Are Dark Patterns Self-destructive for Service Providers?: Revealing Their Impacts on Usability and User Satisfaction

Toi Kojima[1], Tomoya Aiba[1], Soshi Maeda[1], Hiromi Arai[2],
Masakatsu Nishigaki[1] (ORCID), and Tetsushi Ohki[1,2](✉) (ORCID)

[1] Shizuoka University, Hamamatsu, Shizuoka, Japan
{kojima,aiba,maeda}@sec.inf.shizuoka.ac.jp,
{nisigaki,ohki}@inf.shizuoka.ac.jp
[2] Riken AIP, Chuo-ku, Tokyo, Japan
hiromi.arai@riken.jp

Abstract. "Dark patterns," which are known as deceptive designs that intentionally induce users to take actions that benefit the company, have been widely adopted, especially in the field of digital marketing. In recent years, there has been a global increase in efforts to address and regulate dark patterns. The main disadvantages of dark patterns are that waste time as well as money and are addictive. However, there are other possible unintended effects on the user experience. In particular, users who are not deceived by dark patterns understand the methods behind them and may experiences stress and frustration when they spend extra time and effort to avoid them. In this study, we focus on users who are not deceived by these dark patterns and on the detriment to usability caused to these users for avoiding dark patterns. Through this usability study using web pages containing dark patterns, we explored the possibility that the cost incurred by avoiding dark patterns may be a factor that undermines trust in a company.

Keywords: dark patterns · deceptive design · UI · UX

1 Introduction

With the development of the Internet, the availability of online services has increased. However, companies are increasingly using deceptive-design practices that deceive users into taking profit-maximizing actions, particularly in the area of digital marketing. Such designs have been prevalent on the internet for some time. Nevertheless, around 2010, they were labeled as "dark patterns" [4] and became widely recognized.

The problems caused by dark patterns are recognized as concerns worldwide. For example, in India, the government has taken steps to regulate the use of 13 types of dark patterns by e-commerce companies. [1] This reflects a global trend of increasing efforts to address and control such practices in various countries.

F. F.-H. Nah and K. L. Siau (Eds.): HCII 2024, LNCS 14720, pp. 66–78, 2024.
https://doi.org/10.1007/978-3-031-61315-9_5

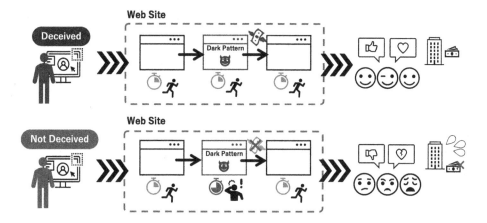

Fig. 1. Reactions to dark patterns from deceived and non-deceived users

Dark patterns are primarily used to maximize corporate profits. However, they can have detrimental effects on users, such as financial loss, exposure of personal information, and addiction to certain services [9]. They can also have unintended effects on users, particularly on those users who are not deceived by the patterns.

Users who are not deceived by dark patterns are assumed to understand the methods employed by dark patterns and to be cautious in their actions. Such users may invest additional time and effort in making decisions based on their intentions when using sites that employ these patterns, and this may lead to stress and dissatisfaction. Therefore, users who are not deceived by dark patterns are expected to experience disadvantages, unlike deceived users.

This chain of events can undermine corporate interests, which is contrary to the intended purpose of the patterns and poses a significant problem for companies as shown in Fig. 1.

This study aims to answer the following research questions:

RQ1 Are there differences in the disadvantages between users who are deceived by dark patterns and those who are not?
RQ2 What factors contribute to users being more susceptible to the negative effects of dark patterns?

This study addresses RQ1 and RQ2 by conducting task-based and question-naire surveys on web pages that contain dark patterns. The goal of this study is to quantitatively analyze the decrease in usability caused by dark patterns and to identify the factors that contribute to this decrease.

2   Related Work

2.1   Taxonomy of Dark Patterns

Existing research on dark patterns has developed primarily as studies focused on its classification. In the early stages of dark pattern research, Brignull [4]

classified and systematized ethically-problematic designs which deceive users. Many subsequent dark-pattern-classification studies have expanded on Brignull's work. For example, Gray et al. [2] extended Brignull's classification by classifying dark patterns into five categories based on strategic motivation. In this study, we selected dark patterns by using the classification prescribed by these studies and conducted our investigation.

2.2   Deceived Users and Methods of Deception

In addition to these classification studies, research has been conducted primarily on the tendencies of users who are deceived by these dark patterns [7]. Studies have also examined the types and persistence of dark patterns and their impact on user decision-making [5]. These studies have focused on users deceived by dark patterns and the deceptive techniques used in them.

Gunawan et al. [3] proposed a methodology to empirically measure the efforts that users may undergo in avoiding dark patterns. Although there are efforts being made to effectively regulate dark patterns, the difficulty of clearly delineating the maliciousness of dark patterns poses a challenge. To address this situation, Gunawan et al. hypothesized that the severity of dark patterns depends, in part, on the severity of the burden users experience when avoiding inducements by dark patterns. The user's effort to avoid such patterns was measured by indicators such as the number of clicks required to access privacy settings or account deletion options, pop-ups encountered, and pages navigated; time taken to complete privacy-related tasks in the presence of dark patterns; or level of navigation required to complete these tasks. They argued that the severity of dark patterns could be quantified by calculating the sum of these indicators.

In this study, we adopt Gunawan et al.'s metric [3] as one of the metrics of the disadvantages that users experience due to dark patterns.

3   Study Design

We conducted a task-based survey on August 5, 2023. Participants were recruited through the crowdsourcing service lancers.jp[1]. They were invited to complete tasks on a survey website created specifically for this investigation and to respond to the associated questionnaires. All participants were provided with an explanation about the research objectives, and their consent to participate in the survey was obtained.

3.1   Participants

We recruited 350 participants, and all of them participated. Of the total, 230 were male, 118 were female, and 2 did not provide gender information. Every participant, who were all Japanese, was paid 165 JPY as a reward for participating in the survey. We set the reward at an amount above the local minimum wage after measuring the median time to complete the task in a laboratory study.

---

[1] https://www.lancers.jp.

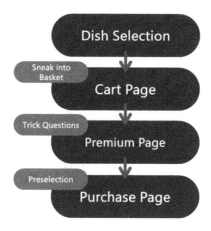

Fig. 2. The process of task-based survey

3.2   Research Method

Participants registered for the survey using their lancers.jp accounts on the dedicated survey request page within lancers.jp. Participants then anonymously performed the task and responded to the questionnaire on the survey webpage. Upon completion of the questionnaire survey, each participant was given a unique password to complete the task. To complete the entire task sequence, participants entered this password on the form created during the task request on lancers.jp. The average time per participant to complete the task survey, from the overview explanation to the display of the task completion password, was approximately 11 min.

3.3   Task-Based Survey

This task-based survey presented a simulation of a scenario, in which participants experienced the process of using a food delivery service on a website that included tasks from selecting dishes to completing an order. All participants were presented with the same website that used dark patterns, and their choices on the website were recorded. In addition, participants were informed of the presence of dark patterns in the post-task questionnaire and were asked whether they had noticed them. The purpose of this approach was to determine whether participants recognized these patterns and knowingly made choices that made them susceptible to these patterns, or whether they were deceived without recognizing these patterns.

The task sequence began with the selection of dishes on the page. After selecting the dishes for the order, participants moved to a cart page that displayed a list of currently-selected dishes. Participants could review the contents of the cart and then select a button to proceed with the order. Next, a page was displayed that promoted membership in a premium subscription. The benefits of becoming a premium member were presented, and the participant had to decide whether or not to subscribe. After they made this choice, the participant was directed to the final purchase page to review the order details. This page included the terms of service, and the participant could finalize the order. This action completed the sequence of tasks shown in Fig. 2. The participant then transitioned to the questionnaire survey. Hereafter, we denote three pages of our main tasks, a page for shopping cart, a page for inviting premium members, and a page for purchasing, as `Cart`, `Premium`, and `Purchase`, respectively.

It should also be noted that the task-based survey only simulated the process of ordering dishes, and there was a potential for behavioral differences between the survey and actual-ordering scenarios. To create a more realistic situation, the survey scenario was designed for a party of six people, and the participants were instructed in advance to order dishes for the group.

3.4 Employed Dark Patterns

The dark patterns used in this survey were selected from various categories elaborated by previous studies [2,4]. In this study, we selected three representative dark patterns that might be expected in the context of delivery services: Sneak into Basket, Trick Questions, and Preselection.

Sneak into Basket: This dark pattern is set in `Cart` and involved adding items to the cart without the consent of the user. This action requires the user to notice the unauthorized items in the cart and manually remove them. In this task survey, users were presented with `Cart` to confirm the items after completing the dish selection. A bottle of iced tea that the user did not order was silently added to the cart. Figure 3 shows the actual dark pattern that was used in the survey.

Trick Questions: This dark pattern is set in `Premium` and guided users to specific choices by using confusing language in questions, making them carefully examine the content of the text to determine the appropriate option. In this task survey, a question on `Premium` encouraging a free trial was phrased as "Wouldn't you not like to try the 3-month premium membership free trial?" This wording aligned "No" with subscribing and "Yes" with not subscribing. The actual dark pattern used in the survey is shown in Fig. 4.

Preselection: This dark pattern is set in `Purchase` and involved "by-default-preselected" options to subscribe to newsletters or paid memberships, which forces users to deselect them if they do not intend to subscribe. In this task survey, a small checkbox for the newsletter was placed on `Purchase`, below the detailed terms and conditions, and was checked by default. The actual dark pattern used in the survey is shown in Fig. 5.

Fig. 3. Example of Sneak into Basket

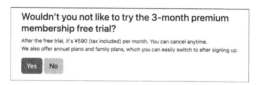

Fig. 4. Example of Trick Questions

3.5   Questionnaire Survey

This study evaluated the impact of dark patterns on usability using both objective and subjective evaluation metrics for RQ1 and RQ2. Objective metrics included task-completion times, whereas subjective metrics involved evaluations obtained through a post-task survey. To objectively assess the effort associated with the user avoiding dark patterns, we measured task completion times using the method proposed by Gunawan et al. [3]. Task completion time was defined as the time between navigating from one page to the next on the website. For example, the time spent on Cart until transitioning to Premium was considered to be the task completion time for Cart.

Table 1 shows the questionnaire items used in the post-task survey. Post-task surveys were used to derive the assessment metrics for evaluating the usability of the service. The questions were adapted from the Web Usability Scale (WUS) [8], which is a well–established scale for quantifying web usability. Participants were asked to rate their perceptions of each screen using a 5-point Likert scale, ranging from "Strongly Agree (5 points)" to "Strongly Disagree (1 point)," without mentioning of the existence of dark patterns. In addition, the Net Promoter Score (NPS) [10], which is a customer satisfaction rating scale, was employed. Participants were asked to rate for each page, on a scale of 0 to 10, how likely they would be to recommend the service to friends or colleagues.

To assess whether participants were deceived by the dark patterns, a set of questions was posed. We informed participants about each dark pattern and asked them to indicate whether they had noticed the tactics. We offered a binary

Fig. 5. Example of Preselection

Table 1. Questionnaire items used in the survey

Category	Content	Number of Questions
Overview of the task survey	Explanation of the study and confirmation of consent to participate	
Subjective metrics evaluation	Questions related to usability assessment of the service (WUS)	63
	Questions related to satisfaction assessment of the service (NPS)	3
Evaluation of deception	Questions about the intention behind the selected actions	3
Literacy assessment	Questions about web access literacy	21
Demographic	Questions about gender, age, and education level	3
Attention test (DQS) [6]	Question to verify whether participants minimize effort	1

choice of "Yes" or "No" in the survey. We also included demographic questions and queries related to internet literacy to analyze the potential influence of individual differences in susceptibility to these tactics. The questions on internet literacy were derived from a subset of the Web Access Literacy Scale, which was developed by Yamamoto et al. [11].

3.6   Methods for Distinguishing Users Deceived by Dark Patterns from Those Not Deceived

In this section, we define the criteria for distinguishing between users who were deceived by dark patterns (D group) and those who were not (ND group). The determination of whether a user was deceived was based on two criteria: whether they made choices that the company was leading them to make on the website and whether they were aware of the dark pattern.

The users were classified into the D group based on this criteria: if a user made choices that the company was leading them to make on the website and answered "did not notice" to the question about whether they were aware of the dark pattern, we classified them as "deceived" and placed them in the D group. Even if a user made choices that did not benefit the company on the website but answered "did not notice the dark pattern" in the questionnaire, we assumed that they accidentally avoided the dark pattern and could still have been deceived, thus meeting the definition of the D group.

Next, the users were classified into the ND group on the basis of not making choices that the company was leading them to make on the website and answered

Table 2. Results of the Mann–Whitney U test for each indicator for each page

Page	Elapsed Time	WUS	NPS
Cart	$p \ll 0.001^{**}$	$p \ll 0.001^{**}$	$p \ll 0.001^{**}$
Premium	$p \ll 0.001^{**}$	0.00144^{*}	0.0384^{*}
Purchase	$p \ll 0.001^{**}$	0.221	0.00260^{*}

"noticed" to the question about the dark-pattern awareness. However, if a user answered "noticed the dark pattern" in the questionnaire but made choices that benefited the company, we assumed that the user found the choice appealing and selected it with intent. Therefore, we excluded such users from both the D and ND groups from the analysis.

4　Result

An analysis was conducted to determine if there were differences in the value of the evaluation metrics between the D and ND groups. Furthermore, we examined the relationships between each evaluation metric and the demographic factors and web literacy. For our analysis, we used the responses from 342 participants (224 males, 116 females, and 2 participants who did not disclose their gender) of the 350 survey respondents. Eight individuals who provided incorrect responses in the attention test, Directed Question Scale (DQS) [6], were excluded from the analysis. With respect to the number of participant in the D and ND groups in each pages, Cart, Premium, Purchase, 141 were classified to the D group as having been deceived by Sneak into Basket in Cart, whereas 29 were classified to the ND group. In the case of Trick Questions in Premium, 246 participants were classified to the D group, whereas 52 were classified to the ND group. For Preselection in Purchase, 200 participants were classified to the D group and 79 were classified to the ND group. We performed Mann–Whitney U test for each group because the Shapiro–Wilk test indicated nonnormality for both groups.

4.1　Task Duration

We tested the task duration against the D and ND groups using Mann–Whitney U test with a significance level of 0.05. If we reject the null hypothesis of no difference between the two groups, this would imply a significant difference in task duration and suggest that the ND group spent significantly more time completing the task. Figure 6(a), 6(b), and 6(c) present the average and standard deviation for the D and ND groups in Cart, Premium, and Purchase, respectively. Table 2 summarizes the p-values obtained from the tests. At a significance level of 0.05, there was a statistically significant difference between the D and ND groups for all pages, where $p < 0.001$ for all tests. This suggests that participants who were not deceived by dark patterns spent significantly more time on each page.

4.2   Usability and Satisfaction Evaluation

We conducted an analysis using the WUS questionnaire responses [8] as indicators and performed the Mann–Whitney U test. As we did with the elapsed time indicator, we developed the null hypothesis that there was no difference between the two groups. Rejecting this null hypothesis would indicate a difference between the two groups and demonstrate that the ND group rated usability of the page significantly lower. The average and standard deviation of the D and ND groups for `Cart` are shown in Fig. 7(a), for `Premium` in Fig. 7(b), and for `Purchase` in Fig. 7(c). The obtained significance probability p is presented in Table 2. Table 2 shows a significant difference between the D and ND groups for `Cart` and `Premium` with significance lefel of 0.05. However, no significant difference was observed for `Purchase`. This indicates that users who were not deceived rated lower usability than the deceived users for `Cart` and `Premium`.

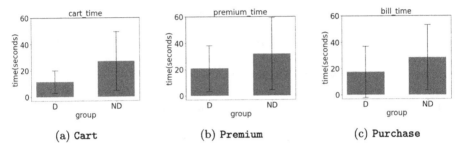

| (a) `Cart` | (b) `Premium` | (c) `Purchase` |

Fig. 6. Mean and standard deviation of the D and ND groups for elapsed time on each page

4.3   Net Promoter Score Evaluation

We conducted an analysis using the responses to the NPS questionnaire [10] as indicators of customer satisfaction. As we did with the elapsed time and WUS indicators, we employed the Mann–Whitney U test, using the null hypothesis that there was no difference between the two groups. Rejecting this null hypothesis would indicate a difference between the two groups, demonstrating that the ND group significantly rated customer satisfaction as lower than those by the D group. The average and standard deviation of the D and ND groups for `Cart` are shown in Fig. 8(a), for `Premium` in Fig. 8(b), and for `Purchase` in Fig. 8(c). The obtained significance probability p is presented in Table 2. At a significance level of 0.05, there is a significant difference between the D and ND groups for all pages, which indicates that satisfaction with each page was significantly lower for users who were not deceived than those who were deceived.

4.4   Individual Differences

We conducted an analysis of the individual differences in the ND group and their susceptibility to disadvantages. First, we performed multiple regression analysis using gender, age, and education level as the independent variables and elapsed time, usability evaluation (WUS), and satisfaction evaluation (NPS) as the dependent variables. However, no significant factors were identified.

Furthermore, we conducted Mann–Whitney U tests between the D and ND groups on the assumption that user awareness of dark patterns is related to internet literacy. The obtained significant probabilities (p) are presented in Table 3. At a significance level of 0.05, no significant differences were observed for any page. Therefore, it can be concluded that there was no difference in internet literacy between users who were deceived and those who were not.

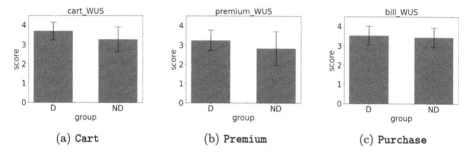

(a) Cart   (b) Premium   (c) Purchase

Fig. 7. Mean and standard deviation of the D and ND groups as per the Web Usability Scale for each page

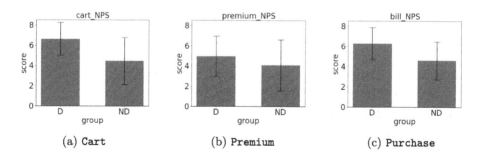

(a) Cart   (b) Premium   (c) Purchase

Fig. 8. Mean and standard deviation of the D and ND groups of the Net Promoter Score for each page

5   Discussion and Limitations

5.1   Discrepancy in Disadvantages Between Users Who Were Deceived and Those Who Were Not

From Table 2, significant differences were observed between the D and ND groups for all items except for the WUS on `Purchase`. Phenomena such as increased task duration, decreased usability, and reduced user satisfaction were observed among users who were not deceived by the dark patterns. These phenomena are considered to be the result of the efforts required to avoid dark patterns. Users who were not deceived by dark patterns likely refrained from making intuitive or suggested choices to avoid them. For example, the Trick Questions used in `Premium` resulted in interfaces that were less intuitiveness and imposed a higher cognitive load, complicating intuitive decision-making and prolonging service use duration. Moreover, users who struggled with intuitive choices challenging generally rated usability lower than those who did not encounter such difficulty. These factors collectively contributed to a decrease in user satisfaction.

In open-ended responses, participants expressed reactions such as "losing trust in the website due to discovery of dark patterns and becoming more skeptical when interacting with websites." The presence of dark patterns contributed not only to decreased usability and satisfaction but also to a diminished trust in the website or service among all users.

Table 3. Results of Mann–Whitney U test for Internet literacy for each page

Page	p-value
Cart	0.813
Premium	0.460
Purchase	0.164

5.2   Factors Contributing to Increased Susceptibility to Disadvantages Caused by Dark Patterns

In Sect. 5.1, a difference between the D and ND groups was confirmed. We hypothesized that a potential factor contributing to this difference is a disparity in internet literacy levels between the D and ND groups. However, as shown in Table 3, no significant difference in internet literacy was observed between users who were deceived and those who were not. Contrary to our initial hypothesis, it implied that users can notice dark patterns independent of their internet literacy skills. In the open-ended sections, several responses were observed indicating that participants were deceived due to a habitual lack of attention or carelessness, while others mentioned that they were able to avoid being deceived because they recognized the tactics from previous experiences. In addition, some responses indicated that complete trust in a website led to deception. These findings highlight the complex interplay between user behavior, experience, and trust in the digital environment's susceptibility to dark patterns.

5.3 Limitations

According to Sect. 5.2, a susceptibility to dark patterns may be influenced by the trustworthiness of websites or services. However, the task-related investigation related to dark patterns conducted in this study involved participants being informed that the tasks simulated a pseudo-service, potentially leading to a perception of trustworthiness that was distinct from that experienced during actual service use. Consequently, if a similar investigation was conducted while participants were engaging with real services, the results could differ. To address this consideration, some existing studies, such as [5], have employed deception by misleading participants into believing that the study involved real services. Therefore, conducting investigations in a "deception study format" may mitigate these influences.

In this study, we employed Sneak into Basket, Trick Questions, and Preselection as dark patterns. However, because there are various types of dark patterns, each with different characteristics in terms of timing, psychological effects, cognitive load, etc., we anticipate that conducting a similar investigation with a different set of dark patterns could yield results distinct from those obtained in this study. Furthermore, the choice of dark patterns may inform the indicators that could be used for validation. Therefore, we believe it is necessary to conduct future validations using alternative dark patterns to comprehensively explore the landscape of deceptive user experiences.

5.4 Ethical Consideration

The questionnaire administered in this study was administered following an assessment of its content and procedures in accordance with the ethical standards established by the ethics committee of the organization to which the correspondence author belong. It was confirmed that the study fell within the required scope and did not require formal ethical review. Before they participated, survey participants were informed about the questionnaire content and voluntarily joined the study by their own free will. In addition, participant incentives were determined by estimating the survey time from lab and pilot surveys and considering the minimum wage in the survey region. The handling of personal information was in compliance with Japan's Personal Information Protection Act.

6 Conclusion

In this study, we conducted both task-based and questionnaire surveys on users who were deceived by dark patterns and those who were not. By analyzing the survey results, we aimed to reveal the presence of specific disadvantages for users who were not deceived by dark patterns by examining the time elapsed during website interactions and conducting usability evaluations of the services.

Our analysis of survey results indicated that, in many cases, users who were not deceived by dark patterns experienced significant disadvantages compared with those who were deceived. This suggests that avoiding dark patterns causes unintended negative consequences for users. The results highlight the specific disadvantages for users who were not deceived by dark patterns and suggest the potential consequences of prioritizing short-term gains in digital marketing, which could lead to long-term risks and loss of trust.

By understanding that the use of dark patterns by companies not only brings short-term benefits but also introduces the risk of customer disengagement, we hope that companies can achieve mutually beneficial service provision for both the company and the users.

References

1. Dipak, K., Dash, A.D.: Government bans e-commerce firms from using 'dark patterns' | india news - times of india (Dec 2023). https://timesofindia. indiatimes.com/india/government-bans-e-commerce-firms-from-using-dark-patterns/articleshow/105668112.cms. Accessed 02 Feb 2024

2. Gray, C.M., et al.: The dark (patterns) side of ux design. In: Proceedings of the 2018 CHI Conference on Human Factors in Computing Systems, pp. 1–14 (2018)

3. Gunawan, J., et al.: Towards an understanding of dark pattern privacy harms. In: Position Paper at the CHI 2021 Workshop: What can CHI do About Dark Patterns (2021)

4. Harry, B.: Deceptive design - formerly darkpatterns.org. https://www.deceptive. design/. Accessed 26 July 2022

5. Luguri, J., et al.: Shining a light on dark patterns. J. Legal Anal. **13**(1), 43–109 (2021)

6. Maniaci, M.R., Rogge, R.D.: Caring about carelessness: Participant inattention and its effects on research. J. Res. Pers. **48**, 61–83 (2014)

7. Nagda, Y.: Analyzing ocular parameters to investigate effect of dark patterns on HCI (2020)

8. Nakagawa, K.: The development of questionnaire for evaluating web usability. Proceedings of the 10th Human Interface Society of Japan, Oct. 2001, pp. 421–424 (2001). https://cir.nii.ac.jp/crid/1571135650780718336

9. Narayanan, A., Mathur, A., Chetty, M., Kshirsagar, M.: Dark patterns: past, present, and future: The evolution of tricky user interfaces. Queue **18**(2), 67–92 (2020)

10. Reichheld, F.F.: The one number you need to grow. Harvard business review **81** **12**, 46–54, 124 (2003). https://api.semanticscholar.org/CorpusID:1681767

11. Yusuke, Y., Takehiro, Y., et al.: Development of a web access literacy scale. Trans. Inform. Process. Society Japan (TOD) **12**(1), 24–37 (January 2019). https://cir. nii.ac.jp/crid/1050001337909767296

Designing Inspiration: A Study of the Impact of Gamification in Virtual Try-On Technology

Sebastian Weber$^{(\boxtimes)}$, Bastian Kordyaka, Marc Wyszynski, and Bjoern Niehaves

University of Bremen, Bibliothekstr. 1, 28359 Bremen, Germany
{sebweber,kordyaka,m.wyszynski,niehaves}@uni-bremen.de

Abstract. Creating innovative, user-centered, compelling services to stand out in the competition is more important than ever today to sustain for service provider. In this line, this study investigates the impact of gamification features in virtual try-on applications on user's motivation in the hairdressing domain. More specifically, it examines how gamification elements influence user motivation in the form of inspiration, employing the Stimulus-Organism-Response (S-O-R) model as a theoretical framework. A mixed-method methodology was adopted, beginning with a qualitative study to identify relevant gamification features in the context of virtual try-on applications followed by an cross-sectional online study to exploratory analyze their effects on inspiration. The cross-sectional online study with 201 participants indicates ambivalent influences of gamification features on inspiration and emphasize the complexity of the underlying psychological mechanisms of the gamification concept. Nevertheless, our findings provide valuable insights that can be used in the design of the development of innovative playful artifacts.

Keywords: gamification · inspiration · service encounter · virtual try-on

1 Introduction

In times of globalization, crises and digitization, competition is constantly increasing. Positive unique selling points are therefore a necessary condition for success in the service industry. The use of innovative technology and user-centered design in service encounters is a vital part of addressing this. Current technologies such as artificial intelligence (AI), virtual reality and augmented reality (AR), for example, offer technological opportunities. Furthermore, so-called virtual try-on (VTO) apps are an interesting possibility in times of ever-increasing online shopping and hybrid services (i.e., complimentary service delivery through conventional and technology-enabled environments) [15,33,39]. With regard to user-centered design, gamification offers a promising approach to successfully design services that motivate users to use the technology continuously [7,13,35].

With regard to technology design, however, it is highly relevant to look closely at how gamification elements affect motivation in detail. A suitable motivational

F. F.-H. Nah and K. L. Siau (Eds.): HCII 2024, LNCS 14720, pp. 79–90, 2024.
https://doi.org/10.1007/978-3-031-61315-9_6

component here is inspiration, as it has a concrete object reference (e.g., testing a new product) [2,31,35]. Research in this regard has been sparse. Hence, this study focuses on this relationship and investigates the consequences of a gamified VTO on users' inspiration. For this we use the service domain of hairdressers, as it is well suited in the context of VTOs due to the visualization options and the potential of a unique selling point. We use the Stimulus-Organism-Response (S-O-R) model as a theoretical framework [16], as it examines technology design (S - Stimulus), psychological processes such as inspiration (O - Organism), and the possible reaction of consumers (R - Response; which we do not research in this study). To this end, we are conducting a cross-sectional study using covariance-based statistics (i.e., regression analysis). In summary, we aim to answer the following research question (RQ):

RQ: What are relevant gamification features to inspire users of a VTO?

With our study, we provide important results for the theoretical context of gamification and inspiration. On a practical level, the results can provide insight into how technology design can contribute to positive unique selling propositions. The rest is structured as follows: First, we describe the theoretical background by describing the context of service encounters and VTOs as well as the theoretical S-O-R framework including gamification and inspiration. Next, we describe our mixed-method methodology. This is followed by Study 1 to identify key gamification elements and Study 2 to investigate the influence of key gamification elements on inspiration. We then discuss the results on a theoretical and practical level and conclude with a short summary.

2   Theoretical Background

2.1   Service Encounter and Virtual Try-On

The design of a service encounter is considered a core issue to meet customers expectation and consequently affects economic outcomes [34]. In this line, the literature provided several definitions. For example, it was defined as a face-to-face interaction between service provider and consumer during the process of service consumption [28]. However, literature also offered a broader view by defining it as "a period of time during which a consumer interacts with a service" [26]which also reflects the other aspects including digital and hybrid service experiences. Based on the changing consumer behavior this viewpoint is needed as a large proportion of users today tend to shop online rather than in a brick-and-mortar stores. In this line, recent technological developments such as Augmented Reality (AR) and Artificial Intelligence (AI) offer new ways to interact with consumers and provide a new service experience that can lead to "spontaneous delights" [1] and as such offer unique selling points in today's competitive environment. Such a service can be vital for both, online as well offline shops. Based on these technologies, big retailers like Ikea or Amazon already offer VTO service experiences [18] to try products virtually (e.g., furniture or fashion) to overcome the so-called fit and match dilemma [20]. However, this dilemma is not only evident in online shopping but also in other service contexts like hairdressing where customers cannot imagine new products (i.e., hairstyles). Several studies already highlighted the potential of VTOs in this realm and studied the adoption behavior [15,22,29,39]. Nevertheless, when service delivery happens increasingly online or hybrid, it is crucial to furthermore understand how service provider can offer a joyful service experience over time to keep consumers and offer them new inspiration for products. In this regard, gamification seems a viable approach for VTOs that has been neglected so far.

2.2   The Stimulus-Organism-Response Model

To explore and understand a gamified VTO, we leverage the S-O-R framework from environmental psychology [16]. This model suggests that specific stimuli can influence a consumer's cognitive and affective processes (referred to as the organism), which subsequently shape the consumers's actions or reactions. In the realm of a gamified VTO, the stimuli on which we focus are the design elements (i.e., gamification features) with which the consumer interacts. These elements serve as external cues that capture users' attention and engage them in the VTO process (e.g., by nudging them to try new hairstyles). The organism, in this context, represents the users' affective and cognitive processes leading to a response. The cognitive aspect involves users' mental engagement and thought processes while the affective component reflects the users' feelings and emotions. The response could manifest through a range of behaviors and attitudes [11]. Adopting the S-O-R framework offers several benefits for our investigation into the impact of gamification on user inspiration within a VTO context. Firstly, it provides a theoretically grounded approach to assess how gamified elements in VTOs act as environmental stimuli. Secondly, it facilitates an analysis of users' psychological processes of the evocation of inspiration based on gamified VTO interactions. Lastly, it underpins a theoretical basis for evaluating the value of a gamified VTO (in future studies) as a result of these processes in the organism.

Gamification as Stimuli. Gamification typically involves integrating game design elements into non-game contexts to enhance user engagement and motivation [5,37]. This approach aims to personalize the user experience by aligning with individual preferences and maximizing engagement opportunities. For the purpose of our paper, we follow the definition of Hamari et al. [7] in which gamification is a process of enhancing services with (motivational) affordances to invoke gameful experiences and further behavioral outcomes of a gamified system [7]. Studies have demonstrated the potential of gamification to elevate user engagement and participation [14,17] in various contexts such as education [4], health [12], marketing [10], and societal impact [6]. Gamification features have typically been grouped into three motivational categories, namely: achievement-related gamification features, immersion-related gamification features, and social-related gamification features [36]. Table 1 defines these dimensions and highlights exemplary gamification features for each dimension based on existing literature [13].

Much literature on gamification has utilized this tri-dimensional structure for categorizing game features [8,9,13,21,23,27,38].

Table 1. Gamification dimensions

Dimension	Definition	Exemplary features
Achievement	Achievement-based features are game design features that are primarily aimed at amplifying the player's feelings of accomplishment and success within the game	Points, badges, or levels
Immersion	Immersion-related features refer to the various components and aspects within a game or interactive experience that are designed to deeply engage and absorb the player, fostering autonomous exploration and curiosity	Avatars, storytelling, or role-playing mechanics
Social	Social-related features in a game are primarily designed to facilitate and enhance user interactions within the gaming community	Multiplayer, networking features, or competition

In the context of our study, there appears to be a gap in understanding how environmental cues like gamification features relate to inspiration. Previous research has shown that gamification platforms can influence users' cognitive and affective system, influencing behavioral outcomes in a desired manner [13]. While there is already evidence in the marketing literature that other stimuli (e.g., advertising) can trigger inspiration [2], there is still a lack of research on how gamification features might evocate the motivational state of inspiration and further responses which is especially true in the context of our study of VTO-based services.

Inspiration as Organism. Within the framework of the S-O-R model, the organism encompasses both the affective and cognitive reactions to a stimulus. Understanding these reactions is crucial as they form a core part of the user's experience in such new service encounters. In our study, we focus on the motivational state of inspiration, which is defined as a motivational state compelling users to realize ideas [19]. This state comprises internal emotional and cognitive response to an external stimulus. Hence, research divides this process of inspiration in "inspired by" and "inspired to" [2,31,32]. As we want to understand the process of beeing inspired, we focus on the state of beeing "inspired by" (the term inspiration refers to "inspired by" from now on). This state of beeing inspired by has an elicitor object which induces a specific emotion, especially a self-transcendent emotion like elevation, admiration, or awe [30]. In this study, we focus on this affective change based on the stimuli of gamification elements which is a crucial first step to understand and explain the possible inspirational value of gamified VTOs.

3   Mixed-Methods Approach

To better understand the relationships between gamification features in our VTO app and inspiration, we used a sequential quantitative design [3]. First, in our Study 1, we reduced a rich list of 46 gamification features in a data-driven manner to identify a set of gamification features relevant to the context of our study the VTO app in the service industry. Second, in our Study 2, we tested the influences of the identified relevant gamification features from Study 1 concerning inspiration in an exploratory manner.

4   Study 1: Identifying Relevant Gamification Features

4.1   Methodology

Data Analysis and Procedure. To identify relevant gamification features in the context of the service industry, we followed a two-step procedure combining literature work and judgments of gamification experts and experienced software developers. Initially, we searched the literature for a study comprising a large portfolio of gamification features (step 1). Following this, in a second step, we used the identified portfolio of gamification features and presented it to a group of four participants asking them to evaluate whether or not the corresponding gamification feature could be relevant in the VTO app (step 2). Based on this evaluation, we shortened the list of gamification features.

Data Collection and Participants. Below, we describe our sampling concerning the empirical part of Study 1 (step 2 of the procedure). To conduct step 2, we ensured that the gender and nation of participants differed. Accordingly, we contacted two male-identifying and two female-identifying participants, where the ages ranged from 28 to 41. As a profession, two participants reported to work as a software developers for several years and the other two participants reported to work as a gamification researcher.

4.2   Results

In the following, we illustrate the identification of relevant gamification features based on the two-step approach described in the following.

Step 1: Finding a Gamification Feature Portfolio. Based on a holistic literature search screening reviews related to gamification features, we selected a study from Koivisto and Hamari [13] proposing a portfolio of 46 gamification features (they called digital affordances) that seemed suitable for our study. Within that study the list of gamification features consisted of (a) ten achievement-oriented features (i.e., points, challenges, badges, leaderboards, levels, performance stats, progress, quizzes, timer, and increasing difficulty), (b) seven social-oriented features (i.e., social networking features, cooperation, competition, peer-rating, customization, multiplayer, and collective voting), (c) five immersion-oriented features (i.e., avatar, narrative, virtual world, in-game rewards, and role play), (d) eight real world-related features (i.e., financial reward, check-ins, motion tracking, physical cards, physical playboards, (e) real world interactive objects, physical objects, and physical dice), and (f) sixteen miscellaneous features (i.e., board games, virtual helpers, virtual currency, reminders, retries, onboarding, adaptive difficulty, game rounds, warnings, penalties, game slogans, funny movies, virtual pets, trading, making suggestions, and virtual objects as augmented reality).

Step 2: Identifying Gamification Features for VTO. Based on the original portfolio of 46 gamification features, we selected the features all four participants agreed upon that were rationale and feasible in a VTO. Only in three cases did different answers occur. We resolved these through a discussion in a joint call following the evaluation, after which no disagreements were left. As a result, we shortened the list to 19 gamification features by excluding 27 of the original portfolio. In particular, the final list of gamification features included (1) Points, Score, XP, (2) Challenges, Quests, Missions, Tasks, Clear Goals, (3) Badges, Achievements, Medals, Trophies, (4) Leaderboards, Rankings, (5) Levels, (6) Performance Stats, Performance Feedback, (7) Social Networking Features, (8) Cooperation, Teams, (9) Competition, (10) Peer-rating, (11) Customization, Personalization. (12) Avatar, Character, Virtual Identity, (13) In-game Rewards, (14) Motion Tracking, (15) Assistance, Virtual Helpers, (16) Virtual Currency, (17) Reminders, Cues, Notifications, Annotations, (18) Making Suggestions, and (19) Virtual Objects as Augmented Reality.

5   Study 2: Analyzing Relationships

5.1   Methodology

Data Analysis and Procedure. To analyze the relationships in our Study 2, we used a cross-sectional survey collecting self-reported data using an online questionnaire. Subsequently, we analyzed the data with covariance-based statistics (i.e., regression analyses) and widespread software applications (i.e., SPSS 28). For this, we tested the exploratory potential of the identified list of relevant gamification features concerning inspiration. For this, we presented the participants in a sequential manner a VTO (https://www.eyeconic.com/help-me/virtual-try-on?start=90), a description and exemplary instantiations of gamification features, and correspondings questions (i.e., importance of gamifications features, inspiration, and demographics).

Data Collection and Participants. To test the relationships, we used a digital questionnaire to collect data from technology application users via the crowd-sourcing marketplace Prolific. After cleaning the data and excluding three cases with missing data the final sample consisted of 201 participants. All participants received USD 1.20 as a reward for participating in our study. On a level of characteristics, 51% of participants identified as female (102), followed by 48%

who identified as male (97), and less than 1% reported other as their identification (2). Additionally, 29.4% of participants were between 36 and 50 years old, 62.3% held (at least) a bachelor's degree (121), and 35% reported their income to be between USD 25.000 and 49.999 a year.

Measurements. Following the best practices of psychometric research, we build a digital questionnaire using empirically validated scales and items from previous research wherever available, asking participants for their self-reported perceptions and behaviors regarding the VTO app.

First, we referred to our list of 19 relevant gamification features of Study 1 to measure gamification. For this, we asked participants, "Please rate the importance of interacting with the gamification feature listed below while using the Virtual Try-On app". For their responses, we provided a scale ranging from 1, "not important," to 5, "very important," in accordance with previous research [36]. Table 2 illustrates the descriptive statistics of all 19 gamification features.

Table 2. Descriptives of Gamification Features.

Number	Gamification features	M	SD
1	Points, Score, XP	2.32	1.32
2	Challenges, Quests, Missions, Tasks, Clear Goals	2.18	1.29
3	Badges, Achievements, Medals, Trophies	1.99	1.26
4	Leaderboards, Rankings	2.12	1.29
5	Levels	2.20	1.32
6	Performance Stats, Performance Feedback	2.86	1.40
7	Social Networking Features	2.37	1.28
8	Cooperation, Teams	2.27	1.23
9	Competition	2.13	1.29
10	Peer-rating	2.59	1.29
11	Customization, Personalization	3.87	1.16
12	Avatar, Character, Virtual Identity	2.81	1.42
13	In-game Rewards	2.50	1.35
14	Motion Tracking	3.91	1.16
15	Assistance, Virtual Helpers	3.50	1.16
16	Virtual Currency	2.06	1.27
17	Reminders, Cues, Notifications, Annotations	2.22	1.26
18	Making Suggestions	3.47	1.04
19	Virtual Objects as Augmented Reality	3.48	1.26

Second, following previous work related to inspiration [31], we measured the reflective scale *inspired by* ($M = 5.50, SD = 1.01, \alpha = .86$) with the aid of five items each asking participants "How much do you agree with the subsequent statements on a scale from 1 "strongly disagree" to 7 "strongly agree" using the arithmetic mean of the scale. The subsequent Table 3 summarizes all wordings of the items and descriptive values of each item.

Table 3. Inspired by items and descriptives.

Item	Wording	M	SD
1	...my imagination would be stimulated for a new hairstyle	5.54	1.28
2	...I would be intrigued by a new hairstyle	5.53	1.17
3	...I unexpectedly and spontaneously would get new hairstyle ideas	4.82	1.54
4	...my hairstyle knowledge would be broadened	5.68	1.20
5	...I would discover new hairstyles	5.96	1.00

5.2   Results

To test the influences of gamification, we conducted a multiple linear regression analysis specifying the 19 identified *gamification features* and the three demographic variables *gender, age, and education* as independent variables to explain the dependent variable *inspired by*. Checking the assumptions of linearity, auto-correlation, and multi-collinearity, neither the scatter plots, nor the *Durbin-Watson statistic* ($DW = 2.05$) seemed to be problematic [24]. However, the *Variance Inflation Factor* of the gamification feature *Badges, Achievements, Medals, Trophies* indicated a concerning value above the recommended threshold of 4 with a value of ($VIFs = 4.95$). After discussing this with the group of authors, we decided to exclude the gamification feature and re-run the analysis with only 18 gamification features. Conducting another multiple linear regression analysis with only 19 *gamification features* and the three demographic variables *gender, age, and education* as independent variables the assumptions of linearity, auto-correlation, and multi-collinearity were met because neither the scatter plots, nor the *Variance Inflation Factors* ($VIFs \leq 3.67$) nor the *Durbin-Watson statistic* ($DW = 2.05$) indicated problematic values [24]. Accordingly, we assumed that our data appeared suitable for regression analysis. The regression equation showed a significant result ($F(21; 179) = 6.86; p < .001$) that explained 38% of the variance of *inspired by*. Furthermore, the four predictor weights of the gamification features *customization, personalization* ($\beta = .17, p < .05$), *virtual currency* ($\beta = -.17, p < .05$), *reminders, cues, notifications, annotations* ($\beta = -.19, p < .05$), and *making suggestions* ($\beta = .22, p < .01$) as well as the two demographic variables *gender* ($\beta = -.13, p < .05$) and *age* ($\beta = .21, p < .001$) played a significant role in explaining *inspired by* (all others $p \geq .06$).

6   Discussion

6.1   Key Findings

We summarize the insights of our study with the subsequent three points:

- First, the gamification features customization and suggestions both had an positive influence on inspired by.
- Secondly, to our surprise, the gamification features virtual currency and reminders both had an negative influence on inspired by.
- Third, female and older participants were more likely to be inspired by gamification design.

6.2   Theoretical and Practical Implications

Based on our results, several implications can be derived that are relevant for existing UX and HCI research on a theoretical level. We will discuss some of them below. First of all, it should be noted that the results of the influence of the gamification feature contain ambivalent results (contrary to our expectations). On the one hand, the two gamification features customization and suggestions showed a positive influence on inspiration, while virtual currency and reminders had a negative influence. This highlights that gamification needs to be tailored and has not always mono-causal influences [25]. In terms of content, we summarize these results in such a way that opportunities to involve potential users in the design process are of particular relevance in the case of novel VTOs, as they can reduce uncertainties with regard to the final service and, thus, foster the inspirational potential intention to try new products. We explain the negative influence of the two gamification features virtual currency and reminders by the hedonic nature of the VTO hairdressing service and the rather utilitarian prompts of the two features. On this base, we see these results as an indication to critically reflect on existing gamification taxonomies' in order to improve the overall user experience and contribute to the success of interactive systems [12]. Furthermore, the findings that female and older participants were more likely to be inspired by gamification design prompt a reexamination of psychological and cognitive development theories to understand why certain gamification elements appeal more to older individuals. This could involve exploring cognitive aging processes and the impact on motivational and inspirational factors.

In addition, our results indicate added value for game and app developers to prioritize and invest in customization features and intelligent suggestion algorithms. This can increase engagement, satisfaction and overall enjoyment of the game or app experience. In summary, the practical implications of these insights are manifold and can be applied across different industries and sectors. Incorporating customization and suggestion features into gamification can lead to engaging, personalized and inspiring user experiences, whether in gaming, education, corporate training, health, marketing or other interactive contexts.

6.3   Limitations and Outlook

As in any empirical study, the procedure in our study was not possible without limitations. We would like to list some of these below in order to give the reader the opportunity to adequately classify our results. Firstly, and this is certainly the most substantial challenge, we are currently in the process of building the technological artifact and had to refer to a hypothetical playful artifact in the context of our study, which naturally limits the empirical insights. Furthermore, we collected our sample via Prolific for reasons of feasibility. Future studies should compare the results of our study with an offline sample with an existing technological artifact. Second, we chose a cross-sectional approach for our study. As a further empirical finding, future studies should look at artifact usage over time. In addition, some limitations arise in connection with our chosen theoretical framework, the S-O-R model. Thus, in the context of our study, we limited ourselves to the relationship between the S in the form of gamification features and the O in the form of psychological inspiration. Further studies can integrate additional components of the psychological processes of the and the resulting consequences of the R in their work, for example with regard to economic consequences such as WoM.

7   Conclusion

In our study, we investigated the relationship between gamification features and inspiration in the context of service industries and VTOs for the first time. Building on an S-O-R framework, our results indicate ambivalent influences of gamification features on inspiration and emphasize the complexity of the underlying psychological mechanisms of the gamification concept. Nevertheless, our findings provide valuable insights that can be used in the design of the development of innovative playful artifacts.

Acknowledgments. This research and development project is funded by the German Federal Ministry of Education and Research (BMBF) within the "The Future of Value Creation - Research on Production, Services and Work" program and managed by the Project Management Agency Karlsruhe (PTKA). The authors are responsible for the content of this publication.

Disclosure of Interests. Nothing to declare.

References

1. Bitner, M.J., Brown, S.W., Meuter, M.L.: Technology infusion in service encounters. J. Acad. Mark. Sci. **28**(1), 138–149 (2000)
2. Böttger, T., Rudolph, T., Evanschitzky, H., Pfrang, T.: Customer inspiration: conceptualization, scale development, and validation. J. Mark. **81**(6), 116–131 (2017)
3. Cameron, R.: A sequential mixed model research design: design, analytical and display issues. Inter. J. Multiple Res. Approach. **3**(2), 140–152 (2009)

4. Caponetto, I., Earp, J., Ott, M.: Gamification and education: A literature review. In: European Conference on Games Based Learning, vol. 1, p. 50. Academic Conferences International Limited (2014)

5. Deterding, S., Dixon, D., Khaled, R., Nacke, L.: From game design elements to gamefulness: defining" gamification". In: Proceedings of the 15th International Academic MindTrek Conference: Envisioning Future Media Environments, pp. 9–15 (2011)

6. Douglas, B.D., Brauer, M.: Gamification to prevent climate change: a review of games and apps for sustainability. Curr. Opin. Psychol. **42**, 89–94 (2021)

7. Hamari, J., Koivisto, J., Sarsa, H.: Does gamification work?–a literature review of empirical studies on gamification. In: 2014 47th Hawaii International Conference on System Sciences, pp. 3025–3034. IEEE (2014)

8. Hamari, J., Tuunanen, J.: Player types: A meta-synthesis (2014)

9. Hassan, L., Rantalainen, J., Xi, N., Pirkkalainen, H., Hamari, J.: The relationship between player types and gamification feature preferences (2020)

10. Huotari, K., Hamari, J.: Defining gamification: a service marketing perspective. In: Proceeding of the 16th International Academic MindTrek Conference, pp. 17–22 (2012)

11. Jacoby, J.: Stimulus-organism-response reconsidered: an evolutionary step in modeling (consumer) behavior. J. Consumer Psychol. **12**, 51–57 (12 2002)

12. Jahn, K., et al.: Individualized gamification elements: the impact of avatar and feedback design on reuse intention. Comput. Hum. Behav. **119**, 106702 (2021)

13. Koivisto, J., Hamari, J.: The rise of motivational information systems: a review of gamification research. Int. J. Inf. Manage. **45**, 191–210 (2019)

14. Landers, R.N., Bauer, K.N., Callan, R.C.: Gamification of task performance with leaderboards: a goal setting experiment. Comput. Hum. Behav. **71**, 508–515 (2017)

15. Lavoye, V., Sipilä, J., Mero, J., Tarkiainen, A.: The emperor's new clothes: self-explorative engagement in virtual try-on service experiences positively impacts brand outcomes. J. Serv. Mark. **37**(10), 1–21 (2023)

16. Mehrabian, A., Russell, J.A.: An approach to environmental psychology (1974)

17. Mekler, E.D., Brühlmann, F., Tuch, A.N., Opwis, K.: Towards understanding the effects of individual gamification elements on intrinsic motivation and performance. Comput. Hum. Behav. **71**, 525–534 (2017)

18. Mohammadi, S.O., Kalhor, A.: Smart fashion: a review of ai applications in virtual try-on & fashion synthesis. J. Artif. Intell. **3**(4), 284 (2021)

19. Oleynick, V.C., Thrash, T.M., LeFew, M.C., Moldovan, E.G., Kieffaber, P.D.: The scientific study of inspiration in the creative process: challenges and opportunities. Front. Hum. Neurosci. **8**, 436 (2014)

20. Pachoulakis, I., Kapetanakis, K.: Augmented reality platforms for virtual fitting rooms. Inter. J. Multimedia Appli. **4** (2012)

21. Peng, W., Lin, J.H., Pfeiffer, K.A., Winn, B.: Need satisfaction supportive game features as motivational determinants: an experimental study of a self-determination theory guided exergame. Media Psychol. **15**(2), 175–196 (2012)

22. Qasem, Z.: The effect of positive tri traits on centennials adoption of try-on technology in the context of e-fashion retailing. Int. J. Inf. Manage. **56**, 102254 (2021)

23. Rohan, R., Pal, D., Funilkul, S.: Mapping gaming elements with gamification categories: Immersion, achievement, and social in a mooc setting. In: 2020 14th International Conference on Innovations in Information Technology (IIT), pp. 63–68. IEEE (2020)

24. Savin, N.E., White, K.J.: The durbin-watson test for serial correlation with extreme sample sizes or many regressors. Econometrica: J. Econometric Soc., 1989–1996 (1977)

25. Schöbel, S., Janson, A.: Is it all about having fun?-developing a taxonomy to gamify information systems. In: ECIS, p. 60 (2018)

26. Shostack, L.G.: Planning the Service Encount in The Service Encounter, pp. 243–254. Lexington Bookss, Lexington, MA (1985)

27. Snodgrass, J.G., Dengah, H.F., Lacy, M.G., Fagan, J.: A formal anthropological view of motivation models of problematic mmo play: achievement, social, and immersion factors in the context of culture. Transcult. Psychiatry **50**(2), 235–262 (2013)

28. Solomon, M.R., Surprenant, C., Czepiel, J.A., Gutman, E.G.: A role theory perspective on dyadic interactions: the service encounter. J. Mark. **49**(1), 99–111 (1985)

29. Tandon, U., Ertz, M.: Modelling gamification, virtual-try-on technology, e-logistics service quality as predictors of online shopping: an empirical investigation. Current Psychol. (2023)

30. Thrash, T., Moldovan, E., Oleynick, V., Maruskin, L.: The psychology of inspiration. Soc. Personality Psychol. Compass **8** (2014)

31. Thrash, T.M., Elliot, A.J.: Inspiration as a psychological construct. J. Pers. Soc. Psychol. **84**(4), 871 (2003)

32. Thrash, T.M., Elliot, A.J.: Inspiration: core characteristics, component processes, antecedents, and function. J. Pers. Soc. Psychol. **87**(6), 957 (2004)

33. Wang T., Keng-Jung Yeh, R., Yen, D.C., Nugroho, C.A.: Electronic and in-person service quality of hybrid services. Serv. Indus. J. **36**(13-14), 638–657 (2016)

34. van Dolen, W., Lemmink, J., de Ruyter, K., de Jong, A.: Customer-sales employee encounters: a dyadic perspective. J. Retail. **78**(4), 265–279 (2002)

35. Weber, S., Klassen, G., Wyszynski, M., Kordyaka, B.: Illuminating the predictive power of gamification to inspire technology users. In: Mensch und Computer 2023 - Workshopband (2023)

36. Xi, N., Hamari, J.: Does gamification satisfy needs? a study on the relationship between gamification features and intrinsic need satisfaction. Int. J. Inf. Manage. **46**, 210–221 (2019)

37. Xi, N., Hamari, J.: Does gamification affect brand engagement and equity? a study in online brand communities. J. Bus. Res. **109**, 449–460 (2020)

38. Yee, N., Ducheneaut, N., Nelson, L.: Online gaming motivations scale: development and validation. In: Proceedings of the SIGCHI Conference on Human Factors in Computing Systems, pp. 2803–2806 (2012)

39. Zhang, T., Dr Wang, W.Y.C., Cao, L., Wang, Y.: The role of virtual try-on technology in online purchase decision from consumers' aspect. Internet Res. **29** (2019)

Virtual Influencers' Lifecycle: An Exploratory Study Utilizing a 4-Stage Framework of Planning, Production, Debut, and Retirement

Joosun Yum🆔, Youjin Sung🆔, Yurhee Jin🆔, and Kwang-Yun Wohn$^{(\boxtimes)}$🆔

KAIST, Graduate School of Culture Technology, Daejeon, South Korea
{yumjoosun,672,jinyuri,wohn}@kaist.ac.kr

Abstract. In this paper, we explored a 4-stage framework(planning, production, debut, and retirement): the life cycle of a 'Virtual Influencer'. To address expected ethical limitations on how VIs are produced and how they communicate with the world, we developed a 4-stage framework for the VI's lifecycle. paper includes a literature review, case studies, and surveys to determine the perception of VIs. We summarized the concepts and characteristics and the market size of VIs. Moreover, we discussed how VIs are created and produced at each stage these days, which is different from human influencers and celebrities. In addition, we adjusted the concept of the retirement of the VIs by suggesting the possible utilities that can be done in the near future. Finally, we identified what social and ethical problems are expected during VI production and activities and proposed solutions such as awareness improvement and preparation of legislation.

Keywords: Virtual Influencer · Virtual Human · Virtual Influencer's Life Cycle · Metaverse · Digital Human

1 Introduction

Planning and Production. Virtual Influencer (VI) is a digital character created using artificial intelligence, computer graphics, or other digital technologies, designed to mimic the role of a real-life influencer. They engage with audiences across social media platforms, promoting products, sharing content, and sometimes even advocating for causes, just like human influencers. Most VIs have detailed backstories, friendly personalities, and attractive appearances to appeal to their followers. The production of VIs presents unique challenges. Recent surveys indicate a nuanced consumer perception towards virtual influencers. While a human resemblance can foster familiarity, making them indistinguishable from real humans may inversely impact their appeal, leading to a decline in favorability. It emphasizes the need for a well-considered approach in the design and presentation of virtual influencers, ensuring a clear distinction between the virtual and real while still harnessing the benefits of human-like relatability.

© The Author(s), under exclusive license to Springer Nature Switzerland AG 2024
F. F.-H. Nah and K. L. Siau (Eds.): HCII 2024, LNCS 14720, pp. 91–107, 2024.
https://doi.org/10.1007/978-3-031-61315-9_7

Debut. Currently, they are stepping out of the fashion&music market and making waves in the tech world, with companies now teaming up to create their virtual personalities. When the decision-makers in the market produce a VI, they need to select the target industry, and the nature of the activities to be pursued is crucial.

Retirement. Like existing graphic artworks, virtual influencers are not immune to unauthorized reproduction and secondary processing, which raises concerns about intellectual property rights and originality. As advancements are made, there is potential for virtual influencers to interact with users through artificial intelligence. It brings forth a critical discussion on the nature of data utilized for training these AI algorithms. It is crucial to ensure that the data is accessible from profanity, hate speech, or biases, especially against marginalized groups such as homosexuals, black individuals, and people with disabilities. The human-like interaction proposed for virtual influencers necessitates stringent measures to avoid biased learning data, ensuring ethical and unbiased engagements. Additionally, the ease of production of virtual influencers could lead to a surge in their numbers, making it imperative to exercise caution to prevent indiscriminate production and potential misuse.

In this paper, we took a closer look at the VI's journey, breaking it down into 4 main stages: planning, production, debut, and retirement. By looking at previous related works, real-world examples, and what people think through surveys, we will explore what the "lifecycle" looks like for virtual influencers (Fig. 1).

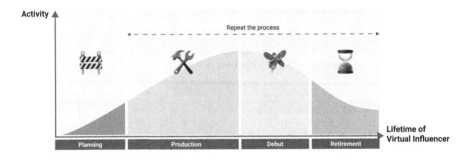

Fig. 1. The 4-stage framework of planning, production, debut, and retirement.

2 Related Works

2.1 Concept and Characteristics of the Virtual Influencer

An influencer is an influential person on a social network service (SNS) who significantly impacts consumers' purchasing decisions for a particular brand or content created through social media. Influencers' activities are not limited to social media. By collaborating with companies and creating advertising content, the boundaries between online and offline activities are more blurred than ever. Recently, due to the COVID-19 pandemic and technological development, virtual influencers not affected by the constraints of reality have begun to appear. Virtual influencers are fictional characters created by combining artificial intelligence and computer graphics, and they are referred to as influencers with great social influence. The technologies utilized include artificial intelligence (AI), augmented reality (AR), and computer graphics (CG) [27].

The virtual influencer market is growing, and the influence is constantly expanding. In particular, the impact of VIs has become more evident with the advent of COVID-19. Influencers gain popularity by actively communicating with the public on social media and earning income through marketing activities such as brand advertising. The market size of many virtual influencers, both domestic and international, is constantly expanding. Therefore, it is expected to have a market size of about 220 million dollars in 2023, and it is expected to grow to a market size of more than 370 million dollars in 2027.

Businesses and consumers have recently embraced VIs for several reasons. First, they are not limited by time and space. Because they're virtual, they can be active anytime, anywhere, and they don't have to coordinate their schedules like celebrities or human influencers do. They can be active when and where they want and for as long as needed. Second, they are easier to control and manage. In the case of humans, when incidents and accidents occur, the advertising brand's image is heavily damaged, and the advertising model has to terminate the contract and pay penalties. However, VIs are less likely to cause incidents and are easier to control and manage, minimizing the damage. Third, VIs have no boundaries to their territory, allowing them to operate freely. While human influencers are limited by their physical characteristics, VIs can work across various content industries, such as singers, YouTubers, actors, models, DJs, etc. [39].

2.2   The Importance of Virtual Influencer Marketing

Virtual influencer marketing has emerged as a compelling strategy, utilizing computer-generated or AI-powered characters to promote products, services, or brands. In 2020, Moustakas et al. conducted an exploratory qualitative study, interviewing six experts to delve into the advantages and disadvantages of using fictional characters in marketing, particularly highlighting VIs' challenges and success factors compared to human ones [34].

Adding to this discourse, Jhawar et al. (2023) set out to explore the rise of virtual social media influencers, recognizing the dearth of research in this area [17]. To address this gap, they proposed a theoretical model to understand social media users' acceptance of VIs, emphasizing the role of parasocial interaction (PSI) in establishing source credibility and fostering acceptance.

Furthermore, Wibawa et al. (2022) conducted a literature review from 2018 to 2022, revealing a positive outlook among digital experts regarding the effectiveness and reliability of using VIs in marketing strategies [44]. They suggest that incorporating virtual characters can bring innovation and depth to global marketing, potentially leading to widespread adoption across industries.

In a related study, Jang et al. (2020) examined the information sources, evaluation criteria, and favored attributes of 3D VIs, providing insights for integrating them into the fashion market and understanding the anticipated effects of their advertisements on brands [16].

Shifting the focus to credibility, Lim and Lee (2023) conducted experiments to investigate the impact of disclosing a VI's origin and employing positive versus negative emotional narratives on their credibility [29]. Their findings suggest that boosting credibility requires disclosing the VI's origin and using positive emotional narratives.

Finally, Franke et al. (2023) delved into the correlation between a virtual influencer's appeal and a consumer's inclination to make a purchase [13]. They found that attractiveness was not directly linked to purchase intention but mediated by mimetic desire and brand attachment.

In conclusion, VI marketing presents a forward-thinking and efficient approach to engaging with audiences in the digital age, offering brands unique opportunities to connect, communicate, and establish a strong online presence.

2.3   Exploring Ethical Terrain on Virtual Influencers

Human influencers naturally have flaws and unique traits, but VIs are artificial creations carefully designed and managed by companies or individual creators. These VIs can be precisely crafted to appeal perfectly to certain audiences without the unpredictability of human behavior.

Sands et al. (2022) examine brands' potential advantages and difficulties in utilizing VIs to connect with and sell to audiences. While VIs offer opportunities for engagement, the authors highlight five main challenges: consumer concerns, unrealistic beauty standards, lack of authenticity, regulatory and ethical issues, and consumer resistance. This substantial difference highlights the need for ethical reflection on using VIs. With their actions and stories tightly controlled, the ease with which VIs can be made into ideal figures raises important questions about truthfulness and manipulation in the digital space. In this research, we investigate how the level of human likeness in a VI's appearance influences consumers' adoption of a recommended stance. Additionally, the study explores whether disclosing the VI's non-real status, despite its humanlike appearance, impacts its persuasive effect. The findings suggest that a more humanlike appearance in VIs increases the sense of social presence, leading to improved perceptions of message quality and brand attitude [34].

As VIs mostly appear on social media platforms, which let users post numerous images(body images [12] and short videos), this content can potentially be repurposed to create deepfakes. Kugler et al.(2021) highlight the growing concern around deepfake technology, which enables the realistic insertion of individuals into videos, leading to a surge in political satire and fake pornography. The research reveals a strong public sentiment against nonconsensual pornographic deepfakes, with a majority favoring criminal sanctions for creators. Though ethical debate around VIs has been focusing on the broader consequences of their use: the possibility of biased audiences [30], and understanding the responsibilities of those who create and control these digital figures [21], a careful and thorough discussion about the ethical aspects of VIs is needed.

In this paper, we highlight the inadequacy of current legal structures in addressing crimes involving VIs (VIs) while also posing critical inquiries: *(1)How should we perceive and handle VIs?*, *(2)Are they to be regarded as entities with human-like attributes or merely as data?* This question emphasizes the need for a carefully thought-out approach, both ethically and legally, to address the growing presence of VIs in the digital world.

3   4-Stage Framework

3.1   Planning

VI becomes a branding tool for companies and products by expressing themselves using their personalities and images. Its extroverted appearance and worldview are important to design a VI that is intimate and attractive to people. A VI is a virtual figure that exists only in the digital world. However, setting up a worldview and persona is essential because it communicates with people and socially affects consumer consumption patterns like a human influencer. First, as an influencer, it is essential to be envied by the public and to communicate in a friendly manner. Second, it should be connected to reality so there is no boundary between virtual and reality. The blurring of the line between virtual and reality, and the more detail, the more people become fans of VIs and go crazy [16].

Background of Persona Formation. Most VIs currently on sale target the MZ generation, which refers to the combination of Millennials and Generation Z (GenZ) as a branding core. Among them, 82.2% watched YouTube in their spare time, and more than half (59.6%) said that there are more than five creators who subscribe to channels such as YouTube, Instagram, and African TV to remember their names and visit them frequently [35]. Therefore, forming a VI image focusing on the MZ generation, which has much contact time with social media, is essential. To target the MZ generation, it is necessary to examine their values. Born in an economic environment with improved national income levels compared to previous generations, the MZ generation forms consumption patterns with leisure, environment, and personality as critical keywords. The VIs currently active have different forms of worldviews centered on the above keywords.

Example of Persona Formation. Lil Miquela is a Brazilian-American woman who is a VI from Los Angeles. About 83 percent of her 2019 Instagram followers are GenZ under 24. As a result, Lil Miquela's worldview has been set as an environmentally friendly, socially participative, and virtual human who speaks his mind like the tendency of the leading SNS follower group. A VI is designed from the overall worldview to personal taste, personality, and daily life. Lil Miquela is forming a familiarity with GenZ followers by sharing her favorite K-pop meeting and parting process through Instagram. In addition, she was set up as a bisexual and added concreteity and reality to the setting of her worldview. She has taken an active stance on human rights issues for the socially disadvantaged, such as sexual minorities and blacks [4]. It is also funded to support the music industry suffering from COVID-19 [10]. Rozy is a VI for Koreans, and she has a world view, including the main interests of GenZ, such as yoga, travel, running, fashion, and eco-life. She has a free-spirited and sociable personality, and her appearance and her enviable appearance will help the MZ generation form a consensus. For example, they frequently post daily photos on Instagram, such as photos taken while traveling abroad or shopping, to form a sense of intimacy with SNS followers. In addition, she is interested in the environment and has a worldview that actively participates in social issues, such as participating in the Zero West Challenge [41]. According to Bibrand, who produced VI Rui, Rui was established as a singer-songwriter by referring to the real model and was produced by embodying the real model's bright and active personality and language. However, even with a high degree of consistency with real people, we must create a new personality that is not in the world. Therefore, when producing VIs, care should be taken to ensure that entertainment companies have detailed differences as if they were managing artists [37].

3.2   Production

VIs are mainly created using deep learning-based graphical and voice technologies. The synthesis of a VI's appearance can be divided into face, body, voice, and facial expression, and fashion is also an important element in completing the VI. When creating a VI face, the key factors are naturalness and freshness. All of the VIs on the market are synthesized faces that don't exist in the real world. A common technique used for face synthesis is the digital double, which is a deep-learning graphics technology that uses a specific actor or model as the primary actor and then applies another virtual face to the real model. Since the face is analyzed and synthesized in 3D, it has a natural advantage over deep fake techniques [43]. In addition, the detailed graphics work realizes the movement and connectivity of all muscles, including skin, bones, and neural networks. Digital doubles can be used to analyze and predict how different muscles around the jawline move when a VI makes various facial expressions or speaks.

Typical VIs who have used this technology are Rozy and Lucy, and in the case of Lucy, the pores and fuzz of her skin are realistic [2]. Rozy was created using 3D modeling technology that captured her face and gestures through a band model with over 800 facial expressions and shapes that Gen MZ preferred [22]. In the case of active VIs, there are no examples of 3D modeling of the entire body, and only the face of the VI is composited with a choreographer as a model [19]. In addition to appearance, voice is a factor that makes VIs realistic and approachable. Most of the VIs that have been released have communicated with users only through on-screen movements or photos without voices, but there are VIs who have produced music with natural pitches and high-quality sound using deep learning-based voice synthesis technology [14]. The voice synthesis technology used to create VIs records authentic human voices over a while, and then the AI analyzes the timbre, tone, and inflection and synthesizes the voice when certain words and sentences are given as input [15]. Recent technological advances have made it possible to create intonation, micro-breathing, and natural-sounding voices to capture emotion and personality in a VI's voice [23]. With this technology, VIs Rozy and Kim Rae-ah released their first singles and debuted as singers. VI facial expression is a depth sensor-based face recognition technology that changes the VI's facial expression when the subject's or model's facial expression changes. The VI's fashion and styling are set to match the persona of each VI. For example, Rozy is styled using street fashion, comfortable and sporty hats and hoodies, and casual jumpers to match her persona of being free-spirited and sociable [25].

User Experience of VI Appearance. A quantitative research method using a structured questionnaire is adopted to grasp the degree of recognition of VIs. The sample group was set at 110 people born in 1987-2003, including Millennials between 1980 and 1994 and Generation Z (Gen Z) between 1995 and 2004. The characteristics of the sample were MZ generation men and women living nationwide, with a gender ratio of male: female = 40.0:59.1, and respondents born in 1995, which is the reference point for generations M and Z, were composed of the average year of birth. The average daily time of the sample group was 1.9 h, and Instagram and Facebook were the most frequently used SNSs. According to the analysis results, 16.4% said they "know well", 61.8% said they "have heard of it" and most respondents recognized VIs. As for whether to visit VI SNS, 27.3% of the respondents had visited (See **Appendix**).

Questionnaire. A quantitative research method using a structured questionnaire is adopted to grasp consumers' perception of the appearance of VI. Evaluation items were organized based on existing prior research to organize the survey items. Um(2023) designed to examine the effects of para-social interaction as relationships between VIs and audiences [42]. This study delves into the effects of perceived human likeness, perceived predictability, and perceived authenticity in evaluating VI advertising. Kim et al.(2023) measured virtuality, expertise, intimacy, and likability [40]. Park Ye-rang et al.(2022) measured attractiveness, reliability, intimacy, expertise, interactivity, and self-consistency in the specialty evaluation of VIs [36]. Hong and Kim (2023) evaluated similarity, attractiveness, reliability, expertise, and interactivity in a study on the preference for VI of the MZ generation [18].

Divide by a numerical scale from 1 to 5, where 1 is very different, 2 is not, 3 is normal, 4 is true, and 5 is very much so. Among the perceptions of VI extroverts, only 37.3% (4: yes, 5: very yes, added up) said they had experienced an unpleasant valley. In addition, 32.7% of respondents said they were "attractive" to VI extroverts (4: yes, 5: very yes). The most noticeable features of the VI's appearance were individual, ideal, and friendly in order, while other responses included "trendy", "ooks popular on social media these da", and "virtual ones stand out". Virtuality is a characteristic that looks real but does not exist, and if it is virtual, it means that it does not look like a real person. Among the extroverted characteristics of VIs, 45.5% (4: Yes, 5: Yes, very yes) accounted for about half of the survey on virtuality awareness. According to the survey analysis results, the part to remember when manufacturing VIs results in problems with the degree of virtuality and the direction of outward formation. From the corporate perspective, after analyzing and discussing how effective it is to make VIs similar to people, VIs should be produced with enough virtuality that consumers can feel comfortable without feeling uncomfortable. From a social point of view, some concerns producing virtual humans could cause social confusion to the point where it is difficult to distinguish them from others.

It is said that people can feel a sense of affinity because they resemble each other, but if they are so similar that it is difficult to distinguish, their likability may decrease (Fig. 2). .

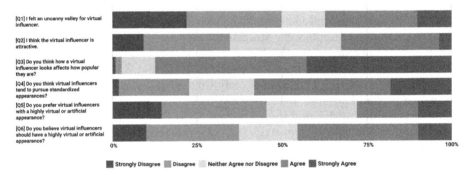

Fig. 2. Results from the survey.

3.3   Debut

VIs are not only in fashion and music but are expanded in the fields of artificial intelligence, with companies creating virtual humans, collaborating, and launching their own VIs. When a company creates a VI, deciding which industry they want to work in and what they want to do is essential. The scope of VIs is broad and unrestricted, so they can become singers, record albums, perform concerts, model products, and promote different brands. They can also become YouTubers, where they can constantly interact with their subscribers. There are two primary industries where VIs are active: the fashion and beauty industry and the game and music broadcasting industry. The media are Instagram, TikTok, YouTube, Twitter, and magazines.

Fashion and Beauty Industry. Many VIs now work as models in the fashion and beauty industry, stimulating consumers' desire to spend. Shudu[1] is a South African virtual model created by British photographer and computer graphics artist Cameron James Wilson based on his experience working in the fashion industry. As one of the first digital models, she rose to fame on social media and was signed by various fashion and beauty brands. She has collaborated with brands such as Calvin Klein and Dior, modeled for Fendi beauty lipsticks, and appeared in commercials for Balmain [11].

Rozy[2] is a VI created by Sydus Studio, a subsidiary of content company Lotus, and designed to represent the preferred beauty standards of Gen MZ. Rozy has an oriental mask and a height of 171cm. She has a free-spirited and sociable personality. Moreover, she modeled for the French luxury brand Celine and participated in a fashion photo shoot for WKOREA. The photoshoot was part of W Korea's 16th annual breast cancer awareness campaign, the "Love Your W" project. Rozy also modeled for W Concept, an online fashion platform, in a photo shoot titled "I am a concept," showcasing different brands' styles [26].

Education Industry Most VIs are active in the entertainment and content industries as influencers, but recently, they have also appeared in the education industry. DA: ON, produced by Minds Lab, a company specializing in AI virtual humans, was a scientific commentator at the online special exhibition'Artificial Intelligence and Art (AIART)' at the National Science Museum in Daejeon. DA: ON is equipped with Wav2Lip, an algorithm-based lip synchronization technology, and deep learning voice generation algorithm (Hi-Fi GAN) voice generation (TTS) technology, enabling free conversation with visitors [8].

Korean Air has recently unveiled a novel inflight safety video that showcases virtual humans. The video features a virtual character named Rina dressed in a Korean Air flight attendant uniform, providing safety instructions in a virtual setting called the "Korean Air Safety Lounge." This simulated space was designed to enhance the clarity of inflight safety rules. Rina, the virtual human, underwent safety training at Korean Air's Cabin Crew Training Center and was honored with the title of an honorary flight attendant. The safety video includes members of MAVE (Members of the Association of Virtual Experience) playing the role of passengers as they follow the safety instructions [1].

Commercial Industry. IMMA[3] is a model for IKEA in Tokyo, created by Japanese 3D imaging startup AWW, and earned 700 million won in 2019 from brand advertising revenue. A video of her spending three days at IKEA's Harajuku showroom and experiencing IKEA products was produced and shown on YouTube and in IKEA stores. In addition, IMMA shared photos from collaborations with Amazon and Porsche on social media [33].

---

[1] https://www.instagram.com/shudu.gram/.
[2] https://www.instagram.com/rozy.gram/.
[3] https://www.instagram.com/imma.gram/.

Music Industry. In the music industry, we typically see VIs as singers, songwriters, DJs, etc. Aespa[4] is the virtual avatar of SM Entertainment's idol group, Aespa, created by Giant Step. Giant Step's U.S. subsidiary provided character design, VFX, real-time content solutions consulting, virtual studio, animation, and character rigging to construct the world of Aespa [38].

IITERNITI[5] is a virtual idol consisting of 11 members in their early 20 s realized with artificial intelligence technology. IITERNITI was developed by Pulse9, a company specializing in artificial intelligence graphics with "deep real" technology, and the 11 members were selected through a referendum. They have released music videos and albums and held fan meetings. The song "Im real" has about 900,000 views, and "No Filter" has about 2.2 million views, showing great public interest in virtual idols. Recently, K/DA organized the opening of the 'Chungcheongnam-do AI Week' online and offline events [32].

K/DA[6] is a fictional four-member girl group created to commemorate the League of Legends 2018 World Championship held in South Korea. K/DA was created by Riot Games, the company responsible for developing and distributing League of Legends (LoL), using the game's characters Akali, Ari, Evelyn, and Kaisa. The virtual girl group performed their new song "POP/STARS" at the 2018 League of Legends World Championship (LOLd Cup) Finals opening ceremony and later released a music video and soundtrack. The music video has received approximately 480,000 views on YouTube [46].

Lill Miquela[7] is a VI created by the American digital character creation company Brud in April 2016 and is a Brazilian-Spanish mixed-race music artist. As a musician, Lill Miquela works in the genres of RB, electronic, and pop, and she released her first single, "Not Mine," on Spotify in August 2017. She has also been featured as an interviewer at music festivals and appeared in music video [9].

Created by dob studio[8], Rui[9] is a virtual cover singer who sings along to famous songs and runs the YouTube channel Rui Coverage. In addition to covering songs, he introduces Korean travel destinations and vlogs about furniture brands from various corners of Korea. Rui is a virtual person who was created by using artificial intelligence technology to acquire the facial data of seven people and create a likeness of a real human being. She has about 40,000 subscribers on YouTube and communicates with the public through YouTube, Instagram, and TikTok [7].

Virtual Influencer Collaboration Activities. VIs work individually and with real people or other VIs to expand the scope of their activities, giving them a more realistic experience. First, examples of collaborative activities with

---

[4] https://www.instagram.com/aespa_official.
[5] https://www.instagram.com/iiterniti.
[6] https://www.instagram.com/kda_music/.
[7] https://www.instagram.com/lilmiquela/.
[8] https://www.dob.world/.
[9] https://www.instagram.com/ruuui_li/.

human influencers are as follows. VI Rozy photographed for the May 2021 issue of WKorea, a high-end fashion magazine, with Irene, a model belonging to the model agency Esteem. Creative content company Esteem and Sidus Studio X, which produces and operates VIs, will jointly manage VI Logistics for the first time in Korea. As a result, Rozy's activity area, which was limited to SNS, has expanded to magazines and broadcasting activities [24]. Secondly, examples of collaborative activities between VIs are as follows. Sidus Studio X and Shudu, the world's first digital model, have launched a photo shoot for Vogue Korea magazine. At the time of the shooting, Rozy raised awareness by filming a fashion pictorial with the world-famous virtual model Shudu because she was a rookie VI. The pictorial was designed with the concept of "Over The Limit" to break down the boundaries in the world and create a world where they coexist. Shudu in Hanbok-style clothes and Rozy in African clothes were posted on Instagram. Within three months, each company created a scene of a VI to promote harmony between two VIs with different images. Taking advantage of VIs, they created a photo taken together in one space without meeting in person [45]. Third, in 2019, Calvin Klein launched the #MYTRUTH campaign featuring supermodel Bella Hadid and VI Lil Miquela with a video showcasing the two endorsers sharing a kiss [5]. Calvin Klein aimed to raise awareness about LGBTQ (lesbian, gay, bisexual, transgender, queer, and questioning or queer) rights through this initiative. On Twitter, the brand explained that the campaign was designed to challenge traditional norms and stereotypes in advertising. Specifically, the video sought to explore the boundary between reality and imagination [6] (Fig. 3).

Fig. 3. List of the VIs.

3.4 Retirement

In this section, we suggest the concept of the "Retirement" of VIs. In general, the word "Retirement" means that an individual gets away from social and economic activities and spends time leisurely. However, unlike humans, VIs do not have a short life span and have no restrictions on their activities. Nevertheless, "retirement" is inevitable even for VIs who are struggling with their fame.

Assuming existing human influencers have the same activity cycle as VIs, the considerable difference would be the concept of "retirement." In the case of VIs, there is no space or time constraint on activities because the concept of "life span" is not applied, but if it does not maintain its popularity like a human influencer, it will disappear [28]. VI is making profits and getting in contact with many consumers. However, if the contents are limited to one-sided interaction, such as text, photos, and videos, the chances of being forgotten by the public can increase. It is difficult to analyze the relationship between industrial changes and idol culture, including the fact that few industrial indicators have been disclosed or it is difficult to secure the reliability of the indicators. Previously, the retirement period was determined by relying on senses, but we would like to establish clear standards based on our understanding of the existing influencer market and utilize them as data for producing new VIs.

There are two types of retirement in the existing entertainment industry: voluntary retirement and involuntary retirement. Since most of the influencer's activities are operated by their agencies, the retirement timing is determined by the agency's decision. The retirement decision relies on the revenue from the market. Influencer activities will naturally decrease and be halted shortly when there is no meaningful profit compared to investment costs. As such, analyzing related data to expect the right timing is crucial. In the case of VIs, the agency can expect the moment to be retired or find the pattern based on the collected data from previous VIs' activities. Based on collected data, we expect that patterns can be set to determine the timing of retirement, such as (1) when no return on investment is expected to be made or (2) when a retirement decision is already made at the early planning/production stage of VI.

It is a new world with virtual humans around, and people might get mixed up if they cannot tell who is real and who is not, which we all need to discuss [34]. Like other digital art, VIs can be copied and messed with, which leads to ownership disputes. With the creation of VIs becoming increasingly fast, potential misuse emerges, such as deepfake pornography [20]. Thus, establishing specific guidelines is crucial.

Data Utilization after Retirement. The data obtained through the activities of VIs includes an understanding of the popularity cycle received by the public. It forms a "popularity scale model during the activity period" by understanding how specific images, personalities, work activities, and communication methods with the public affect popularity. Through the second stage of data collection and utilization, retirement times determined by existing experience and reduction can be determined based on clear data-based standards and can be used to produce next-generation virtual humans.

- **Data utilization** Data values obtained through virtual human activities can be used for training existing entertainment companies. The reality of the influencer market is that it is difficult to make profits unless it gains popularity even after its debut. It is said that about 300 to 400 teams cannot be broadcast every year. To prevent the harmful effects of the current entertainment industry, accumulated know-how will be used to minimize the probability of failure and increase the return on investment the most. In addition, in the case of virtual humans, data is likely to be continuously used in the Intellectual property (IP) industry even after retirement. In some cases, the concepts of "digital human" and "VI" have already been imagined by many people for a long time. Due to the development of technology, the possibility and trust in "digital humans" have been expanded in that they can now work in the advertising and marketing fields like real people [31]. Cyber singer Adam[10] is also known to have worked as a fashion designer to live a second life after retirement. Webcomics containing this worldview were also scheduled to be released in 2016. In the case of VIs, data utilization between platforms is easier than that of human influencers to gain an advantage in the expanding IP industry.
- **Data extension** The life cycle of virtual humans identified in VI can be extended and applied to the virtual human area necessary for society other than the entertainment business. There is a possibility that VI management will try to manage, operate, and design virtual humans first, create databases accordingly, and produce virtual humans in other fields such as teaching and professional. For instance, through VI production, algorithms can identify how VI becomes a person with ethical values and how to produce them so that they do not deviate from the algorithm can be learned in advance. It is expected that VI will be applied to society as a whole as VI expands to other occupational groups. Since virtual humans in augmented reality complement rather than substitute for real things, attempts to fuse reality with metaverse are noteworthy [3]. The agency Gold medalist[11] and digital content developer EVR STUDIO[12] jointly implemented both the actual appearance and facial expression of Kim Soo-hyun[13]. New added value is expected to be created throughout the current VIs industry.

4   Discussion

As the main consumer of VIs is the MZ generation, and influencers' social influence grows, attention must be paid to forming VI personas. Due to the charac-

---

[10] https://kpop.fandom.com/wiki/Adam.
[11] https://goldmedalist.com/en/company.
[12] https://evrstudio.com/.
[13] https://en.wikipedia.org/wiki/Kim_Soo\discretionary-hyun.

teristics of influencers, many followers imitate them, so it is important to design VIs as a personality with a sense of ethics and unique settings. When reflecting current major social issues in the personality of a VI, it should be able to be designed in a positive direction without controversy.

Like existing graphic artworks, VIs are not immune to unauthorized reproduction and secondary processing, which raises concerns about intellectual property rights and originality. As advancements are made, there's potential for VIs to interact with users through artificial intelligence. This brings forth a critical discussion on the nature of data utilized for training these AI algorithms. It is crucial to ensure that the data is free from profanity, hate speech, or biases, especially against marginalized groups such as homosexuals, black individuals, and the disabled. The human-like interaction proposed for VIs necessitates stringent measures to avoid the use of biased learning data, ensuring ethical and unbiased engagements. Additionally, the ease of production of VIs could lead to a surge in their numbers, making it imperative to exercise caution to prevent indiscriminate production and potential misuse. Establishing clear guidelines and ethical frameworks can help maintain a responsible approach toward creating and utilizing VIs, ensuring they serve as a positive extension in the digital realm rather than as a source of misinformation or bias.

We took a close look at the journey of VIs, breaking it down into four main steps: planning, making, introducing them to the world, and what happens when they retire. By looking at past studies, real-world examples, and what people think through surveys, we tried to get a handle on the whole VI scene, including what "retirement" might look like for them and what they could do afterward. It's a new world with virtual humans around, and people might get mixed up if they cannot tell who is real and who is not, which we all need to discuss. Like other digital art, VIs can be copied and messed with, which raises serious questions about who owns what. With the creation of VIs becoming increasingly effortless, a risk of oversaturation or potential misuse emerges; thus, establishing certain guidelines is crucial.

Appendix

A. Participant Demographics

Demographics of survey participants, including gender, birth year, primary SNS used, daily SNS usage time, awareness, and interaction with VIs (Table 1).

Table 1. Survey results.

Question	Answer	Frequency (People)	Percentage (%)
Gender	Male	44	40.0
	Female	65	59.1
	Other	1	0.9
Birth Year	Millennial (1980-1994)	47	42.7
	Gen Z (1995-2004)	63	57.3
Preferred social media (Multiple selections)	Instagram	96	65.3
	Facebook	25	17.0
	Twitter	12	8.2
	Other (Snapchat, Tumblr, TikTok, etc.)	14	9.5
Average screen time on social media per day	0 to <2 hours	61	55.5
	2 to <4 hours	33	30.0
	>4 hours	16	14.5
Awareness of VIs	Very familiar	18	16.4
	Have heard of it	68	61.8
	Do not know	24	21.8
Visit VIs' social media	Have visited	30	27.3
	Have not visited	80	72.7

References

1. Korean Air/ Korean Air. Korean air releases a new safety video featuring virtual humans [online]. 2024. URL: https://www.koreanair.com/id/en/footer/about-us/newsroom/list/240104-new-safety-video (Accessed 10 February 2023
2. Victor Andersson and Tim Sobek. Virtual avatars, virtual influencers & authenticity (2020)
3. Azuma, R.T.: A survey of augmented reality. Presence Teleoperators Virt. Environ. **6**(4), 355–385 (1997)
4. Callahan, K.: Cgi social media influencers: are they above the ftc's influence? J. Bus. Tech. L. **16**, 361 (2021)
5. CalvinKlein/ CalvinKlein. calvinklein (2019). https://www.instagram.com/p/BxhrRE-jbWA/?hl=de (Accessed 19 July 2023)
6. CalvinKlein/ CalvinKlein. calvinklein (2019). https://twitter.com/CalvinKlein/status/1129521041309614085 (Accessed 19 July 2023)
7. Cho, J.: A face that doesn't exist in the world, virtual human louis (2020). https://buybrand.kr/pick/rui/ (Accessed 25 May 2023)
8. Taebeom choi/ Money Today. Ai meets art, 'virtual human science docent' opens at the national museum of science and technology (2022). https://news.mt.co.kr/mtview.php?no=2021110115154398379 (Accessed 10-October-2022)
9. Conti, M., Gathani, J., Tricomi, P.P.: Virtual influencers in online social media. IEEE Commun. Mag. **60**(8), 86–91 (2022)
10. Drenten, J., Brooks, G.: Celebrity 2.0: Lil miquela and the rise of a virtual star system. Feminist Media Stud. **20**(8), 1319–1323 (2020)
11. SaraSemic/ ELLE. Meet the man behind the world's first digital supermodel (2019). https://www.elle.com/uk/fashion/a28394357/man-behind-worlds-first-digital-supermodel/ (Accessed 25 August 2023)

12. Fardouly, J., Vartanian, L.R., Current research and future directions: Social media and body image concerns. Curr. Opin. Psychol. **9**, 1–5 (2016)
13. Franke, C., Groeppel-Klein, A., Müller, K.: Consumers' responses to virtual influencers as advertising endorsers: novel and effective or uncanny and deceiving? J. Advert. **52**(4), 523–539 (2023)
14. Hartholt, A., et al.: All together now. In: Aylett, R., Krenn, B., Pelachaud, C., Shimodaira, H. (eds.) IVA 2013. LNCS (LNAI), vol. 8108, pp. 368–381. Springer, Heidelberg (2013). https://doi.org/10.1007/978-3-642-40415-3_33
15. Higgins, D., Zibrek, K., Cabral, J., Egan, D., McDonnell, R.: Sympathy for the digital: influence of synthetic voice on affinity, social presence and empathy for photorealistic virtual humans. Comput. Graph. **104**, 116–128 (2022)
16. Jang, H., Yoh, E.: Perceptions of male and female consumers in their 20s and 30s on the 3d virtual influencer. Res. J. Costume Cult. **28**(4), 446–462 (2020)
17. Jhawar, A., Kumar, P., Varshney, S.: The emergence of virtual influencers: a shift in the influencer marketing paradigm. Young Consumers **24**(4), 468–484 (2023)
18. Hong, H.J., Kim, S.I.: A study on the virtual influencer preference of mz generation (2023)
19. Kadekova, Z., Holienčinova, M.: Influencer marketing as a modern phenomenon creating a new frontier of virtual opportunities. Commun. Today **9**(2), 22 (2018)
20. Karasavva, V., Noorbhai, A.: The real threat of deepfake pornography: a review of canadian policy. Cyberpsychol. Behav. Soc. Netw. **24**(3), 203–209 (2021)
21. Kim, D., Wang, Z.: The ethics of virtuality: navigating the complexities of human-like virtual influencers in the social media marketing realm. Front. Commun. **8**, 1205610 (2023)
22. Kim, H., Choi, Y.M., et al: A case study of fashion style in accordance with tpo of k-virtual influencer. In: International Textile and Apparel Association Annual Conference Proceedings, vol. 78. Iowa State University Digital Press (2022)
23. Jaehyuck Kim/ Kocca. The secret to dancing models in advertising: the science of creating virtual influencers (2021). https://www.kocca.kr/trend/vol27/sub/s21.html (Accessed 20 April 2023)
24. Koreanupdates/ Koreanupdates. Esteem & sidus studio x (2021). https://twitter.com/KoreanUpdates/status/1386967834471768067 (Accessed 06 August 2023)
25. Lee, S., Park, J., Kim, T., Chun, J.: The effect of virtual fashion influencers' presence on evaluation attributes and relationship maintenance behavior. J. Korean Soc. Clothing Textiles **47**(2), 295–310 (2023)
26. Gyu lee Lee/ The Korean Times. Virtual humans emerge as major trend in burgeoning metaverse (2022). https://www.koreatimes.co.kr/www/tech/2024/01/129_326052.html (Accessed 10 October 2022)
27. Leinatamm, K., Bilali, S.: Virtual avatars rising: the social impact based on a content analysis and a questionnaire in the context of fashion industry (2019)
28. Li, Z.S., et al.: Narratives: the unforeseen influencer of privacy concerns. In: 2022 IEEE 30th International Requirements Engineering Conference (RE), pp. 127–139. IEEE (2022)
29. Lim, R.E., Lee, S.Y.: you are a virtual influencer!": understanding the impact of origin disclosure and emotional narratives on parasocial relationships and virtual influencer credibility. Comput. Hum. Behav. **148**, 107897 (2023)
30. Lou, C.: Social media influencers and followers: theorization of a trans-parasocial relation and explication of its implications for influencer advertising. J. Advert. **51**(1), 4–21 (2022)
31. I Marketing. The top instagram virtual influencers in 2020 (2020)

32. MiranKim. [social archives] interview| park ji-eun, ceo of pulse9 _ why did he create an ai idol?. The Scoop (549), 52–53 (2023)
33. Miyake, E.: I am a virtual girl from tokyo: virtual influencers, digital-orientalism and the (im) materiality of race and gender. J. Consum. Cult. **23**(1), 209–228 (2023)
34. Moustakas, E., Lamba, N., Mahmoud, D., Ranganathan, C.: Blurring lines between fiction and reality: perspectives of experts on marketing effectiveness of virtual influencers. In: 2020 International Conference on Cyber Security and Protection of Digital Services (Cyber Security), pp. 1–6. IEEE (2020)
35. Nisandzic, M.: Are you even real? Virtual influencers on Instagram and the role of authenticity in the virtual influencer consumer relationship. PhD thesis, Master's thesis, Leopold-Franzens-University Innsbruck (2020)
36. Shin Daji Kwon Ji Min Park Jee Young Guo Yijun Yun Jaeyung Park, Ye Rang. A study on user preference based on the characteristics of virtual influencers. Soc. Design Converg. **21**(2), 1–16 (2022)
37. Park, S., Sung, Y.: The interplay between human likeness and agency on virtual influencer credibility. Cyberpsychol. Behav. Soc. Netw. **26**(10), 764–771 (2023)
38. Inc. Priya PR/ Priya PR. Giantstep drops 'savage' vfx and animation for k-pop group aespa music video (2021). https://lbbonline.com/news/giantstep-drops-savage-vfx-and-animation-for-k-pop-group-aespa-music-video (Accessed 05 May 2023)
39. Rodrigo-Martín, L., Rodrigo-Martín, I., Muñoz-Sastre, D.: Virtual influencers as an advertising tool in the promotion of brands and products. study of the commercial activity of lil miquela. Revista Latina de Comunicación Social (79), 70–91 (2021)
40. Won Hyung Choi Eun Kyeong Ko Da Hye Kang Jae Young Yun So Hye Kim, Ha Yeon Shin. A study on the correlation between experience factors and favorability of virtual influencers. J. Korean Soc. Design Cult. **29**(1), 13–23 (2023)
41. Stein, J.-P., Breves, P.L., Anders, N.: Parasocial interactions with real and virtual influencers: the role of perceived similarity and human-likeness. New Media Soc., 14614448221102900 (2022)
42. Um, N.: Exploring the impact of virtual influencer advertising: In-depth interviews with college students. J. DCS **24**(7), 1391–1400 (2023)
43. Um, N.: Predictors affecting effects of virtual influencer advertising among college students. Sustainability **15**(8), 6388 (2023)
44. Wibawa, R.C., Pratiwi, C.P., Wahyono, E., Hidayat, D., Adiasari, W.: Virtual influencers: is the persona trustworthy? Jurnal Manajemen Informatika (JAMIKA) **12**(1), 51–62 (2022)
45. Chang won Lim/ Aju Korea Daily. Digital supermodel shudu to collaborate with virtual social media influencer in s. korea (2021). https://www.etoday.co.kr/news/view/1988477 (Accessed 16 March 2023)
46. Zhe, H., Lee, H.S., Han Zhe and Hyun Seok Lee: The characteristics of user created content (ucc) for virtual band k/da. J. Korea Multimedia Soc. **23**(1), 74–84 (2020)

Exploring the Impact of Virtual Influencers on Social Media User's Purchase Intention in Germany: An Empirical Study

Silvia Zaharia[(✉)] [ID] and Jasmin Asici

University of Applied Sciences Niederrhein, Krefeld, Germany
silvia.zaharia@hs-niederrhein.de

Abstract. Recently, there's been a surge in social media usage, leading companies to employ influencers for product promotion. Alongside human influencers, virtual influencers are now increasingly used for brand and product communication on these platforms.

This research investigates the impact of human virtual influencers (HVIs) on social media user's purchase intention by using an empirical survey. In order to do so a structural model was developed. The determinants believed to positively affect purchase intention include *parasocial relationship, anthropomorphism, credibility* and *authenticity. Age, gender* and *product fit* were examined as moderator variables. Additionally, *anthropomorphism* is assumed to positively influence the parasocial relationship. Data was collected using a quantitative online survey and analysed using multiple regression analysis. The proposed conceptualization and operationalization of the constructs were analysed using exploratory and confirmatory factor analysis.

The most significant and substantial factors influencing customers' intention to buy a product promoted by an HVI proved to be *parasocial relationship, credibility* and a*uthenticity.* The influence of *anthropomorphism* did not prove to be significant. Moreover, *anthropomorphism* has a highly significant influence on *parasocial relationship.*

The examination of moderator effects showed no moderating effects of *age.* In contrast, a significant *gender* moderation effect was observed: the influence of *parasocial relationship* on the *intention to purchase* products promoted by an HVI was significantly stronger for women than for men. Furthermore, moderator effects can be shown with regard to the variable *product fit*: the influence of the *parasocial relationship* and *authenticity* on *purchase intention* is particularly strong when *product fit* is given.

Keywords: human virtual influencers · parasocial relationship · anthropomorphism · credibility · authenticity · expertise · attractivity · trustworthiness · product fit

© The Author(s), under exclusive license to Springer Nature Switzerland AG 2024
F. F.-H. Nah and K. L. Siau (Eds.): HCII 2024, LNCS 14720, pp. 108–126, 2024.
https://doi.org/10.1007/978-3-031-61315-9_8

1 Introduction

In a predominantly digital world, social media has become an essential part of people's lives. As of October 2023, the global number of social media users was 4.95 billion, constituting 61.4% of the world's population (Kepios 2024). The increased use of social media has prompted companies to adapt their customer communication, as social media platforms have a significant influence on users' purchasing decisions (Schmidt/Taddicken (2017). Companies are increasingly using influencers, hereinafter referred to as human influencers (HIs), as an effective communication tool to reach and influence consumers (Ki/Kim 2019). Instagram in particular is the most popular social media channel for influencer marketing (Dencheva 2023).

In addition to the rapid growth of social media platforms and the success of HIs, virtual influencers (VIs) have started to emerge. A VI we define as an artificial representation of a human being. They can be divided into two broad categories, the anime virtual influencers (AVIs) and the human virtual influencers (HVIs). Most VIs used in influencer marketing are human virtual influencers and these also form the focus of this study. HVIs resemble real people and act on social media platforms such as HIs (Wibawa 2022). However, HVIs do not act autonomously, but are controlled by companies and are computer-generated, animated media personalities that have their own social media profile (Hofeditz et al. 2022). A team of developers designs the visual and verbal character of an HVI and its "life" story, considering authentic characteristics (Drenten/Brooks 2020). HVIs do not exist outside of the digital world and are legally required to communicate that they are a virtual entity (Park et al. 2021).

HVIs are particularly well known on image-based platforms such as Instagram, as they attract a lot of attention due to the human traits of their physical appearance, behaviour and personality, as well as through interaction with their followers (Moustakes et al. 2020). Examples of successful HVIs include Lu of Magalu, Lil Miquela and Shudu. They also have millions of followers and have already collaborated with various companies: Lu of Magalu, for example, had 6.4 million Instagram followers in 2023 and has collaborated with Samsung, McDonald's, Vogue, Adidas and Lancome, among others. In that same year, Lil Miquela had 2.6 million Instagram followers and collaborations with Alexander McQueen, Gucci, Prada, Calvin Klein, Karl Lagerfeld, and more. Shudu, the world's first black virtual influencer, had 241 thousand Instagram followers and has worked with luxury fashion brands Karl Lagerfeld and Louis Vuitton as well as with BMW (Storyclash 2023). It can therefore be assumed that HVIs also make a positive contribution to the advertised product with their appearance, their behaviour and the open communication of their virtual entity.

Compared to their real-life counterparts, HVIs have several advantages for companies: HVIs can be customised with their appearance and personality to fit the target group (Kádeková/Holienčinová 2018). Due to their immateriality, they can be used anywhere and at any time (Berrryman et al. 2021). HVIs also offer particular security, as they themselves cannot exhibit unplanned and inappropriate behaviour or have a negative past that could damage the brand (Drenten/Brooks 2020). In addition, companies are perceived as innovative and technically adept if they create or collaborate with an HVI (Conti et al. 2022).

At the same time, when using HVIs, there is a risk that emotional bonds and expressions will only be accepted to a limited extent by users, due to their virtual entity. The appearance and behaviour of an HVI can blur the fictional and human sides and make people feel more fearful and uncomfortable with the apparent perfection of non-human entities (Park et al. 2021). In addition, users may doubt the trustworthiness and credibility of the HVIs when advertising certain products, such as make-up or facial care, as HVIs cannot physically test these types of products (Kádeková/Holienčinová 2018, Conti et al. 2022).

A study conducted in Germany in 2023 found that 19% of all social media users were aware of HVIs. Among 16–29-year-olds, the figure was 44%. 66% of those who are aware of HVIs also follow them (respectively 76% of 16–29-year-olds). 29% of those who follow HVIs and 21% of those who know HVIs want to try out recommendations. The preferred social media channels are mainly the video-based channels Instagram and TikTok (Deckers et al. 2023).

Against this background, this paper aims to answer the following research questions:

– What is the influence of HVIs on the purchase intention of social media users in Germany?
– Which factors influence the intention to buy a product recommended by an HVI?

2   Conceptual Framework

2.1   Literature Review on the State of Research on Human Virtual Influencers

As a relatively recent phenomenon, HVIs have not been extensively studied from a scientific perspective. Drenten/Brooks (2020) address the contradictions between the artificiality and authenticity of HVIs and HIs, arguing that HVIs are no less real than HIs. HIs operate in an online environment of digitally processed images and filters that, like their virtual counterparts, suggest artificiality.

Robinson (2020) argues in his publication on the ontological and ethical question of HVIs that trust and thus transparency can be created by revealing the fictitious form of an HVI. The associated characteristics of the HVI are on a par with the characteristics of an HI and therefore allow products to be advertised equally successfully. Similar to HVIs, HIs are not always authentic and transparent, but create an illusion that does not exist in reality.

Moustakes et al. (2020) showed in their study that HVIs can achieve an emotional bond by elaborating a profound "life" story. Anthropomorphism in particular is important for building a parasocial relationship and can influence the success of product promotions.

Arsenyan/Mirowska (2021) and Park et al. (2021) conducted data analyses of user comments on HVIs and HIs posts. They found that there is a stronger scepticism and aversion towards an HVI. The empirical study by Cheung/Leung (2021) shows a similar result. In this study, the advertising effectiveness of HVIs was analysed in comparison to AVIs by asking test subjects to comment on products advertised by VIs. For this purpose, theoretical constructs of attractiveness, trust and expertise were used from the celebrity endorsement research. This was used to assess whether VIs are attractive, trustworthy and have the professional expertise for advertising. Significant differences were found between the VIs and HVIs in the theoretical constructs of attractiveness and trust, but not in expertise.

Batista/Chimenti (2021) identified five features that characterize an HVI. Firstly, the attractiveness of an HVI and secondly, the authenticity that is attributed to the human traits of the HVI, both in terms of physical appearance and in terms of the conceived character. Another characteristic is anthropomorphism. According to the authors, anthropomorphism can have both positive and negative effects. For example, anthropomorphism can lead to a friendly rapprochement and parasocial relationship between the HVI and its followers. However, anthropomorphism can also be the cause of a negative attitude based on his virtual entity. The final characteristics from the study are controllability and scalability. Lee/Kim (2022) prove in a study the positive effect of anthropomorphism on the parasocial relationship.

Hofeditz et al. (2022) investigated the construct of trust, which helps the HIs to advertise a company's products. Their study shows that the majority of test subjects were unable to perceive the difference between the HI and the HVI. In order to test the theoretical constructs of trust, anthropomorphism and social presence, the identities of the HVI were revealed in the further course of the survey. The results of the study show that all three constructs are perceived significantly lower for an HVI than for an HI.

To summarise, most studies to date have focused on analysing the perception of an HVI. For this purpose, characteristics for the identification of an HVI are established, which can be derived from the appearance and behaviour of an HVI (Batista/Chimenti 2021). The characteristics have been discussed in various papers, which focus in particular on anthropomorphism with its different effects, e.g. with regard to the parasocial relationship (Moustakes et al. 2020, Batista/Chimenti 2021, Hofeditz et al. 2022, Lee/Kim 2022). The authenticity of an HVI is also controversially discussed in research (Drenten/Brooks 2020, Robinson 2020, Park et al. 2021). So far, the research findings on the perception of an HVI presented above have not addressed the influence on user behaviour, such as the intention to purchase an advertised product.

2.2 Theoretical Model and Variables

When developing the research model, in addition to the findings already mentioned in the literature review, effects from research on HIs were transferred to the HVIs research object. Studies were used that deal with the influence of HIs on the purchasing behaviour of social media users (e.g. Djafarova/Rushworrth 2017, Lou/Yuan 2018, Sokolova/Kefi 2020). In the context of this research, the question arises as to what extent HVIs can also influence the purchasing behaviour of social media users. The dependent variable of the research model is the social media user's intention to purchase a product advertised by the HVI.

In the model developed, the purchase intention is influenced by four independent variables (see Fig. 1). These are the parasocial relationship, anthropomorphism, credibility and authenticity. Demographic characteristics such as age and gender as well as product fit were considered as moderator variables.

Several research results show that the *parasocial relationship* between an HI and its followers influences the purchase intention (Lee/Watkins 2016, Hwang/Zhang 2018, Sokolova/Kefi 2019, Hanief et al. 2019). Parasocial relationships arise from previous parasocial interactions (Schramm/ Hartmann 2010). In a parasocial interaction, the interaction is initiated by the media person, in this case the HI, e.g. by the HI directly addressing the audience or allowing them to participate in their lives (Cohen 1999). This creates a simulated mutual exchange (Schramm/Hartmann 2010). The parasocial relationship with the HVI is defined (as with the HI) as a social behaviour characterised by repeated parasocial interaction between HVI and social media users, which creates an interpersonal relationship.

Sokolva/Kefi (2019) showed that a parasocial interaction between the two actors has a significant positive correlation with the followers' intention to purchase. Lee/Watkins (2016) and Hanief et al. (2019) also found that a long-term and intensive relationship between HI and follower increases the likelihood of a repeat purchase. The advertising messages of the HI are particularly effective due to the establishment of parasocial relationships (Hwang/Zhang 2018). These research findings can also be applied to the effects of the parasocial relationship between HVIs and followers on intent. Accordingly, the following hypothesis is derived: *H1: The parasocial relationship has a positive influence on the purchase intention.*

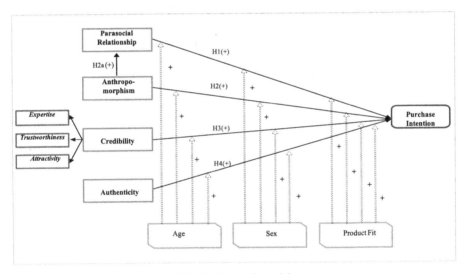

Fig. 1. Research model

Anthropomorphism describes in the present study the human-like appearance of an HVI in facial expressions, gestures and its behaviour towards the social media user. Research in the field of human-robot interaction has intensively investigated anthropomorphism, i.e. the attribution of human characteristics to technology, both from a technological and psychological point of view (Fussel et al. 2008, Bartneck et al. 2008). This research is based on findings from the theory "Computers are Social Actors", according to which people react to computers with social behaviours even though they are aware that they are machines (Nass et al. 1994). People's emotional response to robots initially becomes more positive the more human-like they look. However, this positive reaction reaches a point ("Uncanny Valley") where small but imperfect human-like features lead to the figure being perceived as creepy or scary. Only when the character can no longer be distinguished from a real person does this feeling disappear again (Mori et al. 2012). The phenomenon of the Uncanny Valley also poses a challenge for HVIs.

Several studies suggest that brands employing anthropomorphic advertisers are often viewed more favourably by consumers (Laksmidewi et al. 2017). In addition, they are said to lead to a higher purchase intention than brands that do not use anthropomorphic advertisers (Deligoz/Ünal 2021). Further research has also shown that anthropomorphic entities create increased acceptance and familiarity with the audience and influence purchase intentions (Sheehan et al. 2020). The following hypothesis is derived from this argumentation. *H2: Anthropomorphism has a positive influence on purchase intentions.*

Empirical studies show that anthropomorphism is significantly related to the development of a parasocial relationships between HVIs and social media users. Both qualitative research on HVIs (Moustakas et al. 2020, Batista/Chimenti 2021) and a quantitative study by Lee/Kim (2022) showed that anthropomorphism can strengthen the development of parasocial relationships. Therefore, in addition to the direct effect of anthropomorphism on purchase intention, this study also examines the effect of anthropomorphism on parasocial relationships. *H2a: Anthropomorphism has a positive influence on the parasocial relationship.*

Credibility describes in this study the degree to which social media users are willing to accept the statements of the HVI as valid. Studies in the social media context have already shown that social media users who have an intention to purchase have been influenced by the credibility of the HI (Lee/Kim 2020). Users who perceive an HI as credible are more likely to accept the HI's product promotion, which increases the user's purchase intention (Weismueller et al. 2020, Ata et al. 2022). As HVIs act like their human counterparts, it is assumed that HVIs can also positively influence purchase intentions through their credibility. This results in the following hypothesis. *H3: Credibility has a positive influence on purchase intentions.*

In this paper, credibility is conceptualised as a three-dimensional construct with the dimensions of expertise, attractiveness and trustworthiness of the advertiser (Ohanian 1990).

Expertise refers to the relevant knowledge, experience and skills that a communicator has in relation to a subject matter (Sari et al. 2021). Depending on their area of interest, HIs can be seen as experts in this field. The perceived expertise of the HI is determined by the level of experience and specialist knowledge (Weismueller et al. 2020).

Trustworthiness: Erdogan (1999) defines trustworthiness as the honesty and integrity of an advertiser. In particular, long-term followers who view an HI as trustworthy, caring and professionally competent are more likely to buy the products presented by the HI (Sokolova/Kefi 2019).

Attractiveness: In the context of human advertisers, attractiveness is considered to be the appeal that an advertiser exudes, which attracts the attention of consumers (Sari et al. 2021). In the HI context, the attractiveness of the influencer has an impact on social media user behaviour (Sesar et al. 2022). In particular, the physical appearance of the HI leads to new users following an HI profile on social media (Djafarova/Rushwoorth 2017).

Authenticity is seen as one of the most important factors for the success of HIs (Moore et al. 2018). Lee/Johnson (2021) describe authenticity as a characteristic held by a person who is perceived as sincere and genuine. In terms of HI advertising, authenticity has a positive effect on followers' intention to buy (Pöyry et al. 2019). Authenticity is understood here as the extent to which a social media user perceives the content shown (e.g. the advertising of products) of an HVI as truthful and sincere. Analogue to the findings regarding HI, the following is assumed: *H4: Authenticity has a positive influence on the purchase intention.*

Moderator Variables. Demographic characteristics such as age and gender as well as product fit were considered as moderator variables.

Age. Daily use of social media applications takes place in all age groups, but mainly in the 14 to 29 age group (Beisch/Koch 2021). Studies show that younger users are more influenced by an HI (Skolova/Kefi 2019). There are also differences between 16–29-year-olds and the rest of social media users in Germany in terms of awareness of HVIs and willingness to follow them: while only 19% of all social media users are aware of HVIs, the figure for 16–29-year-olds was 44%. 76% of 16–29-year-olds who know HVIs usually follow them (66% of all social media users) (Deckers et al. 2023). Based on this behaviour, the following moderation hypothesis is formulated: *H-MOD1: The positive influence of the antecedents on the purchase intention is moderated by the age of the user, such that the effect is stronger for younger users.*

Gender: Studies show that women often imitate their favorite celebrities in terms of fashion and lifestyle. This relationship is also evident among HIs and their followers (Djafarova/Rushworth 2017). An analysis of various VI Instagram profiles from 2019 showed that the core target group of HVIs consists of women (Hypeauditor 2019). Based on this behaviour, the following moderation hypothesis is formulated: *H-MOD2: The positive influence of the antecedents on the purchase intention is moderated by the gender of the user, in the sense that the effect is stronger for female users.*

Product Fit: Influencer marketing research examined the fit between the HI and the advertised product. This fit describes the perceived consistency and associative connection between a brand and the advertising (Till/Busler 1998). It has been confirmed that a fit between HI and product can influence the purchase intention of followers (Belanche et al. 2021, Ju/Lou 2022). The following moderation hypothesis is formulated: *H-MOD3: The positive influence of the antecedents on purchase intention is moderated by product fit, such that the effect is stronger when product fit is present.*

Operationalisation. In this paper, existing operationalisations from the literature were adapted to measure the latent constructs and adapted to the object of research. The operationalisation can be seen in Appendix 1. The items were measured using five-point Likert scales (1 = "strongly disagree" to 5 = "strongly agree"). A qualitative (3 interviews) and quantitative pre-test (33 test subjects) took place before the final data collection.

3   Research Design

The data was collected by means of a quantitative survey (standardised online survey). Instagram users in Germany aged between 15 and 60 who use Instagram at least once a week were used as the population for this research project. Instagram was chosen as the social media platform because it is one of the most important social media channels for the use of HVIs (Casaló et al. 2020). The limitation of age is based on the findings on the usage structure by age group on Instagram in Germany (ARD/ZDF-Homepage 2023). Quotas were also set in order to achieve a sufficient number of cases within each gender and age group.

The research object of this paper is the "female" HVI Lil Miquela, a "19-year-old robot living in LA", as stated in "her" bio. Lil Miquela has been active on Instagram since 2016 and had 2.6 million followers on the channel worldwide in early 2024 (@lilmiquella). Lil Miquela behaves like a HI, posts about her leisure activities (e.g. holidays in Barcelona or visits to the hairdresser) and promotes various products and brands. Lil Miquela was created by the design studio Brud from LA (USA). Lil Miquela was chosen because it was not only "the first hyperrealistic virtual influencer to achieve mainstream popularity" and one of the most followed virtual influencers of 2022 (Virtual Humans 2024), but also because it already had advertising partnerships with several world-renowned advertising companies and brands. A selection of these brands includes Samsung, UGG, Calvin Klein, the luxury labels Chanel and Prada, and the car brands Mini and BMW (@lilmiquella).

The survey began with a detailed introduction of the HVI Lil Miquela and its profile on Instagram: Various scenarios in the areas of behaviour, leisure activities and product promotion were shown in the form of photos and videos. The HVI's Instagram profile description and Instagram profile overview were also presented. The questionnaire consisted of 39 questions.

4   Results

4.1   Measurement Validation

The sample comprised 252 participants and was made up of 48% men and 52% women. In accordance with the quota system, two thirds of the test subjects (78%) belonged to the younger age groups (15–39 years). Only 15.5% of the test respondents had experience with HVIs in the past.

Before constructing indices for the subsequent regression analysis, it was first necessary to check the validity of the operationalised latent constructs. The measurement validation took place firstly for the first-order constructs and then for the second-order construct (credibility). This was done in a three-stage procedure according to the approach of Zinnbauer and Eberl (2004). The results (including threshold values) are shown in Table 1.

In the first step, the reliability of these constructs was tested using Cronbach's alpha and item-to-total correlation (ITC). Reliability was confirmed for all constructs. Subsequently, the remaining items were subjected to an unifactorial exploratory factor analysis (with IBM SPSS) to determine whether the indicators represent a single construct (factor). This step also included an assessment of the reliability of the indicators (commonality) and the factor loading. The final step involves carrying out a confirmatory factor analysis with IBM SPSS AMOS. The significance of the factor loadings at the unifactorial level as well as the factor reliability (FR) and the average variance extracted (AVE) are checked first. All constructs met the criteria required in the literature with regard to reliability and convergence validity.

Following the measurement validation of the first-order constructs, the second-order quality test was carried out for the latent construct *credibility* and its dimensions of *expertise*, *trust* and *attractiveness*. The quality tests show that the constructs fulfil the minimum requirements for reliability (Cronbach's alpha: 0.811, ITC > 0.5, commonality > 0.5, factor loading > 0.70) and validity (AVE: 0.843 and FR: 0.907). The significance of the factor loadings of the individual dimensions can also be confirmed.

Subsequently, on a model level, discriminant validity of the 4 antecedents is examined using the Fornell-Larcker criterion: The average recorded variance of a factor should be greater than all its squared correlations with other factors (Zinnbauer/Eberl 2004). As can be seen from Table 2, this criterion is also fulfilled.

The final quality check of the overall model (with AMOS) shows that the required quality measures of the model fit are fulfilled: RMSEA = 0.084 ($\geq$0.08), CMIN/DF = 1.943 ($\leq$2.5) and CFI = 0.939 ($\geq$0.9).

The indexing of the items via the average of the constructs is therefore permitted. The averages (A) of the variables are in the lower range between 1.88 and 2.63. The *purchase intention* of the respondents is rather low (A = 2.04; SD = 1.07). This indicates that the participants have no intention to purchase the product when it is advertised by the HVI.

4.2 Hypotheses Tests

Direct Effects. The hypothesised relationships were examined using multiple regression analysis. The results of the regression analysis are shown in Table 3. The research model makes a highly significant contribution to explaining the purchase intention (adjR^2 = 0.827; F = 162.873).

The *parasocial relationship* (A = 1.88, SD = 0.95) has a strong (ß = 0.468) positive and highly significant (p < 0.001) effect on purchase intention. Accordingly, the hypothesis H1 can be supported.

In contrast, *anthropomorphism* (A = 2.63, SD = 0.97) shows a very weak effect, which is negative contrary to the assumption (ß = - 0.063). However, this result is not significant (p > 0.05). The hypothesis H2 must therefore be rejected.

Credibility (A = 2.33, SD = 1) has a significant medium positive effect (ß = 0.223: p < 0.05) on the purchase intention, the hypothesis can be supported.

The influence of the *authenticity* (A = 2.35, SD = 1.10) shows a highly significant (p < 0.001) medium positive effect (ß = 0.362) on the purchase intention. Hypothesis H4 can be supported.

In addition, the simple regression analysis (adjR^2 = 0.314; F = 62.79) shows a significant influence of the predictor *anthropomorphism* on the *parasocial relationship* (ß = 0.564; p < 0.001). The hypothesis H2a can be supported.

Table 1. Confirmatory factor analysis results including quality criteria

Construct	$\alpha \geq 0.7$	AVE ≥ 0.5	FR ≥ 0.6	Indicator	ITC ≥ 0.5	Sign. FL	FL (CFA) $\geq$ 0.7	Commonality ≥ 0.5
Purchase Intention	0.938	0.855	0.956	PI-1	0.855	***	0.875	0.845
				PI-2	0.823	***	0.833	0.809
				PI-3	0.883	***	0.922	0.878
				PI-4	0.855	***	0.922	0.846
Parasocial Relationship	0.939	0.794	0.953	PR-1	0.803	***	0,838	0,746
				PR-2	0.800	***	0,828	0,742
				PR-3	0.828	***	0,867	0,780
				PR-4	0.832	***	0,860	0,790
				PR-5	0.807	***	0,829	0,753
				PR-6	0.856	***	0,889	0,817
Anthropomorphism	0.871	0.754	0.913	AP-1	0,708	***	0,792	0,709
				AP-2	0,820	***	0,899	0,829
				AP-3	0,762	***	0,820	0,767
				AP-4	0,616	***	0,682	0,589
Expertise	0,925	0,832	0,947	E-1	0,791	***	0,827	0,777
				E-2	0,833	***	0,886	0,825
				E-3	0,848	***	0,878	0,842
				E-4	0,834	***	0,888	0,827
Trustworthiness	0.930	0.840	0.950	TW-1	0,803	***	0,837	0,790
				TW-2	0,821	***	0,880	0,810
				TW-3	0,870	***	0,906	0,865
				TW-4	0,850	***	0,886	0,843
Attractivity	0.898	0.789	0.929	AT-1	0,842	***	0,902	0,845
				AT-2	0,723	***	0,775	0,708
				AT-3	0,821	***	0,845	0,820
				AT-4	0,707	***	0,809	0,691
Authenticity	0.891	0.836	0.933	AU-1	0,814	***	0,884	0,847
				AU-2	0,777	***	0,842	0,811
				AU-3	0,771	***	0,844	0,806
Product Fit	0.919	0.869	0.949	PF-1	0,844	***	0,892	0,869
				PF-2	0,822	***	0,884	0,847
				PF-3	0,840	***	0,892	0,865

*** highly significance on 0.1% level I α = Cronbach´s Alpha I AVE = average variance extracted I FR = factor reliability I ITC = Item-to-Total-Correlation I FL = factor loading

Moderating Effects. The variables *age, gender* and *product fit* were included in the research model as moderators. To investigate the moderating effects, the overall sample was split into groups based on the hypothesised moderators (Urban/Mayerl 2011). The moderator age was split into the two groups younger (15–39 years) and older (40–59 years), the moderator gender was divided into the groups female and male and the

Table 2. Results of the Fornell-Larcker criterion for 2nd order constructs

Fornell-Larcker		PR	AP	C	AU
	DEV	**0,794**	**0,754**	**0,852**	**0,836**
PR	**0,794**	1,00			
AP	**0,754**	0,580	1,00		
C	**0,852**	0,790	0,709	1,00	
AU	**0,836**	0,682	0,604	0,743	1,00

moderator product fit was divided into the groups "non-existing" and "existing" product fit. A multi-group analysis was then carried out at sub-sample level and the difference between the groups was tested for significance using the t-test for independent samples (Table 4):

- No significant group differences can be proven for *age* and the corresponding moderator hypothesis H-MOD1 must be rejected.
- The hypothesis on the moderating effect of *gender* can only be supported for the effect of the parasocial relationship on the purchase intention ($p = 0.033$). For women, the influence of the parasocial relationship on purchase intention is stronger (B = 0.601; $p < 0.001$) than for men (B = 0.440; $p < 0.001$). With regard to the remaining predictors, the H-MOD2 hypothesis must be rejected due to the lack of significance in the group differences.
- The analysis of the influence of the moderating effect of *product fit* shows significant differences in the groups for the parasocial relationship and authenticity ($p = 0.000$).

Table 3. Results of the regression analysis

Dependent Variable: *Purchase Intention* adjR² = 0,827; F = 162,873	Beta (sig.)	adjR² (ordinary linear regression)[a]	Multivariate correlation		Proportion of variance
			Partial	Part	
Parasocial Relationship	**0.468***** (0.000)	0,736***	0,555	0,273	7,45%
Anthropomorphism	−0.063[(ns)] (0.217)	0,314***	−0,108	−0,044	0,19%
Credibility	**0.209**** (0,004)	0,652***	0,247	0,104	1,08%
Authenticity	**0.362***** (0.000)	0,672***	0,476	0,222	4,92%

(ns) = not significant/ *significance on 5%-level/ ** high significance on 1% level/ *** highly significance on 0.1% level

Table 4. Results of hypotheses tests for the moderating effects of Age, Gender and Product Fit

Hypothesis	Moderator		Antecedents of Purchase Intention		
		Parasocial relationship	Anthropomorphism	Credibility	Authenticity
H-MOD1	Age	not supported	not supported	not supported	not supported
H-MOD 2	Gender	supported	not supported	not supported	not supported
H-MOD 3	Product Fit	supported	not supported	not supported	supported

Thus, the influence of the parasocial relationship on the purchase intention is significantly more pronounced with an existing product fit (B = 0.507; p < 0.001) than with a non-existing product fit (B = 0.429; p = 0.003). The influence of authenticity on purchase intention is also moderated by product fit. If product fit is given, there is a greater effect (B = 0.414; p < 0.001) than if product fit is not given (B = 0.264; p = 0.011). With regard to the remaining antecedents, the H-MOD3 hypothesis must be rejected due to the lack of significance in the group differences.

The tested research model can be seen in Fig. 2.

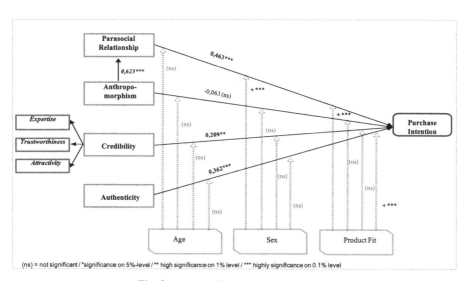

(ns) = not significant / *significance on 5%-level / ** high significance on 1% level / *** highly significance on 0.1% level

Fig. 2. Results of the regression analysis

5 Conclusion

5.1 Discussion and Business Implications

In summary, the present study confirms the model variables that were transferred from the research on HI to the HVI object of investigation. The variables *parasocial relationship, authenticity* and *credibility* have a high to highly significant influence on the *purchase intention*. In contrast, no direct significant influence on the purchase intention could be measured for *anthropomorphism*, the only indicator variable that considers the specifics of HVI´s virtuality and originates from research in the field of human-robot interaction. On the other hand, the indirect effect could be confirmed: As often mentioned in literature, *anthropomorphism* has a direct effect on the *parasocial relationship*. Another result of the present study is that the purchase intention of a product advertised by the analysed HVI is low in Germany.

Based on these findings, recommendations can be made both for the providers of HVIs and for companies wishing to use HVIs as part of their communication strategy.

For Providers of human virtual influencers (HVIs):

- *Focus on parasocial relationships:* Develop strategies that foster a sense of connection between HVIs and their followers. This could involve interactive content or storytelling that resonates with the audience, enhancing the perceived closeness with the HVI. It should be noted that the various measures for building a parasocial relationship cannot be implemented in the short term, however, because a parasocial relationship only develops through repeated parasocial interactions.
- *Leverage anthropomorphism:* Focus on creating HVIs with human-like attributes and the ability to establish strong parasocial relationships. To reduce negative reactions from the Uncanny Valley effect, it's important to clearly represent HVIs as computer-generated entities, emphasizing their virtual nature.
- *Build credibility:* Ensure that HVIs demonstrate expertise, trust, and attractiveness as these attributes significantly boost their credibility. This could involve showcasing their knowledge about the product or industry and creating a trustworthy persona.
- *Ensure authenticity:* Maintain a high level of authenticity in the HVIs' interactions and content. Authenticity significantly impacts purchase intentions, especially when there is a strong product fit.
- *Secure Product Fit:* Align the HVI's persona and the products they promote. A strong product fit amplifies the impact of parasocial relationships and authenticity on purchase intentions.
- *Tailor HVI strategies for different audiences:* Recognize that different audience segments may respond differently to HVIs. Customize content and engagement strategies to cater to these varying preferences and responses. This also means that it is not possible to cover the entire market with a single HVI. It may well make sense to create several HVIs.

Recommendations for companies aiming to use HVIs in their communication strategies to improve engagement and influence purchase decisions:

- *Strategic HVI selection:* Choose HVIs that align with your brand values and the products being promoted, ensuring a strong product fit. The HVIs should have a strong parasocial relationship with their community.

- *Focus on authenticity and credibility:* Utilize HVIs that not only fit your product but also exhibit authenticity and credibility.
- *Consider gender moderation effects:* Tailor strategies for HVIs keeping in mind that women might respond more strongly to parasocial relationships than men.

5.2   Limitations and Future Research

This study focussed on Germany, where there are comparatively few HVIs and therefore consumers have limited experience with them. We would expect significant further insights from testing the developed model in other countries. Further studies should also consider the influence of "experience with HVIs" as an additional factor (either as an antecedent or moderator variable). The subject of this study is the "female" HVI Lil Miquela and "her" profile on Instagram, which may limit generalisability. Future research should include different HVIs across genders, age groups and social media platforms. The influence of anthropomorphism on purchase intention should be further investigated, especially considering the Uncanny Valley effect. Also, the aspect "language and voice" will most probably play a significant role. A methodological restriction of this research is the sample size, indicating the need for larger samples in the future. Finally, an analysis comparing human influencers and virtual influencers would provide us with an indication of what the future upward potential, as well as limitations, of the virtual influencer phenomenon can be.

Appendix 1: Operationalization of the Constructs

Purchase Intention		Source
PI-1	If a product promoted by the HVI were shown to me on Instagram, I could imagine buying it	Hwang/Zhang (2018)
PI-2	I might consider purchasing the product that the HVI has advertised on Instagram	Hwang/Zhang (2018)
PI-3	I would be interested in buying the product that the HVI has promoted on Instagram	Hwang/Zhang (2018)
PI-4	It is likely that I would buy the product that the HVI has advertised on Instagram in the future	Hwang/Zhang (2018)
Parasocial relationship		
PR-1	If an HVI were shown to me on Instagram, I would be pleased to see video or posting content from him	Lee/Watkins (2016)
PR-2	If the HVI were to appear on another social media channel, such as YouTube or TikTok, I would want to view his content	Lee/Watkins (2016)
PR-3	Seeing video or posting content from the HVI would make me feel like part of his friend group	Lee/Watkins (2016)
PR-4	I think it's likely that I would come to view the HVI as an old friend	Lee/Watkins (2016)
PR-5	If I didn't see video or posting content from the HVI for a long time, I would miss them	Perse/Rubin (1989)
PR-6	If I were to see video or posting content from the HVI, I would feel feel like I was among friends	Lee/Watkins (2016)
Anthropomorphism		
A-1	The HVI seems natural to me	Bartneck et al. (2008)
A-2	The HVI seems human to me	Bartneck et al. (2008)
A-3	The HVI looks lifelike to me	Bartneck et al. (2008)
A-4	The HVI seems skilful in its movements to me	Bartneck et al. (2008)
Expertise		
E-1	I have the feeling that the HVI knows a lot about the advertised content/product	Munnukka et al. (2016)

(continued)

(continued)

Purchase Intention		Source
E-2	I think the HVI is competent enough to make statements about the advertised content/product	Munnukka et al. (2016)
E-3	I consider the HVI to be an expert on the advertised content/product	Munnukka et al. (2016)
E-4	I consider the HVI to be sufficiently experienced in his/her field to make statements about the advertised content/the advertised product	Munnukka et al. (2016)
Trustworthiness		
TW-1	I think the HVI seems honest with its followers	Munnukka et al. (2016)
TW-2	I think the HVI is trustworthy	Munnukka et al. (2016)
TW-3	I think the HVI appears credible	Munnukka et al. (2016)
TW-4	I think the HVI comes across as serious	Munnukka et al. (2016)
Attractivity		
AT-1	I find the HVI attractive	Munnukka et al. (2016)
AT-2	I find the HVI very stylish	Munnukka et al. (2016)
AT-3	I find the HVI good-looking	Munnukka et al. (2016)
AT-4	I find the HVI physically attractive	Munnukka et al. (2016)
Authenticity		
AU-1	I feel that the HVI conveys information about the advertised content/the advertised product as authentic	Cabeza- Ramírez et al. (2022)
AU-2	I feel that the HVI stands behind the advertised content/product	Cabeza- Ramírez et al. (2022)
AU-3	I think that the HVI's opinion about the advertised product is reliable	Cabeza- Ramírez et al. (2022)
Product Fit		
PF-1	The HVI and the advertised content/product are a good fit	Belanche et al. (2021)
PF-2	The HVI and the advertised content/product show a high degree of fit	Belanche et al. (2021)
PF-3	The HVI and the advertised content/product are well harmonised	Belanche et al. (2021)
Scale	1 = "strongly disagree" to 5 = "strongly agree"	

References

ARD/ZDF Homepage (2023). https://www.ard-zdf-onlinestudie.de/tabellen-onlinenutzung/soc
ial-media-und-messenger/social-media/. Accessed 02 Mar 2023

Arsenyan, J., Mirowska, A.: Almost human? A comparative case study on the social media presence
of virtual influencers. Int. J. Hum.-Comput. Stud. **155**, 1–16 (2021)

Ata, S., Arslan, H.M., Baydas, A., Pazvant, E.: The effect of social media influencers' credibility
on consumer's purchase intentions through attitude toward advertisement. In: ESIC MARKET
Economic and Busoness Journal, vol. 53, no. 1, pp. 1–21 (2022)

Bartneck, C., Kulic, D., Croft, E.: Measuring the anthropomorphism, animacy, likeability, perceived intelligence, and perceived safety of robots. In: 3rd ACM/IEEE International Conference
on Human-Robot Interaction (HRI), pp. 37–44 (2008)

Batista, A.S.O., Chimenti, P.: Humanized robots: a proposition of categories to understand virtual
influencers. In: Australasian Journal of Information Systems, Vol. 25, September 2021, pp. 1–27
(2021)

Beisch, N., Koch W.: Media Perspektiven (2021). https://www.ard-zdf-onlinestudie.de/files/2021/
Beisch_Koch.pdf. Accessed 28 May 2022

Belanche, D., Flavian, M., Ibanez-Sanchez, S., Arino, L.V.C.: Understanding influencer marketing:
the role of congruence between influencers, products and consumers. In: Journal of Business
Research, vol. 132, no. 2, pp. 1–35 (2021)

Casalo, L., Flavian, C., Ibáñez-Sánchez, S.: Influencers on instagram: antecedents and consequences of opinion leadership. In: Journal of Business Research, pp. 510–519 (2020)

Cheung, F., Leung, W.F.: Virtual influencer as celebrity endosers. In: Advances in Global Services
and Retail Management, pp. 1–7 (2021)

Cohen, J.: Statistical Power Analysis for the Behavioral Sciences, 2nd edn. Lawrence erlbaum
associates, New York (1988)

Cohen, J.: Favorite characters of teenage viewers of Israeli serials. In: Journal of Broadcasting &
Electronic Media, vol. 43, no. 3, pp. 327–345 (1999)

Conti, M., Gathani, J., Tricomi, P.P.: Virtual Influencer in Online Social Media. In: IEEE
Communications Magazine, vol. 60, no. 8, pp. 1–13 (2022)

Deligoz, K., Ünal, S.: The effect of anthropomorphic mascot on the purchasing intention of
consumers: an expertimental study. In: Sosyoekonomi, vol. 29, no. 50, pp. 229–254 (2021)

Dencheva, V.: (2023). https://www.statista.com/statistics/1333431/top-platforms-influencer-mar
keting-us/. Accessed 14 Dec 2023

Djafarova, E., Rushworth, C.: Exploring the credibility of online celebrities' Instagram profiles
in influencing the purchase decisions of young female users. Comput. Hum. Behav. **68**, 1–7
(2017)

Drenten, J., Brooks, G.: Celebrity 2.0: Lil Miquela and the rise of virtual star system. In: Feminist
Media Studies, vol. 20, no. 8, pp. 1–5 (2020)

Drückers, R., Weinand, A.L., Frings, J.: Trend Check Handel, Vol. 7 (2023)

Erdogan, B.Z.: Celebrity endorsement: a literature review. In: Journal of Marketing Management,
vol. 15, no. 4, pp. 291–314 (1999)

Fussell, S.R., Kiesler, S., Setlock, L.D., Yew, V.: How people anthropomorphize robots. In:
ACM/IEE International Conference on Human-Robot Interaction (HRI), pp. 145–152 (2008)

Hanief, S., Handayani, P.W., Azzahro, F., Pinem, A.A.: Parasocial relationship analysis on digital
celebrities follower's purchase intention. In: 2nd International Conference of Computer and
Informatics Engineering, pp. 12–17 (2019)

Hofeditz, L., Nissen, A., Schütte, R., Mirbabaie M.: Trust Me, I'm an influencer!- A comparison
of perceived trust in human and virtual influencers. In: Conference: European Conference of
Information Systems, pp. 1–11 (2022)

Hwang, K., Zhang, Q.: Influence of parasocial relationship between digital celebrities and their followers on followers' purchase and electronic word-of-mouth intentions, and persuasion knowledge. Comput. Hum. Behav. **87**, 1–56 (2018)

Hypeauditor Homepage (2019): The Top Instagram Virtual Influencers in 2019: https://hypeauditor.com/blog/the-top-instagram-virtual-influencers-in-2019/. Accessed 02 Oct 2023

Ju, I., Lou, C.: Does influencer–follower relationship matter? Exploring how relationship norms and influencer–product congruence affect advertising effectiveness across product categories. In: Journal of Interractive Advertrising, vol. 22, no. 1, pp. 1–21 (2022)

Kádeková, Z., Holienčinová, M.: Influencer marketing as a modern phenomenon creating a new frontier of virtual opportunities. In: Communication Today, vol. 9, no. 2, pp. 90–104 (2018)

Kepios Homepage (2024). https://datareportal.com/social-media-users#:~:text=Detailed%20analysis%20by%20the%20team,of%20the%20total%20global%20population. Accessed 02 Jan 2024

Ki, C., Kim, Y.: The mechanism by which social media influencers persuade consumers: the role of consumers' desire to mimic. In: Psychology & Marketing, vol. 36, no. 10, pp. 905–922 (2019)

Laksmidewi, D., Susianto, H., Afiff, A.Z.: Anthropomorphism in advertising: the effect of anthropomorphic product demonstration on consumer purchase intention. In: Asian Academy of Management Journal, vol. 22, no. 1, pp. 1–25 (2017)

Lee, J., Kim, H.: Virtual influencers as friends or rivals: the effects of virtual influencer attributes on intention to imitate and word-of-mouth through parasocial relationship and trust. In: SSRN Electronic Journal, pp. 1–16 (2022)

Lee, S.S., Johnson, B.K.: Are they being authentic? The effects of self-disclosure and message sidedness on sponsored post effectiveness. In: International Journal of Advertising, vol. 41, no. 1, pp. 1–24 (2021)

Lee, J.E., Watkins, B.: YouTube vloggers' influence on consumer luxury brand perceptions and intentions. In: Journal of Business Research, vol. 69, no. 12, pp. 5753–5760 (2016)

Lou, C., Yuan, S.: Influencer marketing: how message value and credibility affect consumer trust of branded content on social media. In: Journal of Interactive Advertising, vol. 19, no. 1, pp. 1–45 (2018)

Miquela (@lilmiquella): https://www.instagram.com/lilmiquela/. Accessed 01 Oct 2024

Moore, A., Yang, K., Kim, H.M.: Influencer marketing: influentials' authenticity, likeability and authority in social media. In: International Textile and Apparel Association Annual Conference Proceedings, vol. 75, no. 1, pp. 1–3 (2018)

Mori, M.: The uncanny valley. In: IEEE Robotics Automation Magazine, pp. 98–100 (2012)

Moustakas, E., Limba, N., Mahmoud, D., Chandrasekaran, R.: Blurring lines between fiction and reality: perspectives of experts on marketing effectiveness of virtual influencers. In: IEEE 2020 International Conference on Cyber Security and Protection of Digital Services (Cyber Secruity), pp. 1–6 (2020)

Munnukka, J., Uusitalo, O., Toivonen, H.: Credibility of a peer endorser and advertising effectiveness. In: Journal of Consumer Marketing, vol. 33, no. 3, pp. 182–192 (2016)

Nass, C., Steuer, J., Siminoff, E.: Computer are social actors. In: Conference on Human Factors in Computing Systems, CHI 1994, Boston, Massachusetts, USA, Conference Companion, pp. 24–28. April: 72–78 (1994)

Ohanian, R.: Construction and validation of a scale to measure celebrity endorsers' perceived expertise, trustworthiness, and attractiveness. In: Journal of Advertising, vol. 19, no. 3, pp. 39–52 (1990)

Park, G., Nan, D., Park, E., Kim, K.J., Han, J., del Pobil, A.P.: Computers as social actors? Examining how users perceive and interact with virtual influencers on social media. In: 15th International Conference on Ubiquitous Information Management and Communication (IMCOM), pp. 1–6 (2021)

Perse, E.M., Rubin, R.B.: Attribution in Social and parasocial Relationships. In: Communication Research, vol. 16, no. 1, pp. 59–77 (1989)

Pöyry, E., Pelkonen, M., Naumanen, E., Laaksonen, S.-M.: A call for authenticity: audience responses to social media influencer endorsements in strategic communication. In: International Journal of Strategic Communication, vol. 13, no. 4, pp. 1–36 (2019)

Robinson, B.: Towards an ontology and ethics of virtual influencers. Australas. J. Inf. Syst. **24**, 1–8 (2020)

Sari, Y.M., Hayu, R.S., Salim, M.: The effect of trustworthiness, attractiveness, expertise, and popularity of celebrity endorsement. In: Jurnal Manajemen dan Kewirausahaan, vol. 9, no. 2, pp. 163–172 (2021)

Schmidt, J.H., Taddicken, M.: Handbuch Soziale Medien, 1st edn. Springer Fachmedien, Wiesbaden (2017)

Schramm, H., Hartmann, T.: Identität durch Mediennutzung? Die Rolle von parasozialen Interaktionen und Beziehungen mit Medienfiguren. In: Mediensozialisationstheorien. Modelle und Ansätze in der Diskussion. 2nd edn. In: D. Hoffmann und L. Mikos. (eds.): VS Verlag für Sozialwissenschaften, Wiesbaden, pp. 201–219 (2010)

Sesar, V., Martincevic, I., Boguszewicz-Kreft, M.: Relationship between advertising disclosure, influencer credibility and purchase intention. In: Journal of Risk and Financial Management, vol. 15, no. 7, pp. 1–21 (2022)

Sheehan, B., Seung, H., Gottlieb, U.: Customer service ChatBots: Anthropomorphism and adoption. J. Bus. Res. **115**, 14–24 (2020)

Sokolova, K., Kefi, H.: Instagram and YouTube bloggers promote it, why should I buy? How credibility and parasocial interaction influence purchase intentions. In: Journal of Retailing and Consumer Services, vo. 53, no. 1, pp. 1–16 (2019)

Storyclash Homepage (2023): Top 10 Virtual Influencers Rocking the Fashion Industry. https://www.storyclash.com/blog/en/virtual-influencers/. Accessed 21 Dec 2023

Till, B.D., Busler, M.: Matching products with endorsers: attractiveness versus expertise. In: Journal of Consumer Marketing, vol. 15, no. 6, pp. 576–586 (1998)

Urban, D., Mayerl, J.: Regressionsanalyse: Theorie, Technik und Anwendung, 4th edn. VS Verlag für Sozialwissenschaften, Wiesbaden (2011)

Virtual Humans (2024). https://www.virtualhumans.org. Accessed 12 Jan 2024

Weismueller, J.: Influencer endorsements: how advertising disclosure and source credibility affect consumer purchase intention on social media. In: Australasian Marketing Journal, pp. 1–11 (2020)

Wibawa, R.C., Pratiwi, C.P., Wahyono, E., Hidayat, D., Adiasari, W.: Virtual influencers: is the persona trustworthy? In: Jurnal Manajemen Informatika (JAMIKA), vol. 12, no. 1, pp. 51–62 (2022)

Zinnbauer, M., Eberl, M.: Die Überprüfung von Spezifikation und Güte von Strukturgleichungsmodellen – Verfahren und Anwendung. In: Schriften zur Empirischen Forschung und Quantitativen Unternehmensplanung (EFOplan), 21 (2004)

Assessing the Influence Mechanism of Media Richness on Customer Experience, Trust and Swift Guanxi in Social Commerce

Kaiyan Zhu[(⊠)] ⓘ, Caroline Swee Lin Tan ⓘ, and Tarun Panwar ⓘ

RMIT University, Melbourne, VIC 3000, Australia
`kelly.zhu@rmit.edu.au`

Abstract. In contrast to traditional e-commerce, social commerce features deliver seller offerings and user-generated content (UGC) to facilitate consumer interactions. Social commerce sellers convey this abundant information using various media types, including livestreams, short videos, and social media channels. However, there is a lack of research on the influence of rich media content on consumers' attitudes and relationships with the seller. This empirical study focuses on live streaming commerce and group buying for fashion products in mainland China, where both social commerce formats are growing fast. Drawing on media richness theory, we propose a conceptual framework to quantify the influence mechanisms of media richness. Our framework encompasses three sets of hypotheses. First, it was proposed that media richness directly contributes to swift guanxi ("guanxi" in Chinese means relationship), customer experience, and trust. Second, to understand how the consumer responds to media richness, we hypothesized that customer experience directly influences swift guanxi and trust. The final hypothesis evaluates whether live streaming commerce, with its rich content akin to face-to-face communication, can positively moderate these established causal connections. Based on an analysis using partial least square structural equation modelling (PLS-SEM) with 694 valid respondents, we found that media richness directly influences swift guanxi and customer experience, but does not directly influence trust. Furthermore, customer experience dominates in transferring the influence of media richness to swift guanxi. Additionally, media richness positively affects trust only by enhancing customer experience. Unexpectedly, live streaming commerce cannot magnify the influence of media richness on the examined psychological factors. This result has managerial implications: live streaming commerce has no advantage over group buying based on its richer media content in forming a more harmonious and trusting relationship with social commerce sellers. Social commerce adopters should focus on the quality and efficiency of media content, rather than information quantity, to facilitate a satisfying consumer experience. We also discuss this study's limitations and offer suggestions for future research.

Keywords: Media richness · Customer experience · Trust · Swift guanxi · Social commerce · Live streaming commerce · Group buying

© The Author(s), under exclusive license to Springer Nature Switzerland AG 2024
F. F.-H. Nah and K. L. Siau (Eds.): HCII 2024, LNCS 14720, pp. 127–142, 2024.
https://doi.org/10.1007/978-3-031-61315-9_9

1 Introduction

Online shopping has changed dramatically since 2016, with online retail businesses evolving from website stores and online marketplaces to mobile commerce [1] and, more recently, to social commerce [2]. Worldwide sales revenue from social commerce skyrocketed to USD724 billion in 2022 and is forecast to surpass USD6 trillion by 2030 [3]. Social commerce includes all e-commerce formats primarily using UGC, such as purchasing on social media [4]. Two social commerce formats, live streaming commerce and group buying, are spreading with astonishing speed in China [5, 6]. The core of group buying is to accumulate order quantities to achieve the seller-requested *tipped point*, after which consumers enjoy discounts [7]. In comparison, live streaming commerce emphasizes the anchors' role in facilitating consumers' synchronous communications with each other and the anchor. Conveying information is critical in either format, represented as efficiently delivering UGC and seller-offered information between consumers and the seller. Social commerce sellers employ various media types, featuring richer media content, to deliver abundant information and attract consumers to spend a longer time interacting with the content. Besides text and static images which dominate traditional e-commerce formats, social commerce has been embedded with GIFs, short videos, live streams, virtual realities, voice, and even reality shows. Live streaming commerce emphasizes synchronous interaction media, such as video livestreams. By contrast, group buying focuses on media tools, such as social media channels and pre-recorded short videos, that can be shared easily with friends, families, and other potential consumers.

Given the importance of media in conveying information, there is a lack of social commerce research focusing on the influence of rich media content, or media richness [8]. The existing limited studies all focused only on live streaming commerce [e.g., 9, 10]. For instance, Chen et al. [9] examined the influence of media richness as one dimension of live streaming services on trust and stickiness. Stickiness is a key aspect of the consumer-seller relationship. It refers to the willingness of users to frequently return to, browse, and use a website. Song and Liu's [10] work identified the causal connection between media richness, consumers' perceived purchase risk and purchase intention. However, the influence mechanism of media richness on consumer psychological factors is still unknown. There is no evidence about the effect of media richness in group buying. As a result, investigations of media richness have not encompassed the broad scope of social commerce.

To contribute insights into this scarcity, we aim to achieve two research purposes in the current empirical study:

1. Quantify the effects of media richness on consumers' primary psychological attitudes, including swift guanxi, customer experience and trust. All these attitudes determine consumers' shopping decisions.
2. Evaluate the difference in the influence mechanism of media richness between live streaming commerce and group buying.

This study starts by introducing the related work on examined variables. It then proposes a conceptual framework and presents the development of the hypotheses. Following this, the paper describes the study method, data collection, and analysis process, before discussing the results and implications. We finish this paper by noting the study's limitations and suggesting future research opportunities.

2 Related Work and Hypotheses Development

2.1 Media Richness

Media richness theory is based on task-media fit [8]. It relates to information transmutability carried by media channels. Media richness is a combination of four criteria: speed of communication, multiple cues, language variety, and degree of personalization. Media richness was termed vividness by Hoffman and Novak [11], and has been used in e-commerce to examine media's informativeness and richness. In the context of online shopping, Klein [12] redefined media richness as the sensory breadth (number of communication channels) and depth (quality within each channel) of the stimuli in the online shopping setting. Media richness links to 5 human senses (i.e., sight, sound, smell, taste, and touch), which have different weights in forming overall experience [9]. Among all media types utilized by social commerce platforms, livestream possesses high media richness, similar to face-to-face communications, and conveys the richest content. Following a sequence from high to low, livestream's media richness is higher than that of videos, voice messages, images, and text [13]. Livestream features real-time communications, text messages, broadcast voice, and videos.

Media richness has been found to impact social commerce consumers' behaviors, such as affecting attraction [14]. With the development of advanced communication technologies and the emergence of numerous affordable mobile telecom devices, more media channels with rich breadth and depth have been applied widely in e-commerce.

2.2 Consumer Psychological Factors

In the current study, we selected two psychological factors, swift guanxi and trust, to represent the relationship between the consumer and the seller. Additionally, a consumer attitude factor – customer experience – has been chosen as the mediatory factor. Based on media richness theory, we propose a conceptual framework, as shown in Fig. 1.

Swift Guanxi. Swift "guanxi" is a concept rooted in a Chinese term referring to "relationship" [15]. Men and Zheng [16] equate swift guanxi with interpersonal relationships between the seller, focal consumer, and other consumers in live streaming commerce. It represents a unique Chinese consumer consumption pattern [e.g., 17, 18, 19]. For example, swift guanxi in live streaming commerce directly impacts consumers' purchase intention [20]. Hossain et al.'s [20] work proposed a conceptual framework involving interactivity, swift guanxi, and consumer purchase intention. Specifically, interactivity impacts swift guanxi positively, and the latter can positively affect consumers' "add-to-cart" behavior in the livestream.

No study has examined swift guanxi in the broader social commerce context, nor examined the relationship between media richness and swift guanxi. A recent live streaming commerce study from Guo et al. [21] recognized that the anchor can initiate swift guanxi because the anchor plays multiple roles in the live chat room: model, shopping

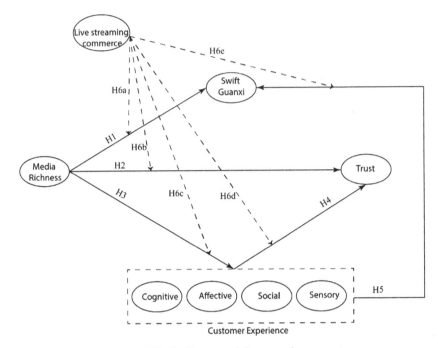

Fig. 1. Conceptual Framework

guide, and seller. During broadcasts, various media types enable the anchor to interact more with the viewer, forming a more harmonious relationship. For example, an elite livestream anchor, *Jiaqi Li*, utilized multiple media types for promoting cosmetics products during 2021's *Double 11* shopping festival, including video clips, social media posts, livestream sessions, and a reality show called *Offers for All Girls*[1] (refer to Fig. 2). He displayed the whole process to his viewers, such as how he negotiated with brands to achieve a product promotion, detailed product information, and backstage stories for each item.

Group buying does not emphasize the livestream; rather it emphasizes social media channels. Take Pinduoduo, a leading group buying platform in China as an example. Consumers can interact with each other by playing games, such as *Price Chop*, through internal social media – *Pinxiaoquan*, and external social media channels (e.g., *WeChat*). *Price Chop* is designed to enable users to invite friends or family members to share products and obtain promoted products at a reduced cost or even for free [22]. In both scenarios of live streaming commerce and group buying, swift guanxi may be formed under the influence of media richness. Therefore, our first hypothesis is:

H1. Media richness in social commerce positively influences swift guanxi.

---

[1] There were 7 episodes in the first season of *Offer for All Girls* in 2021. Its season 2 (2022) and season 3 (2023) also attracted a large number of views. It is the most successful livestream-related reality show in mainland China.

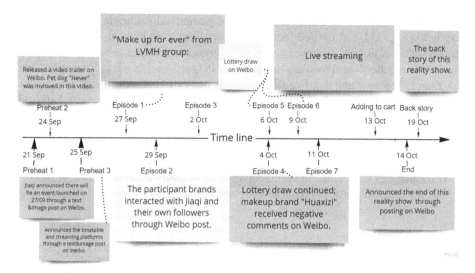

Fig. 2. Context of *Offer for All Girls* reality show, and same topic livestreams, 2021

Trust. As a psychological response of consumers, trust in e-commerce refers to the belief in something or someone based on their characteristics, such as goodness, fairness, honesty, competence, and many others [23]. Trust is a critical factor in internet-based customer behavior [24], such as a consumer purchase stimulator in live streaming [e.g., 25]. In the live streaming commerce research field, trust can be divided into trust in sellers and trust in products [21]. Guo et al. [21] posit that trust is established based on the information conveyed by the anchor, community members, and the product. No research has investigated the role of media richness in group buying. Nadeem et al. [26] found that members in live streaming communities play a critical role in forming and maintaining customer trust. However, richer information conveyed by rich media content can improve consumers' satisfaction based on online consumer research [27]. Hu et al. [28] recognized that the trust relationship between customers and other community members creates a sense of online identification. Hence, we proposed the following hypothesis:

H2. Media richness in social commerce positively influences trust.

Customer Experience. Customer experience originates from various interactions among consumers, and between individual consumer's and the seller's offerings [29]. Klaus and Maklan [30] defined customer experience as a key determinant of consumer behavior. Customer experience is an "anchor point" when studying e-commerce formats from the consumer perspective [27]. Researchers have focused on investigating the influence mechanism from customer experience on consumers' decisions in traditional e-commerce, but researchers thus far have not investigated social commerce. This scarcity of academic inquiry is confirmed by a systematic review undertaken by Dhaigude and Mohan [31]. Their study suggested that future studies should focus on researching the antecedents and outcomes of customer experience. We selected customer experience as an outcome of media richness, following existing empirical studies in the online shopping field [e.g., 11], and posited media richness as the stimulator of swift guanxi and trust. Bilgihan et al. [32] proposed an online customer experience conceptual model, in which media richness and trust are the stimulus and response factor, respectively. Chen and Yang [33] also found that live streaming customer experience directly drives influencer trust. Thus, we hypothesized H3 in the broader social commerce context. No social commerce study yet has measured the customer experience, but consumers' attitudes always form earlier than their relationship formulation, such as swift guanxi and trust. Therefore, we are proposing and examining hypotheses H4 and H5.

H3. Media richness in social commerce positively influences customer experience.

H4. Customer experience in social commerce positively influences trust.

H5. Customer experience in social commerce positively influences swift guanxi.

2.3 Social Commerce

Social commerce emphasizes consumers' social experience during online journeys rather than immediate purchase results [34]. Li [24] claims that comments and reviews, ratings, recommendation lists, tags, and user profiles can encourage consumers to share personal experiences after use. These UGC features may enhance user participation and facilitate Word-of-Mouth (WOM), communication, social interaction, and sharing. Live streaming commerce emerged later than group buying. It was not considered in relevant social commerce studies until recently [e.g., 6, 35], with a pilot study [36] asserting that group buying is a social commerce format. Live streaming commerce emphasizes more real-time interactions than group buying, and therefore live streaming consumers may expect higher informativeness than group buying. Furthermore, face-to-face communication is the richest media type [13] and livestream mimicked face-to-face online. By comparison, information conveyed via group buying is still dominated by the text and static image so that consumers can share it easily. Therefore, we hypothesized H6:

H6. Live streaming commerce enhances at least one of the following relationships within the model:

a) media richness and swift guanxi;
b) media richness and trust;
c) media richness and customer experience;
d) customer experience and trust;
e) customer experience and swift guanxi.

3 Method

3.1 Sample

The partial least square structural equation modelling approach is employed during the statistical analysis procedure, specifically SPSS 28.0 and SmartPLS 4.0. We conducted an online survey among social consumers who recently purchased fashion products from live streaming commerce or group buying in mainland China. A total of 694 valid responses was collected, including 374 live streaming shoppers and 320 group buying participants, as shown in Table 1. Despite the number differences, the ratio of all sample demographics between the two formats is similar.

Table 1. Sample demographics

Measure	Item	*n*	%	LSC	GB
Gender	Male	225	32.4	119	106
	Female	469	67.6	255	214
Age	18–25 years old	111	16.0	46	65
	26–41 years old	534	76.9	303	231
	42–57 years old	39	5.6	20	19
	58–67 years old	10	1.5	5	5
Online shopping frequency	Rarely/I don't remember	9	1.3	3	6
	Once a month	37	5.3	23	14
	Once every fortnight	101	14.6	57	44
	Once a week	193	27.8	96	97
	Between 2–4 times a week	300	43.2	167	133
	5 times or above a week	54	7.8	28	26
Social media usage frequency	Do not/rarely use social media	5	0.7	2	3
	Several times a day	150	21.6	75	75
	1–3 h per day	228	32.9	129	99
	3–5 h per day	190	27.4	106	84
	Almost constantly	121	17.4	62	59

Notes: LSC–live streaming commerce; GB–group buying; *n*–number; %–percent

3.2 Variables

Table 2 reports the adapted measures, which include media richness [10], swift guanxi [21], customer experience [37] and trust [38]. Among these, customer experience is a higher-order construct with four lower-order variables: cognitive, affective, social, and sensory. The 7-point Likert scale was anchored with (1) "strongly disagree" to (7) "strongly agree" for all variables, excepting the control variable–social commerce format.

Table 2. Measurement items

Items	Measure Items
MR_1	Media Richness [10] Social commerce platforms allow me to tailor my messages for others based on my personal requirements (e.g., virtual gifts and emojis)
MR_2	Social commerce platforms can communicate a variety of different cues in the messages (such as emotional tone, attitude, or formality) compared to buying from a website

(*continued*)

Table 2. (*continued*)

Items	Measure Items
MR_3	Social commerce platforms allow all participants to use rich and varied language (e.g., multiple perspectives, text + image; short videos; livestream videos; animated GIFs; catwalk videos; customer review videos; augmented reality product view; avatars; descriptive videos; haptic feedback)
	Swift guanxi [21]
SG_1	The seller/anchor, other consumers and I can understand each other
SG_2	The seller/anchor, other consumers and I treat each other as we treat our friends
SG_3	The seller/anchor, other consumers and I have harmonious relationships
	Customer experience [37]
CCX_1	Information obtained from social commerce platforms is useful
CCX_2	I learned a lot from using social commerce
CCX_3	I think the information obtained from social commerce platforms is helpful
ACX_1	Shopping for products from social commerce platforms is fun
ACX_2	Shopping for fashion products from social commerce platforms is enjoyable
ACX_3	Shopping for fashion products from social commerce platforms is entertaining
OCX_1	There is a sense of human contact on social commerce platforms
OCX_2	There is a sense of human warmth on social commerce platforms
OCX_3	There is a sense of human sensitivity on social commerce platforms
SCX_1	The product presentation on social commerce platforms is lively
SCX_2	I can acquire product information on social commerce platforms from different sensory channels (e.g., imagery, auditory)
SCX_3	Social commerce contains product information exciting to the senses
	Trust [38]
TR_1	I believe in the shopping information that the anchor/ seller provides
TR_2	I do not think that the anchor/seller will take advantage of me
TR_3	I believe that I will be able to use products like those demonstrated on social commerce platforms
TR_4	I believe that the products I receive will be the same as those shown on Social commerce platforms

Notes: MR–media richness; SG–swift guanxi; CCX–cognitive customer experience; ACX–affective customer experience; OCX–social customer experience; SCX–sensory customer experience; TR–trust

4 Results

4.1 Data Analysis and Model Estimation Results

We first conducted a normality test before proceeding with the exploratory factor analysis (EFA). All sample data can be recognized as normally distributed because the highest skewness value is less than the abstract value of 1.3 (item TR_3: -1.246), within the acceptable value of 2 [39]. All measurement items remain according to the EFA results. Notably, customer experience is a higher-order factor. Hence, the EFA of customer experience items was conducted through a separate 4 factor extraction. In the reliability and validity test in PLS-SEM estimation, all variables' factor loadings were examined, and all were greater than the acceptable value of 0.70 at the significant level as all p values were less than 0.05. Table 3 reports composite reliability (CR) values for all variables, ranging from 0.774 for cognitive customer experience to 0.865 for trust, both above the acceptable level of 0.7. Correspondingly, all average variance extracted (AVE) values are also above the cut-off value of 0.5 [39]. Furthermore, we conducted a Fornell-Larcker criterion analysis to check the discriminant validity. The square root of each variable's AVE value is greater than the factor correlations, indicating that discriminant validity is well established. Due to the higher-order construct of customer experience, we referred to Sarstedt et al. [40] who suggest using a two-stage approach to analyze such constructs reliably. The two-stage approach is used to derive latent variable scores for lower-order factors of customer experience, specifically, cognitive, affective, social and sensory dimensions. These scores are then used to compute the higher-order customer experience variable. Therefore, Table 3 does not report factor correlations for lower-order factors of customer experience.

Table 3. Test results for reliability and discriminant validity

Variables	AVE	CR	CX	MR	SG	TR
Customer Experience	.662	.877	.814[a]			
Media Richness	.583	.807	.592	.763[a]		
Swift Guanxi	.620	.830	.626	.441	.787[a]	
Trust	.615	.865	.635	.387	.529	.784[a]
Affective Customer Experience	.555	.789				
Cognitive Customer Experience	.533	.774				
Social Customer Experience	.537	.776				
Sensory Customer Experience	.639	.842				

Notes: [a] –the square root of AVE; AVE–average variance extracted; CR–composite reliability; MR–media richness; SG–swift guanxi; CX–customer experience; TR–trust

This model is valid as its model fitness values derived from the PLS algorithm meet the assessment guidelines suggested by Henseler et al. [41]. Specifically, the saturated model's SRMR value is 0.066, less than the cut-off value of 0.8. Figure 2 reports the model estimation results from the bootstrapping procedure for 5000 samples at a significance level of 0.05. The coefficients of determination, or (R^2) value, represent the explanatory power of the structural model. The R^2 values for all dependent variables are above the acceptable level of 0.25 suggested by Sarstedt et al. [40]: 0.394 for swift guanxi, 0.356 for customer experience, and 0.402 for trust. This result demonstrates that media richness can explain 39.4% of swift guanxi, 35.6% of customer experience, and 40.2% of trust. Figure 3 reports the bootstrapping results for the model estimation. Only the path of media richness→trust shows an insignificant result ($\beta = 0.008$, p = 0.881), thus rejecting H2. The other causal paths are significant. H1, H3, H4 and H5 are supported. Table 4 shows detailed results from examining the hypotheses.

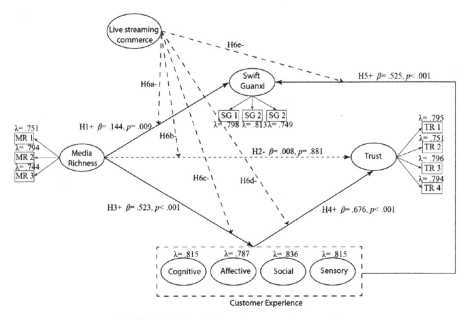

Fig. 3. Model result from bootstrapping procedure

Table 4. Calculation of path coefficient for all hypotheses

	Paths	Coefficients	t-values	p-values	Supported
H1	Media richness→swift guanxi	0.144	8.119	0.009	Yes
H2	Media richness→trust	0.008	13.443	0.881	No
H3	Media richness→customer experience	0.523	1.358	<0.001	Yes
H4	Customer experience→trust	0.676	11.191	<0.001	Yes
H5	Customer experience→swift guanxi	0.525	9.895	<0.001	Yes

4.2 Moderation Analysis of Control Variable

To evaluate whether media richness contributes more strongly to live streaming commerce, we conducted a moderation effect analysis using the social commerce format as the control variable. Live streaming commerce is measured as "1", and group buying format equates to "2". After model estimation, none of the causal relationships is significantly enhanced by live streaming commerce. To be specific, H6a (media richness→swift guanxi): $\beta = -0.085$, $p = 0.313$; H6b (media richness→trust): $\beta = 0.024$, $p = 0.771$; H6c (media richness→customer experience): $\beta = 0.153$, $p = 0.064$; H6d (customer experience→trust): $\beta = -0.106$, $p = 0.227$; H6e (customer experience→swift guanxi): $\beta = 0.080$, $p = 0.383$. Considering the unequal respondent numbers between the two social commerce formats, we also evaluate it by conducting a categorical moderation analysis (PLS-MGA) of 5000 samples in SmartPLS. Table 5 shows that the bootstrap PLS-MGA result. The p-value in Table 3 represents the difference in path coefficients between two investigated social commerce formats. For instance, causal connection of H1 (media richness→swift guanxi) in live streaming commerce is slightly stronger than group buying ($\beta = 0.149 > 0.052$). However, the insignificant p-values ($p = 0.248$) demonstrates that there is no significant difference between two formats after running a 5000-sample estimation. Therefore, the result of PLS-MGA test also supports the moderation analysis finding, as all p-values are insignificant ($>.05$).

Table 5. The results of bootstrap MGA

Path	Path coefficients		STDEV		p-values
	Live streaming commerce	Group buying	Live streaming commerce	Group buying	
H1	.149	.052	.055	.062	.248
H2	.005	.034	.052	.060	.714
H3	.550	.639	.042	.035	.106
H4	.651	.600	.046	.052	.465
H5	.522	.619	.054	.061	.245

Note: STDEV–standard deviation of the estimated parameter over all subsamples

5 Discussion

5.1 Theoretical Contributions

The hypotheses estimation results demonstrate four theoretical findings.

First, media richness directly influences customer experience and swift guanxi. The effect of media richness on swift guanxi is significant, albeit its strength is relatively lower than the other significant path between media richness and customer experience. Both causal connections are empirically confirmed for the first time in social commerce.

This finding corroborates Klein et al.'s [12] work which posits that media richness directly contributes to telepresence in online shopping. Telepresence in the current study belongs to the social customer experience dimension. Specific to social commerce, only two studies relevant to media richness exist for live streaming commerce, while none are present for group buying [9, 10]. Consumer purchase purposes probably explain the reason behind this significance. For example, the initial intent of many focal consumers engaging in social commerce may be to seek entertainment only. By comparison, the shopping journey in traditional e-commerce typically begins with a shopping intent only. For instance, consumers often open a shopping application on their mobile phones because they intend to purchase. Rich media content, such as livestreams, may facilitate delivery of entertaining information from the seller to consumers. Meanwhile, due to the inherent nature of marketing and retailing, the shopping journey in social commerce can be lengthy and more complicated, from initial consumer engagement to the final purchase decision. Multiple media channels and rich media content can help the seller drive a better customer experience. However, media richness is probably only one aspect in consumers forming a positive swift guanxi with the seller. Other factors may count, such as whether the seller conveys a friendly attitude, or the perceived quality of the information provided.

Second, media richness cannot influence trust directly. The reason behind the insignificant linkage between media richness and trust is similar to the low causal connection between media richness and swift guanxi. Compared to swift guanxi, consumers rely on more factors to generate a sense of trust towards the seller or anchor. Social commerce consumers' trust is enhanced only when information provided encompasses characteristics such as "reliability, credibility, trustworthy, and useful" [10]. Social commerce sellers should continuously focus on improving the effectiveness of information rather than focusing only on rich content.

Third, within the influence mechanism of media richness, customer experience is critical to forming a more positive relationship between the social commerce consumer and the seller. It plays the role of a mediatory factor in transferring the influence of media richness in social commerce. Media richness is a significant stimulator of customer experience. However, if consumers do not have a satisfying customer experience, then media richness cannot drive trust and only partially contributes to swift guanxi. Given the R^2 value (0.356) of customer experience in the validated model, other more important stimulator may exist which determine consumers' perceived experience.

Finally, this result reflects that the rich media tools, such as livestreams, do not significantly enhance the effect of media richness compared to lean media tools, such as text and static images. Both social commerce formats show a similar pattern for the media richness influencing mechanism. Richer media convey more deeply engaged information that lean media types cannot (e.g., greetings from the anchor in livestreams). However, the richest information does not equal the most effective for the consumer [13]. Lean media types (e.g., text) also can be the most effective tool for consumers. For example, consumers may prefer to use text and image descriptions over livestreams when they purchase the search goods (e.g., laundry powder) for its efficiency. Further, consumers' media preference varies based on their shopping expectations. For example, live streaming commerce consumers watch livestreams for entertainment purpose. Therefore, they may receive more entertainment content through livestreams. Therefore, what is more important is the "matching" between the extent of richness in the media and the consumers' communication activities within social commerce context.

5.2 Practical Implications

The first managerial implication is that social commerce adopters should emphasize strategies for enhancing consumer experience. Without a positive shopping experience, consumers seem unlikely to generate enthusiasm for swift guanxi and trust. In other words, if consumers have a negative experience during their social commerce shopping journey, they will probably feel the seller/anchor needs to be more trustworthy, and the relationship with the seller/anchor becomes less harmonious. Notably, social commerce sellers can manage CX with relevant marketing strategies that from the consumer perspective improves the task-fit of media richness.

The second managerial implication is that live streaming commerce has no advantage over group buying in forming a more harmonious and trustworthy relationship, even if its media types convey more deeply engaging information. This demonstrates that information quantity does not have a detectable influence on a consumer's shopping attitude and their relationship with the seller. Social commerce adopters should focus on the quality, efficiency, and offer the media content matching with consumers' expectations.

6 Conclusion and Future Research

This study yielded significant results when examining the influence mechanism of media richness on consumers' psychological responses when using social commerce. We have revealed that media richness primarily relies on customer experience to influence two main relationship factors: swift guanxi and trust. We detected no significant difference between live streaming commerce and group buying based on the different media they emphasized. The current study does not lack limitations. First and most significantly, we only focused on Chinese social commerce circumstances, and specifically live streaming commerce and group buying. The established causal relationship and moderation effect may have varied if we conducted this study in other regions or on other social commerce formats. Future researchers can apply the measures in other research settings to evaluate potential variations among different respondent groups and social commerce formats, such as *Facebook Marketplace* in the US. Second, the present study investigated the fashion product segment only. Future studies can select general products to minimize the influence of product segments on the examined model. Finally, we detected the critical role of customer experience in contributing to consumers' decision-making processes. We suggest future social commerce studies could concentrate on empirical studies of customer experience determinants or other customer experience stimulators.

Acknowledgments. This study was partially funded by an RMIT University research grant.

References

1. Bloomenthal, A.: Mobile Commerce. Investopedia (2020). Retrieved 9 November 2023 from https://www.investopedia.com/terms/m/mobile-commerce.asp
2. AP, T.: Influence peddler: How China's Zhang Dayi out-earns Kim Kardashian (2017). https://wwd.com/feature/zhang-dayi-kol-china-influence-peddler-10946898/
3. Yltävä, L.: Social commerce – statistics & facts [Grant] (2024). https://www.statista.com/topics/8757/social-commerce/#topicOverview
4. Wu, S.Y.R., Darpö, O., Zhou, Y., An, L., Zou, T.: Social commerce is remaking online shopping [Grant] (2023). https://www.bcg.com/publications/2023/social-commerce-is-remaking-online-shopping
5. Chow, C.W.C., Chow, C.S.F., Lai, J.Y.M., Zhang, L.L.: Online group-buying: the effect of deal popularity on consumer purchase intention. J. Consum. Behav. 21(2), 387–399 (2022). https://doi.org/10.1002/cb.2013
6. Lin, G.-Y., Lee, M.-X., Wang, Y.-S.: Developing and validating a live streaming social commerce success model. J. Comput. Inf. Syst. 1–19 (2023). https://doi.org/10.1080/08874417.2023.2251417
7. Liu, Y., Sutanto, J.: Buyers' purchasing time and herd behavior on deal-of-the-day group-buying websites. Electron. Mark. 22, 83–93 (2012)
8. Daft, R.L., Lengel, R.H.: Organizational information requirements, media richness and structural design. Manage. Sci. 32(5), 554–571 (1986)
9. Chen, Y.-H., Chen, M.-C., Keng, C.-J.: Measuring online live streaming of perceived servicescape: scale development and validation on behavior outcome. Internet Res. 30(3), 737–762 (2020)
10. Song, C., Liu, Y.-L.: The effect of live-streaming shopping on the consumer's perceived risk and purchase intention in China, Federal Reserve Bank of St Louis, St. Louis. (2021)
11. Hoffman, D.L., Novak, T.P.: Flow online: Lessons learned and future prospects. J. Interact. Mark. 23(1), 23–34 (2009)
12. Klein, L.R.: Creating virtual product experiences: the role of telepresence. J. Interact. Mark. 17(1), 41–55 (2003)
13. Ishii, K., Lyons, M.M., Carr, S.A.: Revisiting media richness theory for today and future. Hum. Behav. Emerg. Technol. 1(2), 124–131 (2019)
14. Dong, X., Liu, X., Xiao, X.: Understanding the influencing mechanism of users' participation in live streaming shopping: a socio-technical perspective. Front. Psychol. 13, 1082981 (2023)
15. Xin, K.K., Pearce, J.L.: Guanxi: connections as substitutes for formal institutional support. Acad. Manag. J. 39(6), 1641–1658 (1996)
16. Men, J., Zheng, X.: Impact of social interaction on live-streaming shopping websites. In: Proceedings of the 18th Annual Pre-ICIS Workshop on HCI Research in MIS, Germany (2019)
17. Lin, J., Li, L., Yan, Y., Turel, O.: Understanding Chinese consumer engagement in social commerce: the roles of social support and swift Guanxi. Internet Res. 28(1), 2–22 (2018)
18. Lin, J., Luo, Z., Cheng, X., Li, L.: Understanding the interplay of social commerce affordances and swift Guanxi: an empirical study. Inf. Manag. 56(2), 213–224 (2019)
19. Mensah, I.K., Zeng, G., Luo, C.: Determinants of social commerce purchase and recommendation intentions within the context of swift guanxi among Chinese college students. SAGE Open 13(2) (2023). https://doi.org/10.1177/21582440231175370
20. Hossain, M. A., Kalam, A., Nuruzzaman, M., Kim, M.: The power of live-streaming in consumers' purchasing decision. SAGE Open 13(4) (2023). https://doi.org/10.1177/21582440231197903

21. Guo, L., Hu, X., Lu, J., Ma, L.: Effects of customer trust on engagement in live streaming commerce: mediating role of swift Guanxi. Internet Res. **31**(5), 1718–1744 (2021). https://doi.org/10.1108/INTR-02-2020-0078
22. Fung.: Group-buying platform – Pinduoduo [Grant]. (2018)
23. McKnight, D.H., Chervany, N.L.: What trust means in e-commerce customer relationships: an interdisciplinary conceptual typology. Int. J. Electron. Commer. **6**(2), 35–59 (2001). https://doi.org/10.1080/10864415.2001.11044235
24. Li, C.-Y.: How social commerce constructs influence customers' social shopping intention? An empirical study of a social commerce website. Technol. Forecast. Soc. Chang. **144**, 282–294 (2019). https://doi.org/10.1016/j.techfore.2017.11.026
25. Xie, Y., Du, K., Gao, P.: The influence of the interaction between platform types and consumer types on the purchase intention of live streaming [Original Research]. Front. Psychol. **13**, 1056230 (2022). https://doi.org/10.3389/fpsyg.2022.1056230
26. Nadeem, W., Khani, A.H., Schultz, C.D., Adam, N.A., Attar, R.W., Hajli, N.: How social presence drives commitment and loyalty with online brand communities? The role of social commerce trust. J. Retail. Consum. Serv. **55**, 102136 (2020)
27. Lemon, K.N., Verhoef, P.C.: Understanding customer experience throughout the customer journey. J. Mark. **80**(6), 69–96 (2016)
28. Hu, M., Zhang, M., Wang, Y.: Why do audiences choose to keep watching on live video streaming platforms? An explanation of dual identification framework. Comput. Hum. Behav. **75**, 594–606 (2017)
29. Gentile, C., Spiller, N., Noci, G.: How to sustain the customer experience: an overview of experience components that co-create value with the customer. Eur. Manag. J. **25**(5), 395–410 (2007)
30. Klaus, P.P., Maklan, S.: Towards a better measure of customer experience. Int. J. Mark. Res. **55**(2), 227–246 (2013)
31. Dhaigude, S.A., Mohan, B.C.: Customer experience in social commerce: A systematic literature review and research agenda. Int. J. Consum. Stud. **47**(5), 1629–1668 (2023)
32. Bilgihan, A., Okumus, F., Nusair, K., Bujisic, M.: Online experiences: flow theory, measuring online customer experience in e-commerce and managerial implications for the lodging industry. Inf. Technol. Tourism **14**, 49–71 (2014)
33. Chen, N., Yang, Y.: The role of influencers in live streaming e-commerce: influencer trust, attachment, and consumer purchase intention. J. Theor. Appl. Electron. Commer. Res. **18**(3), 1601–1618 (2023)
34. Lu, B., Fan, W., Zhou, M.: Social presence, trust, and social commerce purchase intention: an empirical research. Comput. Hum. Behav. **56**, 225–237 (2016)
35. Chen, W.-K., Chen, C.-W., Silalahi, A.D.K.: Understanding consumers' purchase intention and gift-giving in live streaming commerce: findings from SEM and fsQCA. Emerg. Sci. J. **6**(3), 460–481 (2022). https://doi.org/10.28991/ESJ-2022-06-03-03
36. Lai, H., Zhuang, Y.-T.: Comparing the performance of group-buying models – time based vs. quantity based extra incentives. In: Proceedings of the Fourth Workshop on Knowledge Economy and Electronic Commerce (2006)
37. Bleier, A., Harmeling, C.M., Palmatier, R.W.: Creating effective online customer experiences. J. Mark. **83**(2), 98–119 (2019). https://doi.org/10.1177/0022242918809930
38. Zhang, M., Liu, Y., Wang, Y., Zhao, L.: How to retain customers: Understanding the role of trust in live streaming commerce with a socio-technical perspective. Comput. Hum. Behav. **127**, 107052 (2022). https://doi.org/10.1016/j.chb.2021.107052
39. Hair, J.B., Anderson, R., Black, W.: Multivariate Data Analysis (8th edition.). Cengage Learning EMEA (2018)

40. Sarstedt, M., Hair, J.F., Cheah, J.-H., Becker, J.-M., Ringle, C.M.: How to specify, estimate, and validate higher-order constructs in PLS-SEM. Australas. Mark. J. **27**(3), 197–211 (2019). https://doi.org/10.1016/j.ausmj.2019.05.003
41. Henseler, J., Hubona, G., Ray, P.A.: Using PLS path modeling in new technology research: updated guidelines. Ind. Manag. Data Syst. **116**(1), 2–20 (2016)

Artificial Intelligence in Business

ChatGPT and the Medical Industry: A Topic Modeling of Online Discussions by Medical Professionals

Langtao Chen[1](✉) ⓘ, Brenda Eschenbrenner[2], and Youhong Hu[3]

[1] University of Tulsa, Tulsa, OK 74104, USA
`langtao-chen@utulsa.edu`
[2] University of Nebraska at Kearney, Kearney, NE 68849, USA
`eschenbrenbl@unk.edu`
[3] Rolla, MO 65401, USA
`phoenixhu0308@gmail.com`

Abstract. ChatGPT has attracted much attention from industries, such as the medical industry, since its release in late 2022. Being an innovative artificial intelligence (AI) technology that has the potential to reshape the medical industry, ChatGPT has naturally become a hot topic in online communities for medical professionals. Promising opportunities of ChatGPT in transforming various aspects of the medical domain has also come with significant challenges that medical professionals must face and overcome. To understand how ChatGPT has impacted the medical industry as well as issues that have occurred during the adoption and use of the AI tool, this research contributes a systematic topic modeling of ChatGPT-related online discussions by medical professionals. Based on a dataset of 146 posts and 838 comments submitted to a set of online medical professional communities hosted on Reddit, this research identifies 10 important discussion topics. Implications and opportunities for research and practice in both medical and AI fields are highlighted.

Keywords: ChatGPT · Artificial Intelligence · Medical Industry · Medical Professionals · Topic Modeling · Online Communities · User-Generated Content · Latent Dirichlet Allocation

1 Introduction

Chat Generative Pre-trained Transformer (ChatGPT), a generative artificial intelligence (AI) model developed by OpenAI, has gained the attention of many users for both personal and business purposes after its launch on November 30, 2022. In particular, the medical industry and medical professionals have taken an interest in ChatGPT's possibilities to enhance healthcare. For example, medical professionals may view ChatGPT as an additional tool that can assist in medical diagnoses [1]. This interest has also been

---

Y. Hu — Independent Researcher

F. F.-H. Nah and K. L. Siau (Eds.): HCII 2024, LNCS 14720, pp. 145–156, 2024.
https://doi.org/10.1007/978-3-031-61315-9_10

accompanied by concerns over potential issues it may pose, such as accuracy in these diagnoses. Therefore, it will be important to understand the various topics of interest and concern that medical professionals are facing with the use of ChatGPT for it to be successfully leveraged in the medical domain.

ChatGPT relies on generative AI [1, 2]. As a large language model (LLM), Chat-GPT leverages massive datasets to produce natural language responses and content in response to human inquiries. ChatGPT has generated much excitement over its potential to create novel content such as poems. However, concerns have also arisen. For instance, errors in responses or hallucinations, biases (which could stem from the training data being unrepresentative of all members of a population), privacy, security, and insufficient explainability have all been proposed as potential issues with generative AI [1–4].

In consideration of its capabilities, ChatGPT has the potential to enhance healthcare. For example, medical professionals have constraints on their availability and the amount of time they have to serve patients. ChatGPT may be able to save time by answering basic patient inquiries [5]. One study of responses to medical inquiries in an online forum evaluated by healthcare professionals suggested that ChatGPT was more empathetic and provided higher quality responses in comparison to medical professionals' responses [6]. To note, however, ChatGPT responses were typically longer which may have influenced the response evaluations. ChatGPT can also personalize responses to an appropriate literacy level.

Yet, ChatGPT presents challenges that must be addressed for it to be widely adopted by the medical community. For example, not accounting for the diversity in patient populations has been proposed as a potential issue [5]. Datasets used to train ChatGPT may contain biases which, ultimately, produce output with inherent biases. If personal health information is also accessible, this may lead to privacy issues. The lack of regulations and guidelines for appropriate and ethical use of ChatGPT in healthcare settings has presented concerns as well. Because ChatGPT's responses are so fluent, it may instill a false sense of confidence. In addition, ChatGPT responses may not be as effective. For instance, when comparing postoperative care instructions provided by the medical facility versus generative AI applications, those provided by the medical facility rated higher in being comprehensible, actionable, and precise [6].

Medical professionals have been using online communities to discuss various topics of interest, such as emerging technologies. Online communities have facilitated collaboration, enhanced communication, and facilitated knowledge sharing among healthcare professionals [7]. Medical professionals can leverage online communities to distribute best practices and updated information. Emerging technology topics, such as ChatGPT, have emerged in online community discussions. Hence, we utilize online communities for medical professionals to understand the topics of greatest interest regarding ChatGPT within the medical field. By doing so, we contribute a systematic modeling of emerging topics pertinent to the adoption and use of ChatGPT in the medical domain. Drawing from the insights obtained through topic modeling, this research offers valuable implications for both research and practical applications in the realms of medicine, AI, and related fields.

This paper is organized as follows. The next section explains our topic modeling method, followed by results in Sect. 3. In Sect. 4, we discuss the current findings. Section 5 discusses implications and provides directions for future research.

2 Topic Modeling Method

We used latent Dirichlet allocation (LDA) to extract topics from discussions in a set of online communities for medical professionals. The detailed procedures of topic modeling are explained in the following subsections.

2.1 Data

We collected research data from multiple subreddits for medical professionals. Subreddits are communities or forums on the Reddit platform dedicated to specific interests and thematic groups. Users can join these subreddits to participate in online discussions for free. Two primary types of user-generated content in subreddits include posts and comments. Posts are submitted to the online community to initiate a specific discussion thread or share information, while comments are responses to posts or other comments. Figure 1 shows a sample post with one of its comments submitted to the subreddit "r/medicine."

Sample post

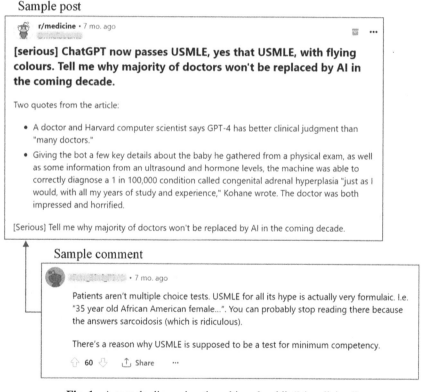

Sample comment

Fig. 1. A sample discussion thread in subreddit "r/medicine."

To collect relevant user-generated content, we used an API (application programming interface) to search posts containing the term "ChatGPT" in subreddits that are dedicated to medical professionals. Next, we obtained comments on those posts. Given that many comments are not relevant to ChatGPT, we only included those comments that contain terms such as "ChatGPT", "GPT", "LLM", "large language model", "AI", "artificial intelligence", and "machine learning." As a result, the raw dataset contains 146 posts and 838 comments submitted to 26 subreddits from December 3rd, 2022 to November 13th, 2023. Table 1 summarizes user-generated content collected for this research.

Table 1. Online posts and comments collected (total 146 posts and 838 comments).

Subreddit	Posts	Comments	Subreddit	Posts	Comments
medicine	43	464	neurology	2	2
nursing	21	68	Cardiology	2	0
pharmacy	13	84	globalhealth	2	0
Radiology	10	21	pathology	1	10
ems	9	78	respiratorytherapy	1	6
publichealth	7	33	Oncology	1	4
emergencymedicine	6	25	nursepractitioner	1	3
medlabprofessionals	6	15	BMET	1	1
healthIT	4	11	BehavioralMedicine	1	1
physicianassistant	4	1	Paramedics	1	1
Psychiatry	3	5	IntensiveCare	1	0
anesthesiology	2	3	optometry	1	0
Dentistry	2	2	surgery	1	0

2.2 Data Preprocessing

In the text mining process, each post or comment was represented as an individual document that contains a bag of words. In total, we obtained 984 documents, each representing a post or comment in the raw dataset. Before being fed into the topic extraction algorithm, these documents were preprocessed using the following steps:

- Remove URLs (uniform resource locators) included in the textual content;
- Remove all punctuation characters such as ",", ".", "?", and "!";
- Remove all numeric digits including possible leading symbols "+" and "-";
- Remove all terms contained in the documents with less than 3 characters;
- Remove all stop words in English such as "a", "an", "the", and "in";
- Change all terms to lowercase;
- Stem terms contained in the documents with the Snowball stemming library. For instance, terms "testing", "tested" and "tests" were converted to their stem "test".

2.3 Latent Dirichlet Allocation

After data preprocessing, all documents were fed into latent Dirichlet allocation (LDA), a popular topic modeling algorithm originally developed by Blei et al. [8]. Prior studies have applied LDA to extract latent discussion topics in a variety of contexts such as online communities [9] and consumer online reviews [10]. The number of topics k is pre-set by researchers. Following a trial-and-error procedure for exploratory studies [9], we found that a 10-topic solution (i.e., $k = 10$) results in a satisfactory set of latent topics that are specific enough to capture primary themes discussed in the raw textual data meanwhile maintaining a sufficient level of thematic abstraction. The output of LDA includes: (1) a table of document-topic probabilities showing the distribution of topics for each document, and (2) a table of topic-term weights indicating the strength of association between each term and each topic.

3 Results

3.1 Latent Topics

Based on the output of LDA, we labeled each topic according to its high-weight terms as well as posts/comments associated with the topic with a high probability. Table 2 presents the 10 topics identified from the online discussion dataset via LDA.

3.2 Topic Dynamics

To investigate the dynamics of online discussion topics, we counted the number of posts/comments with a document-topic probability no less than 0.4058 for each latent topic. The probability cutoff (i.e., 0.4058) was chosen so that 1/10 of the topic-document probabilities would be retained for a 10-topic solution [9, 11].

Figure 2 illustrates the dynamic change of the number of online posts and comments per each topic across 12 months ranging from December 2022 to November 2023. We found that medical professional communities started to discuss ChatGPT related issues right after the AI tool was released on November 30, 2022. The number of such online discussions increased to a very high level between February 2023 and May 2023, followed by a sudden decrease in July 2023. However, September 2023 saw another peak of ChatGPT related discussions. The hottest topics include *T2: ChatGPT mechanism for questioning and answering* (386 documents), *T1: ChatGPT for education and training* (240 documents), *T10: Using ChatGPT for medical diagnosis* (115 documents), and *T8: Human job replacement by AI* (103 documents).

Table 2. Topics extracted from online discussions.

Topic	Label	High-weight terms
T1	ChatGPT for education and training	chatgpt, medic, patient, clinic, human, tool, data, medicin, inform, help, learn, us, train, model, understand, improv, student, healthcar, llm, task, current, health, technologi, potenti, practic, program, decis, idea, languag, provid, nurs, futur, respons, chatbot, set, physician, requir, lot, process, accur
T2	ChatGPT mechanism for questioning and answering	chatgpt, question, gpt, answer, us, chat, write, actual, gener, time, respons, inform, read, check, exampl, sourc, tell, peopl, wrong, note, specif, help, start, output, articl, correct, word, tool, tri, type, studi, letter, stuff, prompt, post, text, explain, impress, try, search
T3	Performance of ChatGPT passing medical tests	test, step, pass, health, usml, perform, peanut, studi, water, intellig, chatgpt, paper, vendor, public, philosoph, marbl, kin, solv, post, world, us, vaccin, question, anim, move, tend, level, field, dean, glass, top, analysi, develop, standard, awar, idea, love, cours, fda, expect
T4	Prompt engineering involving medical emergency scenarios	patient, breath, dose, pain, medic, rate, vital, sign, access, vancomycin, initi, chest, sever, assess, heart, scenario, minut, monitor, pressur, emerg, blood, respiratori, radiat, oxygen, airwai, difficulti, burn, administ, due, includ, histori, sound, provid, appear, report, past, hour, base, elev, cardiac
T5	Using ChatGPT to analyze brains	brain, normal, appear, mass, empyema, ekg, left, caus, space, region, rhythm, tissu, structur, abnorm, ey, pituitari, locat, direct, tumor, procedur, gland, sinus, visual, medial, interpret, chest, involv, shift, size, pleural, cavern, nerv, tempor, hemorrhag, fluid
T6	Using ChatGPT to analyze lab test results	blood, bleed, boolean, system, outprintln, fluid, upper, test, potenti, consid, neg, method, posit, indic, sensit, sourc, valueof, red, statu, volum, abdomen, transfus, measur, trauma, cell, start, varic, liver, diagnosi, stabil, suspect, bacteria, indol, motil, tsi, string, identifi, els, endoscopi, clot

(continued)

Table 2. (*continued*)

Topic	Label	High-weight terms
T7	Using ChatGPT to write professional letters	patient, medic, treatment, letter, write, guidelin, manag, insur, provid, name, alinia, care, recommend, level, emt, cgm, american, infect, glucos, continu, time, request, blastocysti, list, homini, support, diabet, condit, benefit, compani, denial, appeal, monitor, includ, symptom, microbiologi, option, societi, servic, prompt
T8	Human job replacement by AI	job, replac, peopl, human, pharmacist, machin, take, pharmaci, softwar, get, tech, person, worri, lot, career, soon, profess, robot, tell, input, doctor, radiologi, train, happen, cost, advanc, decad, look, chang, school, radiologist, dental, talk, comput, healthcar, anytim, autom, technologi, field, world
T9	Using ChatGPT to generate creative content	nurs, heart, hand, jone, patient, depress, treat, true, team, care, bodi, orthoped, birth, admit, surgeon, hospitalist, uti, labor, friend, staff, rest, fall, oper, thank, hip, elbow, lab, demand, propos, test, vers, dai, pain, world, pin, touch, mlt, tear, soul, choru
T10	Using ChatGPT for medical diagnosis	chatgpt, patient, doctor, diagnosi, diagnos, imag, exam, miss, stori, dai, symptom, report, abl, histori, mri, radiologist, physician, data, care, physic, neurologist, system, list, probabl, chart, radiologi, perfect, reason, month, differenti, mayb, call, potenti, medic, guess, pass, famili, anywai, mean, articl

4 Discussion

Although medicine is a highly regulated industry with significant barriers to entry, we noticed that ChatGPT as an innovative AI tool has been discussed and adopted in a variety of aspects of medicine. In this section, we elaborate on the primary themes of online discussions uncovered from the topic modeling that set forth research directions and practical applications for years to come.

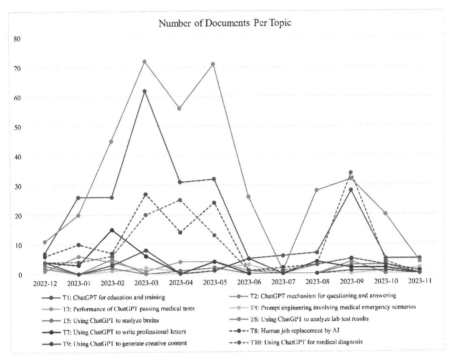

Fig. 2. Number of posts and comments per each topic across months.

4.1 Mechanism of ChatGPT and Hallucinations

The topic with the most documents (i.e., posts and comments) encompasses the use of ChatGPT and potential issues of hallucinations, which have been previously identified as a significant challenge associated with generative AI [1]. Documents included discussions describing how ChatGPT generates responses to questions posed and potential limitations. For example, a user described ChatGPT as:

"You present a question. It runs it through it's complex models and then generates a response based not on the actual CONTENT of what you asked, but on what it predicts are the most likely responses to be given. And it does this letter by letter. So the output you'll get out of a LLM isn't due to it 'thinking it through' but based on what it is guessing is the most probable response letter by letter. Which leads to it doing some really crazy stuff like inventing entire conversations that never happened or giving terrible advice."

Some documents described hallucination issues experienced such as when feasible responses were provided but were actually incorrect. Comments included:

"Unfortunately in my experience it immediately started coming up with completely plausible-sounding answers to my questions that I discovered to be 100% fictional when I referenced the original paper. We still have ways to go with fixing the hallucination problem."

Medical professionals included advice regarding the responsible use of ChatGPT such as:

> *"I try and tell my peers that ChatGPT is fine for general questions, but if you want to use it in any official capacity (including learning clinical and other information), definitely double check what it tells you."*

> *"Basically, AI is awesome, but it's not perfect. We just need to make sure we're using it the right way, and being smart about the info it's working with."*

In summary, online discussions by medical professionals included knowledge sharing to understand how ChatGPT generates its responses, the potential limitations with it, and how to use it responsibly.

4.2   ChatGPT for Medical Education and Training

The second most popular topic in online discussions by medical professionals was using ChatGPT for education and training in the medical field, as well as the training of ChatGPT. For instance, one document suggested that ChatGPT might be utilized for clinical decision-making:

> *"Have students input a patient scenario into ChatGPT and use its responses to help guide their clinical decision-making. This can help students learn how to use technology to aid in their decision-making and critical thinking skills."*

Other discussions addressed the training of ChatGPT and noted positive aspects for the future of ChatGPT in medicine such as:

> *"The current model as far as I know hasn't been trained to any tasks unique to health care. With additional and more specific training, there is potential for much more improvement...I've personally had a lot of success chaining ChatGPT prompts and responses to replicate several common EHR clinical decision support tasks - and this is just me playing around in my spare time. It clearly has limitations but so do existing applications which will not improve as rapidly. I'd be willing to bet ChatGPT with the right prompting will perform better for many clinical decision support tasks within an EHR than traditional systems."*

Based on the discussions posted, medical professionals saw ChatGPT's potential in a variety of education and training contexts. Further, medical professionals shared their perspectives and experiences with ChatGPT, specifically regarding its training. They also offered insights into its ongoing development and use.

4.3   ChatGPT for Medical Diagnoses

The third most prevalent topic addressed the use of ChatGPT for medical diagnoses. Some documents shared positive uses of ChatGPT such as:

"If the timeline in the article is accurate, the patient saw a neurologist early on when the chief complaint was headache. The spine MRI came later, and according to the article, chatGPT was able to suggest tethered cord based on the MRI report. To me, that means that the radiologist report was sufficient for a LLM to interpret it as tethered cord, but obtuse enough that any non-specialists who read the report were unable to make a diagnosis they weren't familiar with."

In addition, another comment noted that ChatGPT may help alleviate some of the strain on medical providers, and this potential has been acknowledged by some in the medical industry as well [5]:

"A half decent AI implementation could reduce a ton of inequities in health care. Think of all the overworked PCPs and frustrated patients. Patients spend 10 minutes EVERY visit recounting their medical history to their PCP because the doctor is too overworked to review the chart and doesn't remember anything about you."

Other documents brought ChatGPT's helpfulness into question, however, such as:

"ChatGPT suggested tethered cord based on the MRI report, which means that the radiologist who read the MRI made the correct diagnosis. This isn't a 'chatGPT saves the day' story, this is a 'rapid fire doctor-hopping leads to communications breakdowns and delay in care' story."

In addition, some discussions highlighted limitations on ChatGPT's application such as:

"Also why would we use it for difficult cases. Thats what we have medical control and physicians to call for. If I call my MC and told them I did something bc ChatGPT said it was a good idea, I would not only expect but hope my license would get ripped away."

To conclude, online discussions indicated that some medical professionals perceive positive outcomes from ChatGPT and its ability to assist time-constrained professionals including primary care providers. Looking forward, some discussions suggested relevant factors to be considered before assuming that ChatGPT is reliable, especially in complex cases.

4.4 Job Replacement by AI

There is an abundance of research across multiple disciplines on human job replacement by AI. Not surprisingly, another topic of discussion focused on ChatGPT's potential for technological unemployment in the field of medicine. Some comments were pessimistic about AI posing a threat to medical professionals. For instance, one document included the following statement:

"If an AI can't be trusted after like 20 years of development to drive trucks/cargo on well defined, largely unchanging highways due to 'human error' of the surrounding cars, it's never going to be able to adjust for patient choices/behavior."

Another comment noted the complexity involved with medicine that could make AI adoption questionable:

"Radiology is so much more complex than just find a bleed or large mass like the AI programs still struggle to do today. If radiology is getting replaced out of the job market, than almost anyone who isn't in manual labor is too."

Others, however, were envisioning AI's existence in certain aspects of medicine. For example, one comment noted:

"The reality is that AI will absolutely replace the endoscopy parts of these jobs. Without question. The problem is perfectly suited to AI - it doesnt get tired and will have the ability to train on massive datasets. Machine vision is rapidly improving and transformer AI systems have changed the paradigm entirely. AI will far exceed the ability of any human endoscopist. Not now. But soon. AI will be cheaper, faster and more accurate. Humans will need to verify results at first but in the long term this may not be the case."

Some online discussions suggested that AI may not be able to replace humans due to factors such as the nuances of treating patients (e.g., patient behaviors) and complexity associated with some medical tasks. Other discussions saw AI fulfilling certain medical functions and outperforming humans, with the acknowledgment that human oversight is currently still needed.

5 Implications and Conclusion

ChatGPT has the potential to enhance the medical communities' ability to properly serve and provide the greatest care possible to patients. It also, however, presents challenges and concerns that need to be addressed for effective usage and adoption to occur. For ChatGPT to succeed in practice, issues surrounding the potential hallucinations will need to be addressed. This may include educating users on ChatGPT's training and opening the "black box" to interpret and understand how it derives a particular response. Medical schools and training programs may consider integrating ChatGPT as part of its educational plans, such as using ChatGPT to assist clinical decision making based on the presentation of a given patient's case. The use of ChatGPT for diagnosing may also be explored further as an assistive tool. Particularly, current medical education programs can be modified to develop appropriate skills to promote effective use of such large language models in the medical field. The potential of ChatGPT creating issues related to technological unemployment should also be addressed to discern areas in which ChatGPT will be most useful.

Future research can explore the human-computer interaction experience to enhance ChatGPT's usage. For instance, studies can focus on prompt engineering to make Chat-GPT the most effective in its responses to given medical inquiries. Future research may leverage the task-technology fit theory [12] to explore mechanisms that will assist medical professionals with integrating ChatGPT responses when making complex medical diagnoses.

This study contributes a systematic analysis of the current adoption and use of Chat-GPT in the medical field by modeling topics embedded in medical professionals' online discussions. Future research can extend this to include patients' use of ChatGPT by analyzing patient discussions on social media platforms such as online health communities [13, 14]. With the capabilities that generative AI presents, understanding all aspects of AI's potential and addressing its challenges may make it another great tool for enhancing patient care.

References

1. Nah, F., Zheng, R., Cai, J., Siau, K., Chen, L.: Generative AI and ChatGPT: applications, challenges, and AI-human collaboration. J. Inf. Technol. Case Appl. Res. **25**(3), 277–304 (2023)
2. Denning, P.J.: Can generative AI bots be trusted? Commun. ACM **66**(6), 24–27 (2023)
3. Nah, F., Cai, J., Zheng, R., Chen, L.: Revealing the dark and bright sides of ChatGPT: an exploratory study on user perceptions. In: Proceedings of the 29th Americas Conference on Information Systems (2023)
4. Nah, F., Cai, J., Zheng, R., Pang, N.: An activity system-based perspective of generative AI: challenges and research directions. AIS Trans. Hum.-Comput. Interact. **15**(3), 247–267 (2023)
5. Diaz, N.: The challenges ChatGPT poses to healthcare, according to CIOs. https://www.bec kershospitalreview.com/healthcare-information-technology/the-challenges-chatgpt-poses-to-healthcare-according-to-cios.html. Retrieved 23 Oct 2023
6. McPhillips, D.: ChatGPT may have better bedside manner than some doctors, but it lacks some expertise. https://www.cnn.com/2023/04/28/health/chatgpt-patient-advice-study-wellness/index.html. Retrieved 23 Oct 2023
7. Kaufman: Online communities for healthcare professionals can help improve communication, collaboration. Formulary **47**(4) (2012)
8. Blei, D.M., Ng, A.Y., Jordan, M.I.: Latent dirichlet allocation. J. Mach. Learn. Res. **3**, 993–1022 (2003)
9. Chen, L.: What do user experience professionals discuss online? Topic modeling of a user experience Q&A community. In: LNCS, vol. 14038, pp. 1–16. Springer Nature, Cham, Switzerland (2023). https://doi.org/10.1007/978-3-031-35969-9_25
10. Kwon, H.-J., Ban, H.-J., Jun, J.-K., Kim, H.-S.: Topic modeling and sentiment analysis of online review for airlines. Information **12**(2), 78 (2021)
11. Chen, L., Baird, A., Straub, D.: An analysis of the evolving intellectual structure of health information systems research in the information systems discipline. J. Assoc. Inf. Syst. **20**(8), 1023–1074 (2019)
12. Goodhue, D.L., Thompson, R.L.: Task-technology fit and individual performance. MIS Q. **19**(2), 213–236 (1995)
13. Chen, L., Baird, A., Straub, D.: A linguistic signaling model of social support exchange in online health communities. Decis. Support. Syst. **130**, 113233 (2020)
14. Chen, L., Baird, A., Straub, D.: Fostering participant health knowledge and attitudes: an econometric study of a chronic disease-focused online health community. J. Manag. Inf. Syst. **36**(1), 194–229 (2019)

Exploring Segmentation in eTourism: Clustering User Characteristics in Hotel Booking Situations Using k-Means

Stefan Eibl[1]([⊠]), Robert A. Fina[1], and Andreas Auinger[2]

[1] University of Applies Sciences Wiener Neustadt, Zeiselgraben 4, 3250 Wieselburg, Austria
stefan.eibl@fhwn.ac.at
[2] University of Applied Sciences Upper Austria, Wehgrabengasse 1-3, 4400 Steyr, Austria

Abstract. In the dynamic field of eTourism, personalization and user segmentation are paramount for enhancing user experience and driving digital platform success. This paper addresses the gap in eTourism research related to understanding consumer behavior through an external lens, due to the limited access to proprietary data from Online Travel Agencies (OTAs). We employ Adaptive Choice-Based Conjoint (ACBC) analysis and k-means clustering on data from a survey (n = 801) based on 346 hotel listings on Booking.com, focusing on Vienna. Attributes such as star category, price, review valence, volume of reviews, scarcity indicators, sustainability cues, and city center proximity were examined to identify consumer preferences. Five distinct consumer clusters were revealed: Cost-Conscious Eco-Bookers, Green-Urban Deal Hunters, Social-Proof Assurance Seekers, Budget-Only Focused Minimalists, and Luxury-Quality Connoisseurs. These clusters vary in their prioritization of hotel attributes and demographics, demonstrating the diverse decision-making criteria within the eTourism market. This paper proposes a foundation for classifying user groups on booking platforms, enabling OTAs and hoteliers to tailor offerings to nuanced consumer segments, thus improving user experiences and potentially increasing conversion rates. The findings offer actionable insights into OTA personalization strategies and contribute to the scientific understanding of consumer behavior in the digital tourism landscape.

Keywords: eTourism Segmentation · k-Means Clustering · Personalization Tactics

1 Introduction

In the rapidly evolving landscape of Online Travel Agencies (OTAs), the significance of user segmentation and personalization strategies cannot be overstated. These approaches have become pivotal in enhancing user experiences and driving business success in the digital tourism domain [1]. However, a critical challenge facing the academic community and external researchers in this field is the limited access to proprietary data held by OTAs. This data, rich with insights on consumer behavior and preferences, often remains confidential and exclusively utilized for internal strategic purposes, thus information

© The Author(s), under exclusive license to Springer Nature Switzerland AG 2024
F. F.-H. Nah and K. L. Siau (Eds.): HCII 2024, LNCS 14720, pp. 157–175, 2024.
https://doi.org/10.1007/978-3-031-61315-9_11

quality may be influenced [2]. Consequently, there is a compelling need to examine user segmentation and personalization from an external perspective, contributing valuable findings to the scientific community. This research aims to fill this gap by exploring user segmentation based on individual preferences, as indicated by utility scores, and correlating them with user demographic data from a sample of Austrian and German respondents (n = 800). This approach is instrumental in creating an external viewpoint on how consumers interact with hotel booking platforms, which is crucial for expanding the existing body of knowledge in eTourism [3]. While OTAs have access to this data for operational and marketing strategies, they often do not share detailed insights publicly, limiting the broader understanding of consumer behavior in this context [4]. By conducting this research, we aim to provide empirical evidence and a theoretical framework for user segmentation that can be utilized by academics and practitioners alike. This is particularly important as it offers an independent analysis of consumer preferences and behavior, which is often shaped by various factors, including psychological, social, and economic elements [5, 6]. Additionally, this research contributes to a deeper understanding of how demographic characteristics interplay with personal preferences in the context of hotel bookings, an area that has seen limited exploration due to the proprietary nature of OTA data [7]. Therefore, our study not only addresses a critical gap in eTourism research but also presents an opportunity for the scientific community to gain insights into consumer segmentation and personalization strategies, which are essential for the continued growth and evolution of the online travel industry [8].

2 Related Work

2.1 User Segmentation in eTourism

User segmentation in eTourism has become increasingly significant as the digitalization of travel and hospitality services continues to evolve. This segmentation process involves categorizing potential customers into distinct groups based on shared characteristics. This categorization is crucial for effective marketing and enhancing user experience on digital platforms [9]. Segmentation theories like the Market Segmentation Theory suggest that distinct groups within a market can be targeted with tailored marketing strategies [10]. The VALS (Values, Attitudes, and Lifestyles) framework, categorizes consumers based on psychological traits and key demographics, which is particularly useful in understanding traveler segments [11]. Segmentation in eTourism is conducted using demographic, psychographic, and behavioral data. Demographic segmentation involves categorizing consumers based on criteria such as age, gender, income, and education [12]. Psychographic segmentation delves deeper into consumers' lifestyles, values, and opinions [13], while behavioral segmentation focuses on purchase history, loyalty, and service engagement [14]. OTAs and hotels apply these segmentation methods to customize recommendations, promotional offers, and design loyalty programs [3, 15]. For example, Gretzel & Fesenmaier [16] showed how understanding travel motivations of different segments enhances marketing effectiveness. The role of data in forming customer segments is particularly significant. The emergence of big data and analytics has provided eTourism platforms with immense information, aiding in more accurate segmentation [17]. This data-driven approach enables a nuanced understanding of consumer behavior and preferences, leading to more effective personalization strategies for companies which have the resources to gather and analyze this data [18].

Building further on the concept of user segmentation in eTourism, it's evident that the effective utilization of segmentation strategies can greatly enhance the personalization of services offered by online travel platforms. The profound impact of tailored marketing and service delivery, based on a deep understanding of different customer segments, is a recurring theme in recent eTourism research [7]. Advanced data analytics techniques have opened new avenues for understanding consumer behavior in the eTourism sector. Data mining and machine learning algorithms, for instance, allow for the extraction of meaningful patterns from large datasets, enabling platforms to identify subtle preferences and behaviors of different segments [17]. These techniques have been instrumental in refining segmentation strategies, allowing for a level of personalization that was previously unattainable [19]. Behavioral segmentation, particularly, has gained traction in the digital era. By analyzing online behavior patterns, such as booking history and interaction with OTA platforms, businesses can gain insights into the preferences and decision-making processes of consumers [20]. This approach aligns with the increasing emphasis on customer experience in the digital marketplace, where personalization is key to customer satisfaction and loyalty [21]. The importance of demographic factors, though traditional, remains significant. Age, income, and education level continue to influence travel preferences and booking behaviors. For example, younger travelers may show a propensity for budget-friendly options and are more influenced by social media marketing, whereas older travelers might prioritize comfort and direct booking experiences [22].

2.2 Personalization Strategies in Online Hotel Booking Situations

Personalization strategies in the online hotel booking sector represent a sophisticated interplay between technology, data analytics, and consumer psychology. In an industry characterized by intense competition and evolving consumer expectations, personalization has emerged as a strategic imperative for enhancing customer satisfaction and loyalty [21]. This approach focusses on tailoring the user experience to individual needs and preferences, often leveraging rich data sets to craft targeted messages and offers [23]. The foundation of personalization lies in the understanding that each traveler's needs are unique [24]. Recognizing and responding to these needs in real-time is the essence of personalization in eTourism [1]. OTAs have been at the forefront of this trend, employing sophisticated algorithms to suggest hotels, special deals, and additional services based on past behavior, search patterns, and preferences [25, 26]. This level of customization is made possible by the immense data users leave as they interact with online platforms, which, when analyzed, can reveal deep insights into consumer behavior [5]. One of the key methods employed by OTAs to achieve personalization is collaborative filtering. This technique uses data from many users to provide recommendations based on similar search and booking patterns [26]. Another method is content-based filtering, which suggests options based on the similarity of items, such as hotels or destinations, to those a user has expressed interest in before [27]. These filtering mechanisms are integral to creating a personalized experience, as they can dynamically adjust the content presented to each user based on their interests and behaviors. The impact of personalization on consumer behavior is great. Studies have shown that consumers are more likely to engage with and purchase from platforms that offer a personalized experience [28]. The personal touch fosters a sense of value and recognition among customers, which, in turn, enhances their loyalty to the platform [29]. In fact, personalization can lead to a virtuous cycle: the more a customer interacts with a personalized service, the more data is generated, which further refines the personalization algorithms, resulting in even more engagement [30]. However, the implementation of personalization strategies is not without challenges. The primary concern is the balance between personalization and privacy. As platforms collect and utilize personal data to tailor experiences, they must also navigate the complex landscape of data privacy regulations and consumer privacy concerns [31]. Transparency in how data is collected, used, and protected is vital for maintaining consumer trust and ensuring the ethical use of personalization technologies [32]. Another challenge is the avoidance of the "filter bubble" effect, where the personalization algorithm over-specializes the content, restricting the diversity of offerings presented to the user [33]. To combat this, OTAs are exploring hybrid recommendation systems that combine collaborative and content-based filtering with techniques that introduce probability and diversity into the recommendations, or by even trying to broaden the segmentation approach [34].

3   Methodology

3.1   Content Mining and Multiple Linear Regression

To analyze which attributes to focus on within the process of analysis, we conducted a web content mining approach using the tool "Octoparse". We also examined the results of the systematic literature review of Eibl & Auinger [8], this aimed to identify factors that influence booking intentions [35, 36]. We selected hotel attributes, focusing on those, visible on the search results page of booking platforms like booking.com. Thus, attributes such as descriptions or room sizes, which are not immediately visible there, were excluded. While images likely influence booking decisions, their analysis was beyond the scope of this web content-focused study. We carried out an analysis of 346 hotel listings on booking.com, with a focus on Vienna within high season, to gather data on various attributes such as star category, price, review valence, the volume of reviews, scarcity indicators, sustainability cues, and distance to the city center. To ensure a comparability across the diverse range of our independent variables, we applied z-standardization, aligning our data on a standardized scale for use in our multiple linear regression model [37, 38]. The results of these two approaches were then incorporated into a conjoint analysis.

3.2   Adaptive Choice Based Conjoint Analysis

In our methodology, the Adaptive Choice-Based Conjoint (ACBC) analysis served as a cornerstone to discern how multiple attributes influence hotel booking decisions on eTourism platforms. A survey administered through Sawtooth Software Lighthouse Studio to a random sampling of individuals from Austria and Germany (n = 800) captured not only demographic information but also the participants' preferences within ACBC scenarios, see Table 1. This process enabled the calculation of utility scores for each hotel attribute, providing a nuanced understanding of the role these attributes play in online hotel booking behaviors [39].

The selection of attributes and levels for the conjoint analysis was a decision informed by the web content mining approach, the application of the multiple linear regression model, and established literature on conjoint analysis by Baier & Brusch [35]. Within an ACBC, it is recommended to have a range of 5 to 12 attributes and each attribute can have between 2 to 12 levels to ensure comprehensive coverage without overwhelming respondents. This range balances detail with manageability, allowing for thorough investigation while maintaining participant engagement and the quality of data collected. [35].

The ACBC analysis, tailored to reflect the intricacies of consumer decision-making, was executed in four structured steps, beginning with a (1) "Build Your Own" phase where respondents were asked to design their ideal hotel by selecting their preferred levels of the presented attributes (like sustainability level or distance to town center). This initial step allows for the identification of each individual's most desired features. Following the BYO, the analysis progresses into a screening phase, where respondents are presented with a series of hotel configurations that are close to their ideal but include some variations. Respondents must decide which of these configurations they would consider acceptable alternatives to their BYO selection. Subsequently, the ACBC approach narrows down the field through Choice Tournaments, where the acceptable configurations compete against each other in head-to-head matchups. In these matchups, respondents are asked to make choices between different sets of alternatives for leisure travel, further refining their preferences.

So, within the survey, participants were presented with a series of hotel options where the identified attributes were displayed side by side, as well as intermixed within each hotel option. This method simulates real-life decision-making by requiring individuals to evaluate and choose between hotels based on a combination of characteristics, such as location and price, without focusing on a single attribute. This approach helps to understand how various factors are weighted against each other in the decision process, reflecting a more realistic scenario where multiple attributes influence the choice of a hotel. [35, 40]. These steps were critical in calculating participants' genuine preferences and allowed us to explore the relationship between consumer demographics and their attribute preferences. Hierarchical Bayesian estimation techniques were employed to calculate utility scores for each attribute level, providing robust, reliable insights into individual preferences and decision patterns [35, 41]. By integrating these steps, we ensured a comprehensive capture of participants' preferences, which are vital in informing the design and personalization of user experiences on hotel booking platforms.

Table 1. Demographics

	N	%		N	%
Age			**Education**		
16–20	19	2.375%	Mandatory school	243	30.375%
21–30	90	11.25%	High school	287	35.875%
31–40	104	13%	Bachelor's degree	91	11.375%
41–50	145	18.125%	Master's degree	155	19.375%
51–60	176	22%	Doctor degree	24	3%
61–70	158	19.75%			
71–80	96	12%	**Net household income per year (EUR)**		
>80	12	1.5%	<19,999	150	18.75%
			20,000–39,999	246	30.75%
Gender			40,000–59,999	166	20.75%
Female	416	52%	60,000–79,999	108	13.5%
Male	382	47.75%	80,000–99,999	65	8.125%
Diverse	2	0.25%	>100,000	65	8.125%
Nationality					
Austria	401	50.125%			
Germany	399	49.875%			
Total	800	100%		800	100%

3.3 K-Means Clustering Approach

Utilizing k-means clustering to analyze similarities and differences between user groups represents a robust method for identifying patterns in consumer behavior. When combined with ACBC results, this approach offers a novel perspective on customer preferences, particularly in the domain of online hotel bookings. This statistical technique is instrumental in segmenting a dataset into a specified number of distinct groups based on inherent similarities within the data, which, in our study, was implemented using the robust capabilities of the XLSTAT software. K-means clustering is a partitioning method that assigns observations into k clusters in which each observation belongs to the cluster with the nearest mean, serving as a prototype of the cluster [42]. The process is iterative and aims to minimize the within-cluster sum of squares, which is essentially a variance measure within each cluster. The methodology rests on defining k centroids, one for each cluster, and then assigning each data point to the nearest centroid based on the Euclidean distance [43]. The centroids are recalculated after each iteration, which results in the reassignment of data points until the within-cluster variation cannot be further reduced, and the clusters become stable [44].

In our research, we leveraged the ACBC individual utility scores, which reflect the relative importance of various hotel attributes to the consumer's decision-making process. These utility scores represent multidimensional data points that the k-means algorithm could effectively analyze to identify coherent clusters of consumers with similar hotel attribute preferences. We chose to focus on five clusters as this number provided the best fit to the data, which was ascertained through the evaluation of several cluster solutions against criteria such as the elbow method and the silhouette score - a measure of how similar an object is to its own cluster compared to other clusters [45]. The selection of five clusters was also validated by the interpretability and managerial implications of the segmentation. It allowed for a detailed differentiation of consumer preference patterns without overcomplicating the model with too many segments, which might have led to an impractical application in a business context [46]. The decision was in line with the parsimony principle, which suggests that models should be as simple as possible, but no simpler - a balance between complexity and practicality [47].

The k-means methodology also was applied to the ACBC utility scores using XLSTAT. The algorithm's application involved several steps: standardizing the utility scores on a scale from 0–100, initializing the centroids, assigning observations to the nearest centroids, recalculating the centroids, and repeating the assignment and recalculating steps until convergence. The standardization of data before clustering is crucial, as it ensures that each attribute contributes equally to the similarity measure and prevents attributes with larger ranges from dominating the distance calculations [48].

Subsequently the sum of squared distances from each point to the centroid of its assigned cluster was minimized, ensuring that the clusters were as compact and separate as possible [49]. Additionally, the silhouette analysis was performed to assess the goodness of fit for each cluster. This involved calculating the average silhouette width for each cluster and for the dataset as a whole, which provided a graphical representation of how well each object lies within its cluster [45].

4   Results

The complex landscape of consumer preferences within the online hotel booking domain presents a multifaceted challenge for market segmentation. Our study's expedition into this domain through the application of k-means clustering has revealed distinct consumer segments, each characterized by unique utility preferences concerning hotel attributes [46].

The five clusters that emerged from our analysis represent distinct archetypes of consumers in the eTourism marketplace. These clusters vary significantly in their valuation of hotel attributes, suggesting differing priorities and decision-making criteria among the groups. The clusters range from price-sensitive consumers to those who prioritize sustainability and luxury, reflecting the diverse nature of the online hotel booking audience. Cross-tabulation was used to further enrich the cluster profiles with demographic data, linking utility preferences to demographic characteristics such as age, gender, income, travel frequency, educational level and marital status [50].

In the subsequent subsections, we will delve into each cluster, outlining their defining characteristics. The detailed breakdown of clusters will provide a rich description of the diverse consumer base that OTAs serve, and how these differences require tailored approaches to marketing and service design. This segment-specific insight is crucial for OTAs and hoteliers aiming to enhance the personalization of their offerings and is consistent with the literature supporting the strategic importance of customized consumer engagement [51]. As we transition to the detailed analysis of each cluster, it is essential to bear in mind that the overarching goal of this segmentation is to identify actionable insights that can inform the personalization strategies of OTAs and hotel-entries, thereby optimizing the consumer experience and driving business performance in the competitive landscape of eTourism [23].

Within the results section of our study, we also present a detailed examination of the k-means clustering analysis applied to various hotel attributes and their influence on user segmentation. The statistical approach undertaken involves an Analysis of Variance (ANOVA), which tests the hypothesis that the means of several groups are equal, see Table 2. This method is instrumental in discerning the significance of each attribute in the formation of distinct user clusters.

The ANOVA results indicate that the majority of hotel attributes have a statistically significant influence on the clustering of user preferences. High F-values and p-values less than 0.05 confirm that attributes such as user ratings, ranging from "9.7" to "6.9", play a pivotal role in segmenting users into distinct groups. Furthermore, proximity to the city center, with varying distances, emerged as a critical factor, with closer distances correlating strongly to user cluster formation.

We also see, that the scarcity cues ranging from 1 to 5 rooms left have a significant effect on building different user clusters, whereas the lack of significance concerning the attribute "only 7 rooms left…" implies that such scarcity cue might not be as influential in shaping user preferences for hotel bookings.

The various price levels, designated as from 84.7 € to 600 € displayed extremely significant p-values, indicating that price is a very important attribute in user segmentation.

These results not only reinforce the validity of our k-means clustering approach but also underpin the significant differentiation among the five user clusters identified in our analysis. The significant variances across key hotel attributes underscore the distinct preferences and decision-making criteria inherent to each cluster. This foundational understanding of the attributes that influence user segmentation allows us to delve deeper into the characteristics of each cluster.

4.1 Cost-Conscious Eco-Bookers (CCEB)

The centroid data for the "Cost-Conscious Eco-Bookers" cluster provides valuable insights into the booking preferences of this group. This cluster is characterized by individuals who prioritize cost-effectiveness but also have an interest in sustainability, as long as it does not involve additional costs. A detailed analysis of the centroid values reveals the following key points about the "Cost-Conscious Eco-Bookers", see Table 3.

The values indicate a clear pattern of price sensitivity, with the highest values associated with the cheapest price options. These consumers are significantly influenced by cost, with the utility scores decreasing as the price increases, showing a clear preference for more budget-friendly options. While this group considers sustainability, it is not their primary concern. The utility scores for sustainability levels are moderate compared to the

Table 2. Analysis of Variances

Attributes	Level	F	Pr > F
Review-Valence	9.7	67.639	<0.0001
	9.0	66.472	<0.0001
	8.3	12.093	<0.0001
	7.6	64.746	<0.0001
	6,9	78.030	<0.0001
Amount of Reviews	5	27.371	<0.0001
	5795	2,767	0.026
	9655	45,534	<0.0001
Hotel star category	3-star	72,981	<0.0001
	4-star	55,223	<0.0001
	5-star	52,195	<0.0001
Scarcity cues	Only 1 room left	30,380	<0.0001
	Only 3 rooms left	16,083	<0.0001
	Only 5 rooms left	15,129	<0.0001
	Only 7 rooms left	1,798	0.127
Distance to town center	0.1 km	90,616	<0.0001
	1.1 km	41,260	<0.0001
	2.1 km	16,211	<0.0001
	6.4 km	84,559	<0.0001
Sustainability level	Level 1	9,917	<0.0001
	Level 2	5,649	0.000
	Level 3	3,988	0.003
	Level 3+	15,802	<0.0001
Price	84.7 €	113,607	<0.0001
	100 €	113,607	<0.0001
	150 €	217,026	<0.0001
	200 €	401,072	<0.0001
	250 €	490,994	<0.0001
	300 €	533,226	<0.0001
	400 €	436,014	<0.0001
	500 €	431,501	<0.0001
	600 €	394,072	<0.0001

scores for lower prices, suggesting that while eco-friendliness is valued, it is subordinate to price considerations. The scores for hotel ratings show a nuanced behavior.

There is an appreciation medium-rated hotels, with a notable peak for "7.6". This might imply a trade-off between quality and cost, where acceptable quality at a lower price is preferred over higher quality at a higher price. The centroid values for scarcity messages indicating limited room availability are relatively high. This suggests that scarcity messages may effectively nudge this cluster towards making a booking decision, possibly due to fear of missing out on a good deal. The closer the hotel is to the city center, the more there is a rather negative stance toward these hotels. The preference for a three-star category over four or five stars suggests a tendency to seek satisfactory accommodations without the need for luxury, aligning with their cost-conscious profile.

This segment shows a high percentage of individuals who did not book any hotels last year, pointing towards a limited need for hotel services. They are primarily in the lower income bracket, suggesting budget constraints influence their booking decisions. Educationally, they span from compulsory to tertiary levels, with a lean towards lower education, which may correlate with their cost-conscious behavior. This group is also marked by a younger demographic, possibly indicating a temporary phase of life with limited financial resources for travel.

4.2 Green-Urban Deal Hunters (GUDH)

Green-Urban Deal Hunters represent a segment of travelers who are looking for more than just the lowest price. They value a good balance between the cost of accommodation and its sustainable credentials, provided the hotel's location allows them to be at the heart of urban life. This cluster might consist of individuals who are environmentally conscious, yet their decisions are also driven by practical considerations of cost and convenience. This cluster has relatively high centroid values for hotels rated as "7.6" and "6.9" which indicates a preference for medium-rated hotels. However, the values for "9.0" and "8.3" are also relatively high, revealing a balanced consideration between quality and affordability. The centroid values for sustainability levels are present but not as pronounced as price indicators, suggesting that while sustainability is a consideration, it does not override the importance of price. The values for proximity to the city center suggest that urban location is important to this cluster, with a desire to be close to city amenities and attractions.

The centroid values across different price points show a downward trend as prices increase, confirming the price sensitivity of this cluster. They are looking for a "deal" that balances cost with the perceived value of sustainability and location. Higher values for larger numbers of reviews indicate a reliance on social proof and the wisdom of the crowd in decision-making, suggesting they seek validation from other travelers' experiences. Lower values for the five-star category and the highest room rates suggest a lesser emphasis on luxury accommodations, indicating a pragmatic approach to booking where excessive spending is avoided.

Green-Urban Deal Hunters are typically married, indicating a potential for family or couple-based travel preferences. They have a balanced booking frequency, reflecting a considered approach to travel, possibly planning around family or work commitments.

Their education levels are quite distributed but show a tendency towards higher education, which may influence their value for sustainability and urban experiences. They book moderately and evenly distributed across income levels, suggesting a conscious balance between quality and cost.

4.3 Social-Proof Assurance Seekers (SPAS)

The "Social-Proof Assurance Seekers" cluster is characterized by travelers who rely heavily on the experiences of others to guide their booking decisions. The prominence of online reviews and scarcity cues in their decision-making suggests that they may seek reassurance from others' endorsements before committing to a booking. They are willing to invest in a higher-rated hotel, provided it comes with a strong backing from many reviews, reflecting a collective confirmation of the establishment's quality. This cluster's centroid data implies a decision-making process highly influenced by social validation and quality assurances, placing significant weight on the opinions and experiences of others. They may perceive a scarce availability as an indication of a hotel's popularity or quality, which can serve as a persuasive factor in their decision-making process. With a significant emphasis on ratings and the number of reviews, this group is likely to seek social validation through the experiences of others. They may exhibit trust in the wisdom of the crowd, using it as a benchmark for their choices.

While they do consider sustainability, it is not the overriding factor in their decision process. However, they appreciate eco-friendly practices as a value-add, especially if such attributes come with strong social proof. This cluster's willingness to pay more for a hotel that has a strong backing in terms of social validation, as reflected in the high ratings and numerous reviews, indicates their preference for assured quality over lower cost. The centroid values for higher ratings categories like "9.7" and "9.0" are notably substantial, suggesting that this cluster places a premium on staying at hotels with excellent reputations. In addition, there is a spread across centroid values for different distances to the city center, which might indicate a certain degree of flexibility in terms of location, if the hotel's quality and social proof are assured.

Social-Proof Assurance Seekers have a higher representation of married individuals, which might be indicative of travel decisions influenced by family considerations or shared experiences valuing social proof. They tend to book hotels with a frequency that suggests regular but not excessive travel. Their education levels skew towards higher education, and they display a broad age range, suggesting diverse life stages from working professionals to active retirees who value others' opinions in their booking choices.

4.4 Budget-Only Focused Minimalists (BOFM)

The "Budget-Only Focused Minimalists" are characterized by their single-minded pursuit of economical options. They exhibit a high degree of price elasticity, responding to cost savings rather than other features such as sustainability, scarcity, or luxury. The limited sensitivity to hotel ratings and the number of reviews indicate that they may rely on basic accommodation standards or are confident in their ability to select suitable accommodations without heavily depending on other travelers' opinions. It is evident that this group prioritizes cost above other attributes when making hotel booking decisions. This cluster demonstrates a strong preference for lower prices, as indicated by the significant utility scores associated with lower price points. The scores across various rating levels do not show a marked preference, suggesting that this group does not weigh ratings as heavily in their decision-making process. There is no significant reaction to scarcity cues such as limited room availability. This group seems to be less influenced by marketing tactics that create a sense of urgency through scarcity. Sustainability levels appear to have little to no impact on their booking decisions.

This segment is characterized by the highest percentages of individuals in the lowest income and education brackets, which directly influences their minimalistic approach to travel. Their booking patterns show a significant number of older individuals. This group's less frequent booking behavior suggest a targeted and essential approach to travel, prioritizing affordability over luxury or brand reputation.

4.5   Luxury-Quality Connoisseurs (LQM)

The "Luxury-Quality Connoisseurs" are sensitive travelers who seek out the best experiences. They are likely to book at well-established, high-starred hotels and may use sustainability as a decider between equally luxurious options. Their booking behavior is motivated by the pursuit of top-quality service, comfort, and an overall luxurious experience. Analyzing the centroid data for this cluster reveals a group that places a premium on high-quality, luxury experiences, and while they have an appreciation for sustainability, it is not their primary concern.

The higher utility scores for top-tier ratings indicate that this group is inclined towards hotels with exceptional reviews. They are likely to seek out establishments that promise an elite experience, denoted by high guest satisfaction levels. While price sensitivity is present, it is not as pronounced as in other clusters. This group is willing to pay more for perceived quality and luxury, as suggested by the balanced utility scores across various price points. Scarcity cues such as "only one room left" may influence their decision to some extent, hinting that while they are looking for luxury, they are also attracted to exclusivity, which scarcity signals can imply. Although sustainability is not disregarded, it is secondary to luxury and quality. They might prefer sustainable options, but not at the expense of comfort or prestige.

As indicated by their income bracket, the Luxury-Quality Connoisseurs show a higher tendency for frequent bookings, emphasizing the importance of travel in their lifestyle. They are often married, which might suggest a preference for shared high-end travel experiences or business travel that allows for more luxurious stays. Their education levels are spread across the spectrum, with a notable percentage holding advanced degrees, possibly reflecting their appreciation for quality and comfort in their travel choices. This group tends to be older, which may correlate with the financial means to prioritize luxury in their bookings.

Table 3. Centroid Data for Clusters

Attributes	Level	Cluster 1 CCEB	Cluster 2 GUDH	Cluster 3 SPAS	Cluster 4 BOFM	Cluster 5 LQM
Review-Valence	9.7	37.919	44.828	55.851	42.137	56.967
	9.0	37.820	46.592	55.616	43.521	58.434
	8.3	44.278	45.536	51.425	47.805	54.220
	7.6	64.280	59.505	48.276	56.938	45.121
	6.9	62.018	52.444	41.993	58.160	40.167
Amount of Reviews	5	64.870	51.179	54.123	57.286	49.080
	5795	48.021	47.687	43.381	46.771	47.674
	9655	44.799	59.763	59.703	53.857	62.030
Hotel star category	3-star	62.985	62.115	49.731	63.153	44.559
	4-star	42.785	44.375	62.270	41.479	53.914
	5-star	30.950	30.756	32.381	31.915	47.734
Scarcity cues	Only 1 room left	52.047	50.268	62.104	48.627	59.288
	Only 3 rooms left	51.609	56.623	45.449	54.888	50.695
	Only 5 rooms left	54.156	54.527	48.634	55.680	46.432
	Only 7 rooms left	48.861	45.064	47.940	48.126	47.414
Distance to town center	0.1 km	39.406	60.244	34.791	47.173	38.121
	1.1 km	51.584	67.725	59.839	51.041	51.181
	2.1 km	50.013	49.945	58.031	46.280	49.080
	6.4 km	61.346	34.075	55.219	57.525	63.155
Sustainability level	Level 1	53.624	55.680	55.733	63.047	57.259
	Level 2	57.913	55.508	63.529	58.785	58.800
	Level 3	48.712	49.172	50.047	43.860	45.972
	Level 3+	47.094	45.615	36.959	36.727	43.531
Price	84.7 €	66.888	36.526	49.460	37.520	46.393
	100 €	66.888	36.526	49.460	37.520	46.393
	150 €	59.014	45.531	53.574	24.895	58.601
	200 €	43.806	49.098	45.326	15.316	61.060

(*continued*)

Table 3. (*continued*)

Attributes	Level	Cluster 1 CCEB	Cluster 2 GUDH	Cluster 3 SPAS	Cluster 4 BOFM	Cluster 5 LQM
	250 €	32.008	45.141	35.036	13.328	57.087
	300 €	28.601	46.484	33.911	12.423	58.122
	400 €	22.583	42.552	28.025	14.331	55.183
	500 €	24.171	43.013	29.367	16.433	57.153
	600 €	23.620	41.529	31.747	18.945	58.467

5 Discussion

The emergence of five distinct consumer clusters - Cost-Conscious Eco-Bookers, Green-Urban Deal Hunters, Social-Proof Assurance Seekers, Budget-Only Focused Minimalists, and Luxury-Quality Connoisseurs - reflects a spectrum of prioritization across multiple hotel attributes, from price and location to sustainability and social proof. The significant implications of these clusters for OTAs and hoteliers lie in their potential application for precision-targeted marketing strategies and the enhancement of the personalization of services. This insight aligns with the works of Gretzel et al. [1] and Xiang et al. [23], which emphasize the need for a deep understanding of consumer behavior to drive personalization in the digital tourism sphere. Our findings mirror the shift in market segmentation theories, moving beyond demographic data towards a richer psychographic and behavioral understanding as outlined by Weinstein [12] and Smith [10]. The statistical validation of our clustering approach, evidenced by the ANOVA results, resonates with the importance of varied attributes in influencing consumer preferences. This relates closely to the findings of Buhalis & Law [18] and Li et al. [5], who highlighted the role of big data in enabling nuanced market segmentation in tourism. Moreover, the utility scores for attributes like review valence and scarcity cues support the perspectives of Morrison [7], indicating the ongoing significance of consumer-perceived value and urgency in booking decisions. Our research contributes a significant layer to the body of eTourism literature by proposing a novel model for user segmentation based on direct preferences for hotel attributes. This model has the potential to bridge the gap identified by Kotler & Keller [9] and Plummer [11], where the interplay of consumer psychology and market segmentation has been a enduring focus.

Cost-Conscious Eco-Bookers. The Cost-Conscious Eco-Bookers (CCEB) cluster exemplifies a segment balancing financial caution with environmental concerns. This group's price sensitivity echoes Kotler & Keller's [9] emphasis on cost-effective marketing strategies tailored for budget-aware segments. The moderate interest in sustainability aligns with Bahja et al.'s [6] findings that ecological concern influences consumer choices in hospitality. Yet, for CCEB, environmental friendliness is secondary to affordability, suggesting a need for competitively priced eco-friendly options. OTAs and hoteliers can target CCEB with value-oriented eco-friendly packages that do not compromise on cost. This strategy could be augmented by leveraging scarcity cues, as this group shows responsiveness to such marketing tactics. Given their preference for quality at reasonable rates, OTAs should present them with transparent review-based quality indicators, aligning with the social proof concept highlighted by Jamal et al. [22]. Hoteliers can emphasize their sustainable practices without additional costs, potentially appealing to the CCEB segment's eco-awareness. Scarcity-based promotions can also be effective, nudging this cost-sensitive segment towards quicker booking decisions.

Green-Urban Deal Hunters. Green-Urban Deal Hunters (GUDH) prioritize sustainability but not at the expense of convenience or cost. This reflects the VALS framework's principles, where values like environmentalism coexist with pragmatic purchase behaviors [11]. Their urban-centric preferences suggest a lifestyle-oriented segmentation approach, as discussed by Weinstein [12]. For OTAs, the strategy should focus on well-rated, centrally-located hotels with clear sustainability features, providing a mixture of urban experience and eco-consciousness. Offering dynamic pricing and limited-time offers could effectively target GUDH, appealing to their deal-seeking nature without abstaining to their green values. Hoteliers can attract GUDH by showcasing their sustainable credentials and proximity to urban attractions, potentially incorporating flexible pricing strategies that reflect the value of their location and green initiatives.

Social-Proof Assurance Seekers. Social-Proof Assurance Seekers (SPAS) are heavily influenced by the experiences of others, as seen in their reliance on reviews and ratings. Their behavior underpins the theories of social validation and assurance in consumer behavior [16]. The cluster's willingness to pay more for socially validated quality points to the trust economy's impact highlighted by Komiak & Benbasat [30]. OTAs should implement reputation-based recommendation systems, highlighting hotels with high ratings and numerous reviews. Personalized marketing communications that cite customer testimonials and ratings can resonate well with SPAS, reinforcing the quality assurance they seek. Hoteliers should encourage satisfied guests to leave positive reviews and can design experiences that are likely to be shared on social media, leveraging the power of user-generated content to build trust and influence booking decisions.

Budget-Only Focused Minimalists. The Budget-Only Focused Minimalists (BOFM) cluster's focus on cost above all reflects the Market Segmentation Theory's cost-focused consumer group [10]. Their limited interest in ratings or sustainability cues suggests a functional approach to booking, consistent with Morrison's [7] discussion on budget-driven travel behavior. For OTAs, this indicates the necessity of a stripped-down, price-focused marketing approach. Highlighting the lowest available prices and basic amenities could effectively capture this segment. Bundling options are less likely to appeal to BOFM unless they present clear cost-saving opportunities. Hoteliers can satisfy BOFM by offering basic accommodations and transparent pricing, ensuring that guests don't pay for unnecessary extras, thereby aligning with their budget-focused values.

Luxury-Quality Connoisseurs. Luxury-Quality Connoisseurs' (LQM) preference for high-quality, luxurious experiences aligns with the psychographic segmentation that associates lifestyle and luxury [14]. Their appreciation for sustainability when choosing between high-end options reflects a premium consumer's sophisticated decision-making process, as evidenced by the work of Apostolakis et al. [24]. OTAs targeting LQM should focus on curating a selection of premium, high-starred hotels that highlight both luxury and sustainability. Personalized high-touch services, loyalty rewards, and exclusive offers can cater to their expectations for a tailored experience, reinforcing the importance of a customer-centric approach as discussed by Paluch & Tuzovic [29]. For hoteliers, this means providing flawless service and high-quality amenities. They could also create exclusive sustainable programs that appeal to LQMs, offering a synthesis of luxury and environmental responsibility.

6 Concluding Remarks and Limitations

The practical contribution of our research lies in its application to the eTourism industry, providing businesses with a nuanced understanding of customer segments through the clustering of user characteristics. This segmentation enables the creation of tailored marketing campaigns, such as targeting eco-conscious travelers with green travel packages, thereby enhancing the precision and effectiveness of marketing efforts. Moreover, our findings inform the customization of booking platforms to align with specific consumer preferences, thereby elevating the user experience and potentially increasing customer loyalty. The insights also aid eTourism companies in making strategic decisions, optimizing resource allocation based on the attributes most valued by their clientele. This approach not only streamlines operations but also furnishes eTourism operators with a competitive edge by facilitating the delivery of personalized customer experiences.

The theoretical contribution of this paper to the scientific community, particularly within the domain of eTourism, lies in its comprehensive analysis of consumer behavior during the hotel booking process. Traditionally, research in this area has predominantly focused rather on internal factors of OTAs, such as hotel attributes and how they influence customer decisions. However, our study extends this perspective by integrating external factors, specifically the diverse characteristics of user groups, into the evaluation process. By employing an approach that considers both the attributes presented by OTAs and the distinct preferences of various user clusters, our research highlights the multi-dimensional nature of the booking process. This methodology may enrich the current understanding of how internal factors, like hotel attributes, impact consumer choice and also how these choices are nuanced by the external factors, such as the socio-demographic profiles of the users and their unique travel motivations.

The primary limitation of this study is the reliance on self-reported data of respondents, which may introduce bias. Additionally, the study's focus on a specific demographic within Austria and Germany limits its generalizability to other regions and cultures. Future research should explore the applicability of the proposed segmentation model across diverse global markets and investigate the impact of real-time data analytics on the accuracy of user segmentation. Our study centers on the overarching findings of the k-means clustering approach, and as such, we do not delve into the detailed outcomes of the web content mining and multiple linear regression analyses. The specific regression results fall outside the scope of this paper, with our focus being on the superior insights derived from the conjoint analysis and k-means clustering. We also recognize that selecting a five-cluster solution based primarily on fit metrics may raise concerns about overfitting and limit the robustness and applicability of our findings across different datasets, which we identify as a limitation of our study. Moreover, longitudinal studies could provide insights into the stability of the identified segments over time, and experimental designs could test the effectiveness of tailored marketing strategies derived from the segmentation model.

Acknowledgments. We would like to express our gratitude for the analytical tools provided by Sawtooth Software Lighthouse Studio, which significantly contributed to the data analysis and insights presented in this paper.

Disclosure of Interests. The authors have no competing interests to declare that are relevant to the content of this article.

References

1. Gretzel, U., Sigala, M., Xiang, Z., Koo, C.: Smart tourism: foundations and developments. Electron. Mark. **25**, 179–188 (2015)
2. Xiang, Z., Du, Q., Ma, Y., Fan, W.: A comparative analysis of major online review platforms: implications for social media analytics in hospitality and tourism. Tour. Manage. **58**, 51–65 (2017)
3. Law, R., Qi, S., Buhalis, D.: Progress in tourism management: a review of website evaluation in tourism research. Tour. Manage. **31**(3), 297–313 (2010)
4. Buhalis, D., Foerste, M.: SoCoMo marketing for travel and tourism: empowering co-creation of value. J. Destin. Mark. Manag. **4**(3), 151–161 (2015)
5. Li, J., Xu, L., Tang, L., Wang, S., Li, L.: Big data in tourism research: a literature review. Tour. Manage. **68**, 301–323 (2018)
6. Bahja, F., Cobanoglu, C., Berezina, K., Lusby, C.: Factors influencing cruise vacations: the impact of online reviews and environmental friendliness. Tourism Rev. **74**(3), 400–415 (2019). https://doi.org/10.1108/tr-12-2017-0207
7. Morrison, A.M.: Marketing and Managing Tourism Destinations. Routledge, London (2013)
8. Eibl, S., Auinger, A.: On the Role of User Interface Elements in the Hotel Booking Intention: Analyzing a Gap in State-of-The-Art Research. In: International Conference on Human-Computer Interaction, pp. 170–189. Springer (2023). https://doi.org/10.1007/978-3-031-35969-9_12
9. Kotler, P., Keller, K.L.: Marketing Management (15th global ed.). England: Pearson, pp. 803–829 (2016)
10. Smith, W.R.: Product differentiation and market segmentation as alternative marketing strategies. J. Mark. **21**(1), 3–8 (1956)
11. Plummer, J.T.: The concept and application of life style segmentation: the combination of two useful concepts provides a unique and important view of the market. J. Mark. **38**(1), 33–37 (1974)
12. Weinstein, A.: Handbook of market segmentation: Strategic targeting for business and technology firms. Routledge (2013)
13. Wind, Y.J., Green, P.E.: Some Conceptual, Measurement, and Analytical Problems in Life Style Research. Marketing Classics Press (2011)
14. Kahle, L.R., Chiagouris, L.: Values, Lifestyles, and Psychographics. Psychology Press (2014)

15. Sigala, M.: Integrating customer relationship management in hotel operations: managerial and operational implications. Int. J. Hosp. Manag. **24**(3), 391–413 (2005)

16. Gretzel, U., Fesenmaier, D.R.: Experience-based internet marketing: an exploratory study of sensory experiences associated with pleasure travel to the Midwest United States. In: ENTER, pp. 49–57 (2003)

17. Fotaki, G., Spruit, M.R., Brinkkemper, S., Meijer, D.: Exploring big data opportunities for online customer segmentation. Int. J. Bus. Intell. Res. **5**, 58–75 (2014)

18. Buhalis, D., Law, R.: Progress in information technology and tourism management: 20 years on and 10 years after the Internet—The state of eTourism research. Tour. Manage. **29**(4), 609–623 (2008)

19. Gan, M.Y., Ouyang, Y.: Study on tourism consumer behavior characteristics based on big data analysis. Front. Psychol. **13**, 876993 (2022). https://doi.org/10.3389/fpsyg.2022.876993

20. Sigala, M.: Social media marketing in tourism and hospitality. Inf. Technol. Tourism **15**(2), 181–183 (2015). https://doi.org/10.1007/s40558-015-0024-1

21. Kwon, K., Kim, C.: How to design personalization in a context of customer retention: who personalizes what and to what extent? Electron. Commer. Res. Appl. **11**, 101–116 (2012)

22. Jamal, S., Newbold, K.B.: Factors associated with travel behavior of millennials and older adults: a scoping review. Sustainability **12**(19), 8236 (2020)

23. Xiang, Z., Magnini, V.P., Fesenmaier, D.R.: Information technology and consumer behavior in travel and tourism: insights from travel planning using the internet. J. Retail. Consum. Serv. **22**, 244–249 (2015). https://doi.org/10.1016/j.jretconser.2014.08.005

24. Apostolakis, A., Jaffry, S., Kourgiantakis, M.: Examination of individual preferences for green hotels in Crete. Sustainability **12**(20), 8294 (2020). https://doi.org/10.3390/su12208294

25. Merinov, P., Massimo, D., Ricci, F.: Sustainability driven recommender Systems. In: Italian Information Retrieval Workshop (2022)

26. Ricci, F., Rokach, L., Shapira, B.: Recommender Systems: Introduction and Challenges. Recommender Systems Handbook, pp. 1–34 (2015)

27. Lops, P., De Gemmis, M., Semeraro, G.: Content-based recommender systems: State of the art and trends. Recommender Systems Handbook, pp. 73–105 (2011)

28. Dzulfikar, M.F., et al.: Personalization features on business-to-consumer e-commerce: review and future directions. In: 2018 4th International Conference on Information Management (ICIM), pp. 220–224 (2018)

29. Paluch, S., Tuzovic, S.: Persuaded self-tracking with wearable technology: carrot or stick? J. Serv. Mark. **33**(4), 436–448 (2019)

30. Komiak, S.Y., Benbasat, I.: The effects of personalization and familiarity on trust and adoption of recommendation agents. MIS Quarterly, pp. 941–960 (2006)

31. Gal-Or, E., Gal-Or, R., Penmetsa, N.: The role of user privacy concerns in shaping competition among platforms. Inf. Syst. Res. **29**, 698–722 (2018)

32. Martin, K.D., Murphy, P.E.: The role of data privacy in marketing. J. Acad. Mark. Sci. **45**, 135–155 (2017)

33. Pariser, E.: The Filter Bubble (2012)

34. Praditya, N.W.P.Y., Erna Permanasari, A., Hidayah, I.: Designing a tourism recommendation system using a hybrid method (Collaborative Filtering and Content-Based Filtering). In: 2021 IEEE International Conference on Communication, Networks and Satellite (COMNETSAT), pp. 298–305 (2021)

35. Baier, D., Brusch, M.: Methoden - Anwendungen - Praxisbeispiele. In: Baier, D., Brusch, M. (eds.) Conjointanalyse (2009). https://doi.org/10.1007/978-3-662-63364-9

36. Schegg, R.: European Hotel Distribution Study 2022. University of Applied Sciences and Arts of Western Switzerland, Statista (2022)

37. Field, A.: Discovering Statistics using IBM SPSS Statistics. Sage (2013)

38. Tabachnick, B.G., Fidell, L.S., Ullman, J.B.: Using Multivariate Statistics. Pearson Boston, MA (2013)
39. Orme, B.K.: Getting started with conjoint analysis: strategies for product design and pricing research (2006)
40. Hair, J.F.: Multivariate data analysis (2009)
41. Allenby, G.M., Rossi, P.E.: Hierarchical bayes models. In: The Handbook of Marketing Research: Uses, Misuses, and Future Advances, pp. 418–440 (2006)
42. Jain, M., Kaur, G., Saxena, V.: A K-Means clustering and SVM based hybrid concept drift detection technique for network anomaly detection. Expert Syst. Appl. **193**, 116510 (2022)
43. Steinley, D.L.: K-means clustering: a half-century synthesis. Br. J. Math. Stat. Psychol. **59**(Pt 1), 1–34 (2006)
44. MacQueen, J.: Some methods for classification and analysis of multivariate observations. In: Proceedings of the Fifth Berkeley Symposium on Mathematical Statistics and Probability, vol. 1, no. 14, Oakland, CA, USA, pp. 281–297 1967
45. Rousseeuw, P.J.: Silhouettes: a graphical aid to the interpretation and validation of cluster analysis. J. Comput. Appl. Math. **20**, 53–65 (1987)
46. Wedel, M., Kamakura, W.A.: Market segmentation: Conceptual and methodological foundations. Springer Science & Business Media (2000). https://doi.org/10.1007/978-1-4615-4651-1
47. Kass, R.E., Raftery, A.E.: Bayes factors. J. Am. Stat. Assoc. **90**(430), 773–795 (1995)
48. Milligan, G.W., Cooper, M.C.: A study of standardization of variables in cluster analysis. J. Classif. **5**, 181–204 (1988)
49. Ja, H.: A k-means clustering algorithm. JR Stat. Soc. Ser. C-Appl. Stat. **28**, 100–108 (1979)
50. Malhotra, N.K.: Marketing Research: an Applied Prientation. Pearson (2020)
51. So, K.K.F., King, C., Sparks, B.A., Wang, Y.: The role of customer engagement in building consumer loyalty to tourism brands. J. Travel Res. **55**, 64–78 (2016)

Keywords Effectiveness in Textile Product Sales Performance: A Case Study of the Shopee Website

Pei-Hsuan Hsieh$^{(\boxtimes)}$ (iD) and Ambrose Phong

National Chengchi University, Taipei 116, Taiwan
hsiehph@nccu.edu.tw

Abstract. Online shoppers usually conduct product searches using one or more keywords, in addition to applying filters, to find their favorite products on an e-commerce platform. However, it remains uncertain which combinations of keywords are effective for online sellers to include in their titles to ensure that their listed products match buyers' search keywords and appear in the search results. The purpose of this study is to identify the keyword strategies that sellers use when listing their products on an e-commerce platform. The study further explores how sales performance can be improved by employing specific strategies for keyword combinations. In this study, web scraping techniques were employed twice to obtain two 30-day product sales datasets from Shopee. One dataset was validated by analyzing the other. The Term Frequency-Inverse Document Frequency statistical method, in conjunction with the Beautiful Soup library, was used for data analysis. The results show that there is indeed a higher sales volume when using the 10 most frequently used keywords in sellers' product titles. In both datasets, these top 10 keywords consistently appearing in product titles include two textile products (i.e., T-shirt and underpants) and seven adjectives (i.e., breathable, cooling, customized, group, rayon, in stock, Taiwan). The results reveal that sellers can increase their sales by strategically combining these high-frequency keywords. By identifying these keywords, textile product sellers can better optimize their product titles, increase visibility, and attract more interest from potential buyers.

Keywords: Keyword Search · Sales Performance · e-Commerce · Beautiful Soup · TF-IDF (Term Frequency-Inverse Document Frequency)

1 Introduction

The COVID-19 pandemic has driven a surge in online shopping as consumers seek to avoid the risks associated with traditional retail. This surge has endowed e-commerce platforms with significant development potential. Shopee is one of the most vital e-commerce platforms in Southeast Asia, including Taiwan, with 70 million monthly visits in these regions. In Taiwan, Shopee is especially popular, establishing itself as the most widely used local online shopping platform [1]. Similar to other e-commerce platforms in Taiwan (such as Yahoo!Kimo, Ruten, Momo, Books, PChome), Shopee

© The Author(s), under exclusive license to Springer Nature Switzerland AG 2024
F. F.-H. Nah and K. L. Siau (Eds.): HCII 2024, LNCS 14720, pp. 176–186, 2024.
https://doi.org/10.1007/978-3-031-61315-9_12

organizes various promotional activities. On this platform, buyers can enjoy a convenient shopping experience while exploring a wide range of products, and sellers can expand their product sales market. Given its large user base and diverse marketing tools, Shopee was selected for this case study to research the effective use of search keywords in online sales.

All online products can be searched using one or more keywords on a search engine or through paid/sponsored search advertising [2, 3]. However, the effectiveness of keywords in sales relies on advertisers' grouping strategies, especially in uncertain environments [2, 4, 5]. To increase the likelihood of actual transactions, buyers must first be directed to an e-commerce platform such as Shopee and then use certain keywords to find more products on the platform. The keywords, coupled with filter options (e.g., brand, price, listing date) provided by Shopee, are typically designed to identify product categories. Some products have specific keywords or tags, making them easily searchable for buyers. These keywords, which incorporate concise product descriptions or product features, are directly integrated into the product name, creating a more informative product title. This helps enhance the visibility of products and attract more potential buyers. However, it is uncertain which combinations of keywords are more effective in capturing consumer attention and influencing purchasing decisions, which is what motivated the current study.

The purpose of this study is to deepen our understanding of how sellers optimize their product titles by strategically selecting and combining keywords to enhance product visibility and boost clothing sales. The results of this study are intended to assist textile product sellers in gaining insights into their target buyers and improving sales efficiency. Sellers of other types of products can also derive valuable marketing strategies and decision support from this study.

2   Literature Review

A consumer who wishes to search for a product online might simply use a search engine or enter a search term on an e-commerce platform. Sellers using keyword advertising expect that the buyers' keyword searches will easily lead them to their products. However, a study has revealed that, in reality, buyers tend to avoid keyword advertisements [6]. The experimental study measured the eye movements and number of clicks of 451 volunteer searchers and found that the searchers clicked to see more organic (i.e., unpaid) search results than keyword advertisements. Moreover, the placement of the keyword advertisement did not affect the number of observations the searchers made. The study suggested advertisers focus not only on attracting consumer attention but also on increasing the effectiveness of keyword advertising. The study also stated that while the searchers tended to avoid clicking on keyword advertisements, they did glance at the sponsored area. This means that advertisers should focus on presenting concise, clear focal messages in their advertising copy instead of encouraging clicks.

Product brand names or categories without any specifications are often used by buyers as general keywords when searching for digital products. General keywords have positive and significant effects on sales performance; however, they bring less total product sales than specific keywords [7]. (Note that contrary results were found when buyers searched for non-digital products, e.g., clothes). Another study proposed an automatic keyword generation approach using the internal search engine of online stores [3]. This approach involved extracting keywords, particularly from consumers' goal-directed search patterns (as opposed to exploratory ones), and it proved effective in expanding the number of profitable keywords.

A prior study suggested that managerial insights that are based on business goals are critical for advertisers when strategically grouping the keywords in sponsored search advertising, especially under uncertain environments like budget increases and profit considerations [5]. Advertising with more keywords and allocating more budget to keyword advertising do not necessarily result in higher profits [4, 8]. Sellers need to take into account other variables, such as competitive market dynamics influencing buyers' decisions and specific brands that they follow [4].

The literature also suggests using various keyword portfolios to enhance risk-adjusted return and sales performance [7–9]. A study found that mobile advertising can increase the variety of keyword portfolios than traditional PC online search advertising [8]. The study suggested using keyword portfolios to target the goal-oriented buyer's search behaviors in product titles and subcategories (e.g., product features or adjective keywords) of the products, thereby increasing direct sales (vs. focusing on indirect sales with subcategories). The study finally suggested using different keyword portfolios for different sales channels to attract different types of consumers.

As discussed in the literature, advertisers should strategically use various keywords or establish different keyword portfolios to achieve the business goal of boosting product sales performance. This study is therefore motivated to explore the specific strategy of keyword combination, hoping to provide concrete suggestions to sellers for crafting textile product titles.

3   Methodologies

3.1   Data Sources and Data Collection

This study focused on the top-selling clothing products on Shopee, including men's and women's clothing. Product information on men's and women's clothing was collected using the Python beautifulsoup4 library as it had been proven as a successful web scraping tool [10]. We gathered product title, price, location, and monthly sales from Shopee.

To ensure the accuracy of data collection, two rounds of data collection were conducted. The first rounds of data collection took place on September 14 (Time A), and the second round, on October 23 (Time B). Each data collection targeted Shopee's "men's clothing" and "women's clothing" categories to include monthly sales data and full product titles for the 30 days preceding the data collection day. All products on the first four web pages of the Shopee platform were identified as top-selling and were scraped (as buyers are less likely to click beyond these pages). As each Shopee web page displays 60 entries, 480 entries were retrieved in each data collection round (240 for men's clothing and 240 for women's clothing), resulting in a total of 960 product information entries. The results of the second-round data analysis validated those obtained in the first round, which confirmed that there were no data collection errors and established the reliability of the data analysis at a later stage.

3.2 Data Cleaning and Organization

After the data were collected, they were cleaned by removing items that were not textile products, such as accessories and shoes, as they were unrelated to the analytical goals of this study. Next, term frequency-inverse document frequency (TF-IDF) was used to organize the data [11]; specifically, a tool in the Python programming language called "jieba" was used to handle Chinese characters in text segmentation. In addition, emoticons, English letters, mathematical symbols, and punctuation found in product names/titles were removed to facilitate text analysis at a later stage. The screenshot of the original web-scraped data points is shown in Fig. 1. The data cleaning and organization resulted in an analyzable dataset of 942 data points (472 for Time A and 470 for Time B).

Product Titles	Price	Sales	Location
【全店免運 免費排版設計】客製化衣服 長袖POLO衫 排汗POLO衫 吸濕 透氣 工作服 企業LOGO 上衣 一件起訂	$250	月銷量 1.1萬	臺北市南港區
客製化t恤訂製班服衣服工作服客製圖體服客制工作衣訂做排汗衫印刷電繡製作印製浸金工作上衣團服廟會宮廟服上班女團體男印字設計	$221	月銷量 8,688	桃園市桃園區
♥ Vivy peach ♥ 附發票 現貨秒出 陽離子 短袖 寬心 涼感衣 吸濕排汗 抗UV 機能運動背心 男生 FILA	$79 - $129	月銷量 1,434	新北市三重區
快速出貨 訂製衣服 客製化t恤 水洗 復古 重磅 客製化 100%精梳純棉 寬鬆 落肩 厚挺 質感超棉 短袖t恤 上衣	$78	月銷量 9,831	桃園市蘆竹區
有大呎碼【XS-7XL】專櫃團體服客製化t恤袖t恤 來圖印製純棉t恤 刺繡 排汗衫班服 親子裝t恤 活動服印字	$10 - $50	月銷量 9,358	桃園市蘆竹區
客製化衣服t恤訂製班服客製tshirt印花短袖訂做團體服排汗衫印刷圖服電繡團體製作印製上衣燙金logo設計印字印廟會服裝	$270	月銷量 8,802	桃園市桃園區
⊛ 冰絲涼感 ⊛ 高彈力 ⊛ 現貨透氣褲 長褲 涼感褲 工作褲 優跑褲 素面長褲 直筒褲 素面長褲 直筒褲 健身褲	$86	月銷量 12,076	臺南市安南區
臺頭客製化馬甲背心 印製文字 圖案 電繡 選舉服 志願者 班服 公司服飾 義工服飾 廟會背心 團體服	$179	月銷量 7,020	桃園市蘆竹區
訂製翻領短袖成人男女同款夏季polo衫訂製圖體化印製印logo工作服裝潮流上衣桃POLO衫DIY	$279	月銷量 6,900	桃園市蘆竹區
☺ 28s.studio ☻ 素面 買起 超挺 棉短褲 男女皆可 棉褲 抽繩棉褲 短褲	$199	月銷量 1,794	臺南市北區
☂ 限時免運 ⚡ 英國代購正品極厚外套 男生女生 經典款 衝鋒衣 三層拉鍊 防風衣 防潑水 防寒 連帽外套 保暖夾克	$1,280 - $1,580	月銷量 1,590	嘉義縣大埔鄉

Fig. 1. Examples of the original web-scraped data points collected on September 14, 2023. Note: The listed price is in New Taiwan Dollars (NTD), with an exchange rate of approximately 1 USD = 32 NTD. The sales column displays the number of items sold monthly, ranging from 1,434 to 12,076 (for the 11 examples in this figure only). The number of items sold monthly for the full dataset (472 data points) ranges from 301 to about 29.9 thousand.

3.3 Data Analysis Procedures

The cleaned and organized data from Time A and Time B were analyzed separately using TF-IDF, with a specific focus on the product titles. According to the literature, TF-IDF can analyze a substantial amount of data by calculating word frequencies to identify consumers' preferred search keywords for online shopping products [11]. The process of using TF-IDF involved extracting keywords in the product titles and calculating the frequency of these keywords. Those extracted high-frequency product title words that correlated with high sales performance reflect a good match between sellers' descriptions of the products and buyers' needs. In other words, a higher frequency use of certain product title words can contribute to increased overall product sales.

The analysis results yielded 70 keywords, which were sorted from high to low based on word frequency. A comparison between Time A and Time B found that the same top-frequency keywords appeared in both sets. This pattern affirmed the accuracy of data collection, cleaning, and organization processes in both timeframes. The Time B dataset subsequently underwent additional analysis to identify two or more integrated keywords for optimizing product titles and achieving top sales performance on Shopee. It is important to note that, even though the data for men's clothing and women's clothing were collected separately, the additional analysis for Time B combined them into one dataset of 470 product items. This decision was influenced by the common seller's practice of cross-posting products under different categories to attract more buyers across all genders. Besides, the product titles made it challenging for the researchers to differentiate one-category from two-category products.

The final analysis for the Time B dataset proceeded as follows: The number of products sold and their respective locations were identified by checking for the presence of two or more high-frequency, top-ranking integrated keywords in product titles. Simultaneously, the calculation of the average number of products sold in Time B was conducted. Bar graphs were then used to identify the optimizing strategies for crafting product titles to achieve higher sales performance. The process of data collection, cleaning, organization, and analysis is illustrated in Fig. 2.

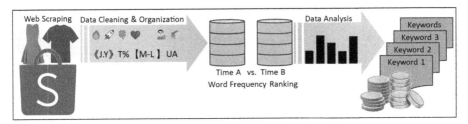

Fig. 2. The process of data collection, cleaning, organization, and analysis.

4   Results

The top ten keywords consistently appearing in the product titles in both Time A and Time B are shown in Table 1. All keywords have been translated from the original traditional Chinese characters presented in parentheses. It was found that two types of textile products (i.e., T-shirts and underpants) and seven adjectives/keywords (i.e., breathable, cooling, customized, group, rayon, in stock, Taiwan) consistently ranked high. The keywords that described the texture of textile products had higher sales performance, as demonstrated by the analysis results presented in the next section. Adjectives that ranked lower included collarless, sport, comfortable, disposable, etc.

The above initial findings confirmed the accuracy of the data collection method. Thus, the second dataset (i.e., Time B) was worthy of further analysis to identify one or more integrated keywords associated with the best sales performance. The final section of the analysis results presents the extra findings on the influence of product locations on sales performance.

4.1   Sales Performance of T-shirts

The monthly sales performance of T-shirts in Time B is shown in Fig. 3. Each frequently used keyword for T-shirts resulted in a higher average monthly sales than two randomly selected, infrequently used keywords (e.g., "sport" ranked 22 + "collarless" ranked 26).

Table 1. Top 10 keywords consistently used in product titles on Shoppe.

Rank	Frequency and Keyword in Time A	Frequency and Keyword in Time B
1	127 Underpants (內褲)	140 Customized (客製化)
2	109 In stock (現貨)	127 Underpants (內褲)
3	95 Breathable (透氣)	104 Taiwan (台灣)
4	81 Rayon (冰絲)	92 In stock (現貨)
5	81 Cooling (涼感)	88 Breathable (透氣)
6	75 Taiwan (台灣)	77 Tailor-made (訂製)
7	71 Customized (客製化)	74 T-shirt (T恤)
8	66 T-shirt (T恤)	70 Rayon (冰絲)
9	62 Pure cotton (純棉)	62 Group (團體)
10	58 Group (團體)	59 Cooling (涼感)

Clearly, the keywords "T-shirt + Taiwan," followed by "Customized," "Breathable," "Group," "In Stock," "Rayon," and "Cooling," were effective in achieving higher sales performance for this type of products.

4.2 Sales Performance of Underpants

The monthly sales performance of underpants in Time B is shown in Fig. 4. Except for two keywords (i.e., customized and group), product titles that integrated any of the top five frequently used keywords had higher sales than those that used infrequently used keywords. Clearly, products with the keywords "Underpants + Rayon," followed by "In Stock," "Taiwan," "Cooling," and "Breathable," had higher product sales performance.

4.3 Influence of Product Locations

In large cities (e.g., Taipei, New Taipei, Taichung, Tainan, and Kaohsiung, also known as special municipalities in Taiwan), more items are sold because sellers living in these cities put more items for sale. As shown in Table 2, the product locations were dominated by six special municipalities (Time A: 84.96%; Time B: 85.11%). The average number of items sold in Time A and Time B was 11,927 and 20,894, respectively. However, aside from these larger cities, the manufacturing factories located to produce products also play a significant role in sales performance. In Keelung City, Hsinchu County, Changhua County (only in Time B), and Pingtung County, sellers listed fewer products but still achieved sales of more than 10,000 items. Therefore, an additional noteworthy finding is the significant impact of product locations on sales performance. On average, sales performance was much better than in these larger cities. In Time A, the highest average items sold were 8,925.67, achieved by only three product listings from Hsinchu County. In Time B, the highest average items sold were 10,960.00, achieved by only five product listings from Pingtung County on the Shopee platform.

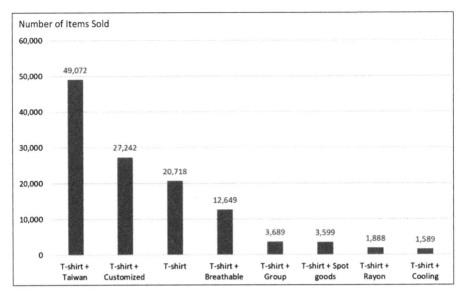

Fig. 3. Sales performance of T-shirts with different keyword combinations in their product titles, arranged from high to low.

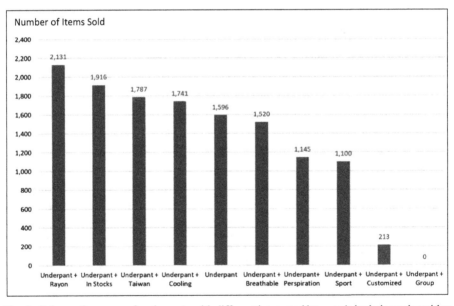

Fig. 4. Sales performance of underpants with different integrated keywords in their product titles, arranged from high to low.

Table 2. Product location and monthly sales performance

Location	Number of Products Listed (Time A Time B)		Number of Items Sold (Time A Time B)		Average (Products Listed / Items)	
Taipei City	23	26	22,998	63,974	999.91	2,460.54
New Taipei City	107	104	11,576	41,463	108.19	398.68
Taoyuan City	86	109	13,125	10,029	152.62	92.01
Taichung City	90	81	12,625	12,500	140.28	154.32
Tainan City	21	33	12,713	11,381	605.38	344.88
Kaohsiung City	45	47	9,699	10,708	215.53	227.83
Keelung City	8	5	27,375	28,857	3,421.88	5,771.40
Hsinchu County	3	5	26,777	10,960	8,925.67	2,192.00
Changhua County	29	13	6,508	14,934	224.41	1,148.77
Pingtung County	7	5	24,345	54,800	3,477.86	10,960.00

Number of products listed in other cities and counties (in alphabetic order): Chiayi City (6, 4), Chiayi County (8, 5), Hsinchu City (4, 7), Hualien County (0, 2), Mainland China (17, 6), Miaoli County (5, 5), Nantou County (3, 2), Unknow (1, 2), Yilan County (2, 3), Yunlin County (7, 6)

5 Discussions and Conclusion

Recognizing significant keywords in e-commerce is essential for all sellers. This study identifies two textile products and seven adjectives that consistently appeared in product titles on Shopee. The analysis of sales performance revealed that these frequently used integrated keywords were effective. For online sellers, the results of this study can serve as important references for designing a keyword portfolio specifically for textile products during the warmer seasons in Taiwan. The influence of product location also indicates that cities or counties with more sellers or where textile manufacturing factories are located are likely to generate a large number of sales. Novice sellers can collaborate with the factories in these areas or consider shipping items directly from these locations, combined with the keyword analysis results, as a valuable reference for starting their businesses.

5.1 Academic Implications

To identify effective keywords, this study used the beautifulsoup4 library, followed by the TF-IDF, to extract high-frequency words used as keywords in product titles [10, 11]. The beautifulsoup4 library was employed to scrape details from the Shopee platform across various categories of clothing. This technique for data cleaning and organization can be further enhanced by considering the native languages used by different people groups [e.g., 12]. For example, keywords like "harvest festival (豐年祭)" used by the

Amis and "grateful to you (承蒙您)" used by Hakka speakers may be used by Taiwanese sellers on the Shoppe platform.

In addition, this study utilized "jieba" to segment words in textile product titles. Different word segmentation tools, such as Ansj, HanLP, and Word, can be employed to find similar results [13]. Other word segmentation tools include CKIP, Ekphrasis, Linguistica, and Stanza. Advanced natural language processing techniques like Stanford Core NLP, TextBlob, SpaCy, and GenSim NLTK processing corpus can also be considered. Moreover, the automatic generation of keywords by continuously retrieving and updating algorithms to develop a variety of keyword portfolios for advertisers, producing the most effective keywords for buyers in real time, relies on the advancement of computing speed in machine learning [3].

5.2 Practical Implications

Advertisers are advised to develop strategies aligned with business goals in managing keyword portfolios to achieve high sales performance [4, 8]. In addition, the keywords used to describe products need to be continuously updated since modern terminology appears for a short period but creates strong spiral effects in the marketplace and easily influences buyers to change their search behaviors. For example, the term "rayon (冰絲)" shares the same meaning as "cooling (涼感)"; however, the traditional Chinese characters of the former provide a fresh descriptor to attract buyers' attention when searching in product titles on an e-commerce platform.

Some customers do not have specific goals when browsing web pages or conducting a search. Thus, enriching product titles with additional keywords is always strategically beneficial to reach potential buyers [6, 8]. It is still important for sellers to understand their target customers, as this information is crucial for designing a keyword portfolio that aligns with the customers' search preferences. Both practices increase not only the chances of products being discovered through search queries but also the overall relevance of product listings.

Finally, we suggest considering developing universal keywords with broad customer appeal. These are keywords that transcend specific product categories and are likely to capture the attention of more customers. "Free Shipping and Delivery (免運)," for example, can be regarded as a universal keyword, and this keyword is a value proposition that tends to attract customers across diverse product categories. "Limited Time Offer (限時優惠)" creates a sense of urgency, encouraging customers to make quicker purchase decisions. "Exclusive (獨家)" refers to the uniqueness and specialty of the products, increasing interest among potential buyers. "Best Seller (暢銷商品)" is also a universal keyword that highlights the popularity and trustworthiness of a product, instilling confidence in prospective customers.

Overall, by strategically integrating such universal keywords into product titles and descriptions, sellers can potentially boost click-through rates and conversions. It is also essential to adapt these recommendations when designing keyword portfolios or finding more universal keywords while considering the cultural and linguistic nuances of the target customers to maximize the effectiveness of keywords in different marketplaces.

5.3 Limitations and Future Research Direction

Since this study was conducted during a warmer season in Taiwan, product sales may be influenced by seasonal variations [14]. Different seasons might exhibit distinct consumer behaviors and preferences. The study found that a specific textile product type (i.e., T-shirt) and some adjectives, such as cooling and rayon, were frequently used keywords. If a study is conducted using the same process during other seasons, the results may be entirely different. Therefore, the findings of this study cannot be directly adopted in different situations. Extending the study to include data from various seasons for comparison would provide a more comprehensive understanding of this topic.

Future studies are suggested to span across multiple seasons, years, and diverse periods to examine how consumers' search behaviors differ over time. The dynamics of buyer search behaviors influenced by different factors such as changing fashion trends, economic conditions, and brand promotion activities can also be understood [4]. In other words, a longitudinal study can reach a thorough understanding of the advantageous relationship between keywords and sales performance.

Finally, this study did not consider the significant impact of product images on sales. Previous studies have highlighted the importance of product images, emphasizing the interplay between keywords in different channels [8, 15–17]. The quality or style of product images, combined with seemingly inefficient keywords, might still result in increased sales due to visually appealing or well-presented photos. High-quality and visually appealing photos can enhance consumer perceptions and drive sales, even when the keywords are less efficient. Thus, future studies could examine the effect of visual elements on customer engagement, search behaviors, or purchasing decisions while also comparing them with different integrated keywords and/or different marketing channels.

Disclosure of Interests. The authors have no competing interests to declare that are relevant to the content of this article.

References

1. Buii, M.: Top 10 Taiwan online shopping sites 2024. EcomEye (2024). https://ecomeye.com/top-10-online-shopping-ecommerce-websitesapps-in-taiwan/
2. Erdmann, A., Arilla, R., Ponzoa, J.M.: Search engine optimization: the long-term strategy of keyword choice. J. Bus. Res. **144**, 650–662 (2022). https://doi.org/10.1016/j.jbusres.2022.01.065
3. Scholz, M., Brenner, C., Hinz, O.: AKEGIS: automatic keyword generation for sponsored search advertising in online retailing. Decis. Support. Syst. **119**, 96–106 (2019). https://doi.org/10.1016/j.dss.2019.02.001
4. Kim, A.J., Jang, S., Shin, H.S.: How should retail advertisers manage multiple keywords in paid search advertising? J. Bus. Res. **130**, 539–551 (2021). https://doi.org/10.1016/j.jbusres.2019.09.049
5. Li, H., Yang, Y.: Optimal keywords grouping in sponsored search advertising under uncertain environments. Int. J. Electron. Commer. **24**(1), 107–129 (2020). https://doi.org/10.1080/10864415.2019.1683704

6. Lo, S.K., Hsieh, A.Y., Chiu, Y.P.: Keyword advertising is not what you think: clicking and eye movement behaviors on keyword advertising. Electron. Commer. Res. Appl. **13**(4), 221–228 (2014). https://doi.org/10.1016/j.elerap.2014.04.001

7. Lu, X., Zhao, X.: Differential effects of keyword selection in search engine advertising on direct and indirect sales. J. Manag. Inf. Syst. **30**(4), 299–326 (2014). https://doi.org/10.2753/MIS0742-1222300411

8. Cao, X., Yang, Z., Wang, F., Lu, C., Wu, Y.: From keyword to keywords: the role of keyword portfolio variety and disparity in product sales. Asia Pac. J. Mark. Logist. **34**(6), 1285–1302 (2021). https://doi.org/10.1108/APJML-02-2021-0145

9. Symitsi, E., Markellos, R.N., Mantrala, M.K.: Keyword portfolio optimization in paid search advertising. Eur. J. Oper. Res. **30**(2), 767–778 (2022). https://doi.org/10.1016/j.ejor.2022.03.006

10. Sagarino, V.M.C., Montejo, J.I.M., Ceniza-Canillo, A.M.: Sentiment analysis of product reviews as customer recommendations in Shopee Philippines using hybrid approach. In: Proceedings of 7th International Conference on Information Technology and Digital Applications, pp. 1–6. IEEE (2022). https://doi.org/10.1109/ICITDA55840.2022.9971379

11. Xiao, Y., Qi, C., Leng, H.: Sentiment analysis of Amazon product reviews based on NLP. In: Proceedings of 4th International Conference on Advanced Electronic Materials, Computers and Software Engineering, pp. 1218–1221. IEEE (2021). https://doi.org/10.1109/AEMCSE51986.2021.00249

12. Kunilovskaya, M., Plum, A.: Text preprocessing and its implications in a digital humanities project. In: Proceedings of the Student Research Workshop Associated with RANLP 2021, pp. 85–93. INCOMA Ltd. (2021). https://aclanthology.org/2021.ranlp-srw.13

13. Zhang, X., Wu, P., Cai, J., Wang, K.: A contrastive study of Chinese text segmentation tools in marketing notification texts. J. Phys. Conf. Ser. **1302**(2), 022010 (2019). https://doi.org/10.1088/1742-6596/1302/2/022010

14. Laik, J., Mirchandani, P.: Effect of seasonality, sales growth rate, and fiscal year end on cash conversion cycle. Decis. Sci. **54**(1), 43–63 (2021). https://doi.org/10.1111/deci.12545

15. Liu, L., Wu, S., Cai, G.: Impact of online product presentation on sales: the effects of text-image introductory information and celebrity endorsements. J. Prod. Brand Manage. **32**(8), 1220–1232 (2023). https://doi.org/10.1108/JPBM-08-2022-4109

16. Wang, X., Ding, Y.: The impact of monetary rewards on product sales in referral programs: the role of product image aesthetics. J. Bus. Res. **145**, 828–842 (2022). https://doi.org/10.1016/j.jbusres.2022.03.052

17. Xia, H., Pan, X., Zhou, Y., Zhang, Z.: Creating the best first impression: Designing online product photos to increase sales. Decis. Support. Syst.. Support. Syst. **131**, 113235 (2020). https://doi.org/10.1016/j.dss.2019.113235

Predictive Analysis for Personal Loans by Using Machine Learning

Hui-I. Huang[1], Chou-Wen Wang[1], and Chin-Wen Wu[2][(✉)]

[1] National Sun Yat-Sen University, Kaohsiung, Taiwan
[2] Nanhua University, Chiayi County, Taiwan
cwwu@nhu.edu.tw

Abstract. This study adopts five common machine learning algorithms for predicting consumer personal loan uptake, including Logistic Regression, Support Vector Machine, Multilayer Perceptron, Gradient Boosting Decision Trees Catboost, and Xgboost. The research utilizes data from Thera Bank available in the public database Kaggle, featuring fields like age, work experience, income, family size, average credit card expenditure, education level, home loans, securities account, deposit account, and internet banking usage. The study addresses the issue of imbalanced data using the SMOTE (Synthetic Minority Over-sampling Technique) method and compares the accuracy and stability of predictions using the five models with three different sampling rates to identify the optimal model and key factors. Empirical results show that the Gradient Boosting Catboost model and the Support Vector Machine model perform with stability and precision across different sampling ratios, making them the best models. Moreover, through the Gradient Boosting Xgboost model, the study identifies key features such as educational factors, income, family size, the existence of a deposit account, and annual credit card spending. The findings of this research can provide crucial factors for financial institutions when formulating marketing strategies for personal loans.

Keywords: Bank · Machine Learning · Personal Loans · Support Vector Machine · Gradient Boosting Model

1 Introduction

This research delves into the application of machine learning models in the domain of personal consumer loans and their impact on the decision-making processes of financial institutions. In the context of rapidly evolving financial technology, traditional evaluation methods face increasing challenges, not only from the rapidly changing market environment but also from the need to process large volumes of complex financial data and make precise predictions about market strategies. Therefore, the integration of machine learning models to enhance prediction efficiency and analyze the crucial factors affecting loan applications has become a priority.

The core purpose of this study is to use machine learning models to predict consumers' acceptance of personal consumer loan activities and explore which models are most effective in this field. Additionally, we investigate which factors decisively influence consumers' acceptance of loan activities from financial institutions, aiding these institutions in optimizing their marketing activities. The main objectives of the study are:

1. Which machine learning model can most effectively predict customer applications for personal loans?
2. What key factors are most likely to influence customer applications for personal credit loans?

The personal loan case study in this research is derived from the public data in Kaggle's database, provided around 2019 by Walke in the Thera Bank case. The data, covering 5,000 records with 14 variables, include demographic information (such as age, income), the customer's relationship with the bank (e.g., mortgage loans, securities accounts), and the customer's response to the last personal loan activity (whether they accepted the personal loan).

Through various machine learning models, the study explores which models can effectively predict whether customers will accept the bank's marketing activities, enabling financial institutions to understand customer needs better and adjust their financial service strategies accordingly.

In summary, the study found that machine learning models, particularly the Catboost and SVM models, excel in predicting whether customers will apply for credit loans. The SVM model showed higher accuracy in predicting customers who have already applied for loans. However, after a comprehensive comparison, the Catboost model outperformed in overall performance (including predictions for both types 0 and 1), indicating that financial institutions can leverage these efficient models to adjust marketing strategies, target different customer groups, and enhance the accuracy of predictions and loan application rates. For institutions with larger data volumes, the Catboost model is recommended; otherwise, the SVM model should be the primary choice.

Further analysis of feature importance from the Xgboost model revealed that education, income, and family situation are key factors affecting loan decisions. This suggests that financial institutions should pay more attention to these factors in customer credit assessments, assigning them higher weights. Additionally, institutions can delve deeper into the nuances of these features to better understand customers' credit risk and develop more targeted loan products.

2 Literature Review

2.1 Application of Machine Learning Models in the Finance

In recent years, the development of machine learning in the financial area has been rapid, with a wide range of applications. For instance, in financial credit scoring, machine learning can more finely evaluate the risk of borrowers; in fraud detection, it can effectively prevent fraudulent activities through the analysis of large amounts of data. Automated trading systems utilize machine learning to process and analyze market data, making real-time trading decisions. Furthermore, financial institutions leverage machine learning to optimize customer service, such as developing intelligent investment advisors and personalized banking services, demonstrating the significance of machine learning in the innovation of financial products and services. These advancements not only revolutionize traditional banking and financial services but also bring a more data-driven, efficient, and customer-centered development direction to the finance industry.

Huang, J., Chai, J., & Cho, S. (2020) comprehensively evaluated the application of deep learning (DL) models in the financial and banking sectors, systematically assessing these models, including data preprocessing, input data, and model evaluation, and exploring the factors affecting the outcomes of financial deep learning models. The study provides observations on the latest applications of deep learning in the financial and banking sectors, highlighting the rapid development trend of deep learning in these industries, pointing out the shortcomings of existing literature, and directing future research in this field.

Zhu, Zhou, Xie, Wang, and Nguyen (2019) proposed a new hybrid ensemble machine learning method to improve the accuracy of predicting credit risk for small and medium-sized enterprises (SMEs) in supply chain finance. The study utilized random subspace and MultiBoosting techniques, analyzing data from 46 SMEs and 7 core companies listed on the Chinese stock market from 2014 to 2015, to verify the feasibility and effectiveness of the new method. The results showed that this new approach performs well in handling small sample data, providing practical insights for improving the financing capabilities of SMEs.

Tax, K. J., Dosoula, Smith, and Bernardi (2021) explored how to use supervised machine learning algorithms to improve fraud detection performance in anti-money laundering. The study tested four different machine learning algorithms: logistic regression, SVM, random forest, and artificial neural networks (ANN), using a simulated dataset for experimentation. The study found that, among these techniques, random forest performed best in terms of accuracy, while the accuracy of artificial neural networks was the lowest. This research provides new insights into the anti-money laundering field, showcasing the potential application of machine learning technologies in detecting money laundering activities.

Ma and Sun (2020) delved into the application of machine learning and artificial intelligence in market marketing, reviewing machine learning tasks and methods, and discussing AI-driven marketing trends and practices. The article proposes a unified conceptual framework and a multifaceted research agenda, exploring how machine learning methods can be integrated into market marketing research, highlighting the advantages of machine learning methods in handling unstructured data, making predictions, and extracting features, and discussing the potential applications of these methods in market marketing.

The aforementioned literature studies on various financial aspects illustrate that machine learning is ubiquitous. Through machine learning models, financial institutions can comprehensively review and strategize in marketing, trading, fraud, risk management, and other areas, simultaneously improving accuracy and efficiency. In the next section, we will explore the literature on personal loans combined with machine learning models.

2.2   Combining Machine Learning Models with Personal Loan Prediction

Arun, K., Ishan, and Sanmeet (2016) delved into machine learning's role in streamlining the personal loan approval process. By employing algorithms such as decision trees, random forests, and SVMs, their research aimed to mitigate risk in loan vetting, seeking the most accurate predictive model. Eletter and Yaseen (2017) applied Linear Discriminant Analysis, MLP, and CART to evaluate loan default probabilities, aiding credit decisions in Jordanian banks. Their comparison highlighted MLP's effectiveness in minimizing misclassification costs and identifying potential defaulters.

Ibrahim, Ridwan, Muhammed, Abdulaziz, and Saheed (2020) investigated the Catboost model's application in predicting loan approval and employee promotions, highlighting Catboost's superior performance over other classifiers. Their research points to the significance of feature engineering in enhancing model accuracy. Nosratabadi et al. (2020) reviewed various data science methodologies, including deep learning and hybrid models, across economic sectors. Their systematic review underscored the advanced performance and future potential of hybrid deep learning models in economic analyses. Zhang, Gong, Yu, and Wang (2020) combined factor analysis with Xgboost models to predict investor lending willingness in P2P platforms under incomplete information scenarios. Their method improved prediction accuracy, informing better investment decisions in the P2P sector.

Sreesouthry, S., Ayubkhan, Rizwan, Lokesh, and Ra (2021) explored logistic regression within machine learning to forecast loan approvals. Their work, leveraging Kaggle's public dataset, emphasized machine learning's capability to refine credit scoring and risk assessments, showcasing the predictive strength of logistic regression in banking. Akça and Sevli (2022) demonstrated the use of Support Vector Machines (SVM) with various kernel functions to predict bank loan application outcomes, finding the polynomial kernel to be most effective with an impressive 97.2% accuracy rate. This study underlines the importance of choosing the right kernel function and the impact of data imbalance on model outcomes.

Anand, Velu, and Whig (2022) employed a variety of machine learning models to predict personal loan behavior, contributing to safer credit lending practices. Their comprehensive approach, assessing models from logistic regression to Gaussian Naive Bayes and SVMs, demonstrated the effectiveness of machine learning in identifying loan default risk. Li, Zhang, Qiu, Cai, and Ma (2022) focused on predicting loan defaults in P2P lending using models like logistic regression, random forest, and Catboost, enhanced by blending techniques. Their findings suggest that such integrated approaches can significantly reduce credit risk for online lending platforms.

Collectively, these studies not only showcase the broad applicability and potential of machine learning models in financial services, particularly in personal loan approval and risk assessment, but also offer insights into the methodological advancements and data strategies that enhance predictive accuracy and reduce risk.

3   Research Methodology

3.1   Data Source

The personal loan case comes from the public data repository on Kaggle, provided by Walke around 2019, featuring a real-case scenario of Thera bank. However, the data release description does not record the period of data acquisition; it is inferred that the data published in 2019 were obtained from the 2018 database. Therefore, it is understood that the data represent 5,000 customers who applied for business due to personal loan marketing activities in the previous year. The original dataset contains a total of 5,000 entries and covers 14 variables, including demographic information of the customers (such as age, income, etc.), the relationship between customers and the bank (e.g., mortgage loans, securities accounts, etc.), and the customers' response to the last personal loan activity (whether they accepted a personal loan or not). Through different machine learning models, we will delve into which model can effectively predict whether customers will accept the bank's marketing activities, thereby allowing financial institutions to deeply understand customer needs and adjust corresponding financial service strategies accordingly.

3.2   Feature Preliminary Analysis

In the case study, we examine a growing customer base of T Bank, which primarily comprises depositors with varying balance sizes and a limited proportion of loan customers. T Bank aims to increase its loan customer ratio by leveraging its existing depositor base to introduce loan services, aiming to generate more interest income. Analysis of the previous year's data on depositors converting to loan customers showed a successful conversion rate of over 9%, prompting the personal loan department to devise more targeted activities to increase this ratio under minimal budget constraints. The bank provided data on 5,000 customers, of which only 480 (9.6%) accepted the personal loans offered in the previous activity.

Table 1 presents the data features of T Bank's case, including mean, standard deviation, maximum and minimum values, and types. The data, free from missing or duplicate values and devoid of character-type variables, had 56 outliers removed across variables like age, experience, income, family size, average credit card spending, education level, mortgage, securities account, CD account, online banking, and credit card ownership, ensuring no significant bias in assessment results. Credit card average spending (CCAvg) was annualized to align with income measurement standards.

Data Source: Compiled for this Study. This customer data table offers an in-depth analysis of variables including age, work experience, income, family status, education level, and other financial-related variables. These insights provide objective references for banks or financial institutions to understand their customer base, develop more effective marketing strategies, and offer suitable financial products (Table 2).

1. Age and Experience: The wide age distribution, with an average age of 45 years, indicates the bank's appeal to various age groups. The average work experience of

Table 1. Feature Description

ITEM	VARIABLE	QUANTITY	MEAN	STANDARD DEVIATION	MINIMUM	MAXIMUM
1	Age	4944	45.35	11.46	23	67
2	Experience	4944	20.35	11.25	0	43
3	Income	4944	72.85	45.27	8	218
4	Family	4944	2.40	1.15	1	4
5	Education	4944	1.88	0.84	1	3
6	Mortgage	4944	54.13	97.68	0	617
7	Personal Loan	4944	0.09	0.29	0	1
8	Securities Account	4944	0.10	0.30	0	1
9	CD Account	4944	0.05	0.22	0	1
10	Online	4944	0.59	0.49	0	1
11	Credit Card	4944	0.29	0.45	0	1
12	ann_CV	4944	22.80	20.45	0	120

Table 2. Comparison of Model Results at Different Sampling Ratios

Model	Accuracy		
Training and Testing Sets	10%	20%	30%
Logistic Regression	89.09%	89.56%	88.85%
SVM	98.11%	98.17%	97.44%
Xgboost	98.99%	98.79%	98.52%
Catboost	99.19%	98.58%	98.58%
MLP	98.38%	97.44%	97.44%
All Zeros Model	90.11%	90.90%	90.11%

Data Source: Compiled for this study.

20.35 years suggests a customer base with considerable professional tenure, informing age-related financial products and services like retirement plans or long-term investments.

2. Income and Family Status: An average income of 72.85% shows stable annual earnings among customers, though a high standard deviation indicates significant income disparity. An average family size of 2.40 suggests most customers have smaller families.

3. Education Level: Customers' education levels range from below college (1) to master's (2) and doctorate (3) degrees, with an average level of 1.88, indicating a higher education level among most customers. This may affect their financial understanding and risk tolerance, necessitating consideration in financial education and communication.

4. Mortgage: The average mortgage amount and its high standard deviation indicate a segment of customers with substantial mortgage needs, suggesting targeted financial planning and services for this group.

5. Personal Loan: The low average uptake of personal loans (0.09) from the previous activity indicates relatively low demand, necessitating a deeper understanding of customers' attitudes and needs regarding loan products.

6. Financial Products (Securities Account, CD Account, Online Banking, Credit Card): These binary variables reveal customer participation in these financial products or services, suggesting the need for better promotion or the development of new financial services to attract customers.

7. Credit Card Spending (ann_CV): Variability in credit card usage among customers, indicated by the average and standard deviation of credit card spending, suggests the potential for targeted rewards programs and promotions for specific customer segments.

Based on these analyses, banks can tailor financial products like retirement plans and family loans to customers' diverse age, income, and family situations. For customers with higher education levels, more in-depth and professional financial education can enhance their understanding of financial products. Banks could offer customized financial and loan solutions for those with significant mortgages and strengthen the promotion of financial products like securities accounts, online banking, etc., to attract more participation. For frequent credit card users, launching corresponding rewards programs could increase loyalty and potentially encourage personal loan applications.

3.3 Brief Review of Machine Learning Models

This study divides the dataset into training and testing sets with ratios of 90%/10%, 80%/20%, and 70%/30%, respectively. We utilize five machine learning models for training: Logistic Regression, Support Vector Machine (SVM), Xgboost, Catboost, and Multilayer Perceptron (MLP). Here's a concise overview of each model:

1. Logistic Regression (LR): A well-known classification algorithm, particularly effective for binary classification problems. Despite its name, LR is a classifier that predicts the probability of an instance belonging to one of two classes using a logistic function. It assumes a linear relationship between input features and the log odds of the output.

2. Support Vector Machine (SVM): Developed by Vapnik, Boser and Guyon in 1992, SVM works well for both linear and nonlinear problems, particularly in high-dimensional spaces. It aims to find a hyperplane that best divides a dataset into two classes, maximizing the margin between the closest points of the classes, which are known as support vectors.

3. Catboost (Categorical Boosting): A gradient boosting decision tree algorithm optimized for categorical data without the need for extensive data preprocessing like one-hot encoding. Introduced by Yandex in 2017, Catboost excels in handling categorical features directly, improving efficiency and performance in large datasets.
4. Xgboost (eXtreme Gradient Boosting): An efficient and scalable implementation of gradient boosting framework by Tianqi Chen in 2014. Xgboost is known for its speed and performance, especially in structured data competitions. It uses advanced regularization (L1 and L2), which helps in reducing overfitting and improving model performance.
5. Multilayer Perceptron (MLP): A type of artificial neural network with multiple layers, including input, hidden, and output layers, where each layer is fully connected to the next. MLP can model complex nonlinear relationships through its network structure and is particularly useful for deep learning tasks.

Each model offers unique advantages in handling specific types of data and learning tasks. Logistic Regression is straightforward and interpretable, making it suitable for simpler binary classification problems. SVM is powerful for datasets with a clear margin of separation, even in high-dimensional spaces. Catboost and Xgboost are advanced tree-based algorithms that perform well on structured data, with Catboost being particularly adept at handling categorical features directly. MLP, representing deep learning approaches, is versatile in modeling complex patterns but requires substantial data and computational resources.

In summary, the selection of a machine learning model depends on the specific characteristics of the dataset, including the linearity of relationships, presence of categorical features, and the size of the data. This study aims to compare these models' performance in predicting customer responses to bank loan marketing activities, providing insights into the effectiveness of each model in financial services applications.

3.4   Assessing Model Predictive Metrics

Following the introduction of the five models in the previous section, this section explains how we use evaluation metrics such as the confusion matrix, accuracy, precision, recall, F1 score, and ROC curve to thoroughly analyze and assess the performance of the models. These metrics will help reveal the strengths and limitations of the models in various aspects, providing a more compelling interpretation and discussion of the research findings.

Confusion Matrix is a table used to evaluate the performance of classification models, particularly in binary classification. The confusion matrix categorizes predictions into True Positives (TP) and True Negatives (TN) for correct predictions, and False Positives (FP, Type I error) and False Negatives (FN, Type II error) for incorrect predictions, based on the actual observation being positive (P) or negative (N).

Using the confusion matrix, we can calculate various evaluation metrics, including accuracy, precision, recall, and F1 score, offering a comprehensive assessment of the model's performance in different aspects. Common evaluation metrics include Accuracy, which assesses overall model performance by the proportion of correct predictions; Precision, indicating the accuracy of positive predictions; Recall (Sensitivity), measuring the model's ability to identify all positive cases; and F1 Score, a balance between precision and recall, with values closer to 1 denoting higher model accuracy and effectiveness.

4 Research Methodology

This chapter delves into the predictive factors of personal loan uptake using statistical analyses and machine learning models. Initially, a correlation heatmap revealed associations between variables like age, work experience, and the positive correlation between income and personal loan applications. Scatter plot analyses showed that higher-income individuals are more likely to apply for loans, while age has a lesser impact; however, higher annual credit card spending and certain levels of work experience influence loan applications, with mid-experience individuals less likely to apply.

In terms of machine learning model outcomes, the study addressed data imbalance using the SMOTE technique and employed various models including Logistic Regression, SVM, Xgboost, Catboost, and MLP for prediction. These models demonstrated varying degrees of accuracy and stability across different sampling ratios, with Catboost and Xgboost showing superior performance in accuracy, precision, and recall, particularly in handling class imbalance. The SVM model excelled in predicting Type 1 outcomes. Furthermore, an importance analysis of factors in the Xgboost model highlighted education, income, and family size as key determinants of loan uptake, with education's importance slightly increasing with higher sampling ratios, underscoring its pivotal role in prediction. Overall, this chapter thoroughly examines various factors affecting personal loan uptake and identifies the most effective predictive models, offering strategic insights for financial institutions in formulating personal loan strategies.

4.1 Model Comparison

In this study, six models were employed to tackle a binary classification problem, including a simplistic model that predicts all zeros, Logistic Regression, SVM, Xgboost, Catboost, and MLP. These models exhibited distinct performances and characteristics when applied to the data. Here is a concise comparison of their results for training and testing sets with ratios of 90%/10%.

1. All Zeros Model: Predicts all instances as class 0, achieving an accuracy rate of approximately 89%, indicating a bias in handling imbalanced datasets with poor recognition of class 1.
2. Logistic Regression: Exhibits high computational efficiency with training time around 4.4 s and almost instant prediction, reaching an accuracy of 90.2%. The model shows balanced performance across classes with precision, recall, and F1 score close to 0.9 for both classes, validated by a cross-validation score of about 89.8%.
3. SVM: Requires longer training time around 2 min but demonstrates exceptional predictive capability with an accuracy of roughly 98.1%. It shows almost perfect precision and recall for both classes, slightly lagging in cross-validation scores compared to Catboost and Xgboost.

4. Xgboost: Features quick training (10 s) and prediction times, achieving an impressive accuracy of approximately 98.99%. It shows high accuracy for both primary and secondary classes with a cross-validation score of about 98.58%, indicating robustness and efficiency.
5. Catboost: Stands out for its training efficiency and remarkable accuracy of around 99.19%, demonstrating superior performance in correctly predicting both classes and handling data imbalance. It slightly leads in performance metrics, especially in recall rates for less prevalent classes, with a cross-validation score of about 98.70%.
6. MLP: Despite longer training times, it excels in prediction accuracy (approximately 98.38%) and showcases high precision, recall, and F1 scores for both classes, supported by a cross-validation score of around 97.68%, proving its consistency and robustness.

For Comparative Summary, to clearly compare the models, Catboost and Xgboost outshine others in balance across accuracy, precision, recall, and stability in cross-validation, showing superior generalization capabilities and effectiveness in managing class imbalance. SVM excels in predicting class 1 but slightly falls short in cross-validation performance compared to Catboost and Xgboost. Catboost marginally leads across most performance indicators, demonstrating its strong predictive accuracy and reliability in identifying less common classes, an essential factor in model performance evaluation.

4.2 Robustness Check

For datasets split into 80% training and 20% testing, a comparative analysis shows that the Xgboost model excels with the highest accuracy at 98.79%, followed by Catboost at 98.58%, and SVM at 98.17%. These models demonstrate high consistency and reliability across metrics for class 0, with Xgboost and Catboost also showing superior cross-validation scores, indicating their robustness and generalization ability. In contrast, the MLP and Logistic Regression models have slightly lower overall accuracy but still perform well above 97% and 89%, respectively. Notably, the all zeros strategy, while achieving a 90.90% accuracy rate due to class imbalance, fails entirely to identify the minority class, highlighting its ineffectiveness in imbalanced datasets.

For datasets split into 70% training and 30% testing, all five models display high accuracy rates in personal loan prediction tasks, with Xgboost and Catboost slightly outperforming others, achieving over 98.5% accuracy. These models particularly excel in predicting class 0, with near-perfect precision and recall. SVM and Xgboost show higher precision for class 1, while Catboost's lower recall indicates a more cautious approach in identifying class 1. The MLP model reveals some weaknesses in recall for class 1 but still maintains commendable overall performance. Compared to the all zeros strategy, which reaches 91% accuracy but fails to identify class 1, resulting in significantly lower macro-average metrics, SVM shows standout performance in class 1 predictions but slightly lags behind Catboost and Xgboost in cross-validation scores. Overall, Xgboost and Catboost provide the most balanced and reliable performance for this task.

For class 0 predictions, nearly all models achieve or approach perfect scores, likely due to the larger sample size of class 0. In predicting class 1, accuracy and recall vary across models, but Catboost and Xgboost generally provide higher values. Considering the average cross-validation scores, an important indicator of model generalization ability, both Catboost and Xgboost excel, particularly Catboost, which reaches 98.70% at a 10% sampling ratio.

Comparing all metrics across sampling ratios, Xgboost and Catboost exhibit superior and consistent performance. However, given Catboost's slightly higher accuracy at a 10% sampling ratio and its relatively better performance in predicting class 1, Catboost emerges slightly ahead among these models. Nonetheless, selecting the best model also depends on the data characteristics in practical applications, including the uniformity of data distribution and class balance. If class 1 prediction is equally critical, a model with a higher F1 score for class 1 might be preferred. Additionally, training and prediction times are practical considerations in model selection. While the SVM model leads in class 1 metrics, it slightly lags behind Catboost in class 0 predictions, impacting its average cross-validation score. Overall, considering stability and performance across different sampling ratios, the Catboost model is identified as one of the best models.

4.3 Key Factors for Applying Personal Credit Loans

The analysis of key factors influencing customer applications for personal credit loans reveals that education, income, and family size are paramount. Education stands out as the most significant predictor across various sampling ratios, suggesting that individuals' educational backgrounds play a crucial role in their likelihood of applying for credit loans. This importance slightly increases with larger sample sizes, highlighting the growing relevance of education in the predictive models as more data is considered.

Income follows as the second most influential factor, indicating that higher earnings are associated with a greater propensity to apply for loans. This relationship is consistent across all sampling ratios, underscoring the importance of financial stability in loan application decisions.

Family size ranks third in importance, maintaining a stable position across different sampling ratios but showing minor fluctuations compared to education and income. This suggests that while family size has a steady impact on loan application likelihood, its influence is somewhat less pronounced than that of education and income.

Additionally, the feature importance of having a deposit account shows more variation across sampling ratios but does not emerge as a top predictor, indicating its lesser, though still present, contribution to predicting loan applications. Overall, education, income, and family size are key to understanding and predicting customer behavior regarding personal credit loan applications, with education being the most consistent and significant factor.

5   Conclusion

From the findings and analyses of this study, several conclusions can be drawn. Firstly, machine learning models, especially Catboost and SVM, have shown excellent performance in predicting whether customers will apply for credit loans, with the SVM model being particularly accurate for customers who have already applied for loans.

This aligns with Akça and Sevli's recommendation for financial institutions to use the SVM model for predictions. However, after a comprehensive comparison, the Catboost model performed better overall, indicating that financial institutions can utilize these efficient models to refine marketing strategies and tailor plans for different customer segments to enhance prediction accuracy and loan application rates. For large datasets, the Catboost model is recommended; otherwise, the SVM model may be preferable.

Secondly, the feature importance analysis from the Xgboost model further reveals education, income, and family situation as key factors influencing loan decisions, suggesting financial institutions should pay more attention to these factors and assign them higher weights in customer credit evaluations. Additionally, institutions could delve deeper into these features' nuances to better understand customer credit risk and develop more targeted loan products.

Furthermore, financial institutions should enhance their data analysis of these features to optimize loan products and services, including designing customized loan plans for different customer groups and introducing more flexible loan products to adapt to changing market demands.

Lastly, this study suggests financial institutions explore using machine learning models for deeper customer behavior analysis, monitoring credit histories and behavior patterns, and considering external environmental factors. This will help institutions better understand customer credit needs and risks and more accurately tailor marketing strategies and risk management measures.

In summary, the insights from this study aim to improve loan marketing strategies and services, enhance the likelihood of individual loan applications, leverage machine learning models, focus on important features, and deepen customer behavior analysis. This approach can help financial institutions maintain a competitive edge in the financial market, offer better loan products and services, reduce the risk of bad loans, and contribute to sustainable growth and customer satisfaction.

The original data obtained for this study did not include the loan amounts approved by financial institutions for customers who consented to personal loans, preventing an analysis of whether customers with good educational backgrounds and income levels are more likely to receive bank loans or whether higher-educated and higher-income customers prefer applying for personal loans. Additionally, if financial institutions can obtain sufficient information internally, they could explore loan applications, credit records, and approved amounts by age group, for instance, to tailor marketing strategies for target customer groups, thereby making marketing more efficient and precise. Building on this study's foundation, it aims to assist financial institutions in constructing effective machine learning models and implementing precise marketing strategies.

References

1. Akça, M.F., Sevli, O.: Predicting acceptance of the bank loan offers by using support vector machines. Int. Adv. Res. Eng. J. **6**(2), 142–147 (2022)
2. Agarwal, K., Jain, M., & Kumawat, A.: Comparing classification algorithms on predicting loans. In: Information Systems and Management Science: Conference Proceedings of 3rd International Conference on Information Systems and Management Science (ISMS) 2020

(pp. 240–249). Springer International Publishing (2022). https://doi.org/10.1007/978-3-030-86223-7_21

3. Amari, S.: A theory of adaptive pattern classifiers. IEEE Trans. Electron. Comput. **3**, 299–307 (1967)

4. Anand, M., Velu, A., Whig, P.: Prediction of loan behaviour with machine learning models for secure banking. J. Comput. Sci. Eng. (JCSE) **3**(1), 1–13 (2022)

5. Arun, K., Ishan, G., Sanmeet, K.: Loan approval prediction based on machine learning approach. IOSR J. Comput. Eng **18**(3), 18–21 (2016)

6. Boser, B.E., Guyon, I.M., Vapnik, V.N.: A training algorithm for optimal margin classifiers. In Proceedings of the Fifth Annual Workshop on Computational Learning Theory, pp. 144–152 (1992)

7. Cox, D.R.: The regression analysis of binary sequences. J. R. Stat. Soc. Ser. B Stat Methodol. **20**(2), 215–232 (1958)

8. Cramer, J.S.: The origins of logistic regression (2002)

9. Eletter, S.F., Yaseen, S.G.: Loan decision models for the Jordanian commercial banks. Global Bus. Econ. Rev. **19**(3), 323–338 (2017)

10. Huang, J., Chai, J., Cho, S.: Deep learning in finance and banking: a literature review and classification. Front. Bus. Res. China **14**(1), 1–24 (2020)

11. Ibrahim, A.A., Ridwan, R.L., Muhammed, M.M., Abdulaziz, R.O., Saheed, G.A.: Comparison of the CatBoost classifier with other machine learning methods. Int. J. Adv. Comput. Sci. Appl. **11**(11) (2020)

12. Li, X., Ergu, D., Zhang, D., Qiu, D., Cai, Y., Ma, B.: Prediction of loan default based on multi-model fusion. Procedia Comput. Sci. **199**, 757–764 (2022)

13. Ma, L., Sun, B.: Machine learning and AI in marketing–Connecting computing power to human insights. Int. J. Res. Mark. **37**(3), 481–504 (2020)

14. Nosratabadi, S., et al.: Data science in economics: comprehensive review of advanced machine learning and deep learning methods. Mathematics **8**(10), 1799 (2020)

15. Prasad, K.G.S., Chidvilas, P.V.S., Kumar, V.V.: Customer loan approval classification by supervised learning model. Int. J. Recent Technol. Eng. **8**(4), 9898–9901 (2019)

16. Sreesouthry, S., Ayubkhan, A., Rizwan, M.M., Lokesh, D., Raj, K.P.: Loan prediction using logistic regression in machine learning. Ann. Romanian Soc. Cell Biol. **25**(4), 2790–2794 (2021)

17. Tax, N., et al.: Machine learning for fraud detection in e-commerce: a research agenda. In: Deployable Machine Learning for Security Defense: Second International Workshop, MLHat 2021, Virtual Event, August 15, 2021, Proceedings 2 (pp. 30–54). Springer International Publishing (2021). https://doi.org/10.1007/978-3-030-87839-9_2

18. Zhang, D., Gong, Y., Yu, L., Wang, X.: P2P online loan willingness prediction and influencing factors analysis based on factor analysis and XGBoost. J. Phys.: Conf. Ser. **1624**(4), 042039 (2020). IOP Publishing

19. Zhu, Y., Zhou, L., Xie, C., Wang, G.J., Nguyen, T.V.: Forecasting SMEs' credit risk in supply chain finance with an enhanced hybrid ensemble machine learning approach. Int. J. Prod. Econ. **211**, 22–33 (2019)

UX-Optimized Lottery Customer Acquisition Processes Through Automated Content Creation: Framework of an Industry-University Cooperation

Diana Kolbe[1,2]([✉]) [ID], Andrea Müller[1], Annebeth Demaeght[1], and Barbara Woerz[2]

[1] Hochschule Offenburg – University of Applied Sciences, Badstrasse 24,
77652 Offenburg, Germany
`diana.kolbe@hs-offenburg.de`
[2] Burda Direct GmbH, Hubert-Burda Platz 2, 77652 Offenburg, Germany

Abstract. Artificial intelligence (AI) and Machine Learning (ML) are rapidly turning from trending topics to requirement for competitiveness for enterprises. For marketing departments, AI and ML offer potential for improvement of their processes such as optimizing user experience and personalizing campaigns for selected audiences. Nevertheless, the integration of new technologies such as AI and ML into the existing marketing mix portfolio means a great challenge for marketing managers as their implementation requires new skills and knowledge which is not always already developed. The objective of the paper is to demonstrate how an industry-university cooperation (IUC) can enable the adaptation to new business contexts. Thus, this paper proposes a framework on IUC involving different project phases. It describes the process for placing AI-generated individual content, recommendations and references for specific interests.

Keywords: AI-Generated Content · AI-Implementation · Climate Lottery Marketing · Personalized Social Media Communication · Industry-University Collaboration

1 Customer Acquisition Processes in the German Social Lottery Market

Digitalization requires new skills such as expertise in the collection and evaluation of data, social media communication, user experience (UX) design and content production [1]. The promotion of lotteries on social media is a key trend in the gambling industry. Practices such as the posting of winning prizes on social network websites are becoming a popular instrument [2]. Nevertheless, German lotteries are lacking to attract younger audiences. Partly, this is due to the restrictions that mark the gambling law but on the other side there have been few innovations that might attract the younger clientele. In 2022, the first German climate lottery ClimaClic gGmbH was founded which has the focus to attract younger audiences through a digitalized product, means the ticket can only be bought online, and social media communication. As the brand is relatively new, it has the aim to stabilize its position in the market through an enhanced customer acquisition process including personalization and automation. AI is becoming more and more prevalent in marketing activities and the administration of marketing resources. One key feature of AI that sets it apart from traditional advanced analytics is the automation of feedback loops and improvement, so-called machine learning (ML), the process by which a system learns how to operate better and then improves itself [3].

Recent investigations highlight to study topics such as social media marketing, personalization, AI, and ML conjointly [4]. The constraints to do so, are from industry-side the lack of knowledge in the implementation of AI or ML technologies and from university-side the access to information and databases. IUC is a popular instrument to overcome these barriers. In the specific case of a climate lottery (ClimaClic) and the Offenburg University (HSO), the collaboration through a postdoctoral tandem program on UX-Optimized Customer Acquisition Process through Automated Content Creation is exemplified.

Universities and businesses have different goals and face with different challenges; this difference may increase the benefits of cooperation, but it can also lead to difficulties [5]. Through IUCs, companies can benefit from highly qualified people resources, such as researchers or students [6]; they get access to technology and expertise [7]; and they can use expensive research infrastructure [8]. On the other hand, universities can benefit from improved access to industry equipment, patents and licenses [7]. Industrial partnerships have become an essential component of university funding, and contributions from businesses and international organizations for research and development in the higher education sector represent an important source [9]. Given these positive effects and their financial significance, it's critical to handle IUCs successfully to guarantee the benefits for both parties.

The present IUC has the objective to evaluate the potential of AI and ML in automated content creation to enable customer acquisition, retention and revitalization for the specific case of a climate lottery. As cold leads, these special-interest target groups generate extremely poor responses due to the inhomogeneous target groups. With the help of AI or ML, suitable customer profiles are to be identified, selected and customers contacted with personalized content. To be able to implement effective and efficient response – i.e. feedback from the target group – in a more targeted and cost-effective manner and thus to be able to offer content individually tailored to the expectations of the person, various tools and platforms are already being developed to support better automated dialogue marketing communication. However, these are usually tied to platforms, e.g. with a sub-function for target group selection, as is the case of Google, Facebook or Instagram and are not yet able to develop specific topics, such as climate protection, with machine-generated personalized content.

This paper aims to propose a framework which might be helpful in the implementation of AI projects. We propose a concept which includes the creation, testing and optimization of AI or ML-based algorithms, which are empirically and quantitatively evaluated in the context of representative UX and A/B testing. This requires the extensive iterative primary data collection of the selected target groups, which are supported by quantitative multivariate analysis methods, such as analysis of variance, regression analysis, factor analysis and cluster analysis.

The remainder is as follows: first, the specific marketing challenges are discussed such as market conditions and specific requirements for climate lotteries. Afterwards, the channel and content needs are presented in order highlight the need for the industry-university collaboration. Third, the framework for the collaboration with respect to the AI-based social media content automation is presented and lastly, some recommendations are given to further develop the thematic.

2 Chances and Risks (not only) for Starts-Ups in the German Gambling Market

2.1 The German Gambling Market

In Germany, the gambling market can be differentiated between the regulated market, i.e. games of chance such as lotteries or slot machines in gambling halls, and the non-regulated market, which includes online casinos or online poker. The regulated gambling market generated $9.4 billion in gross gaming revenue in 2021 [10]. The state lotteries accounted with 43 percent the largest share in the regulated gambling market in Germany. Class, social, and savings lotteries only account a market share of 12%. Nevertheless, online lotteries are becoming increasingly popular among consumers [11].

Due to the federal structure of the country, the legislation on gambling law in Germany is a matter for the federal states. Each federal state has its own regulations and laws, nevertheless, these are based on the Interstate Treaty on Gambling (GlüStV) which was concluded between all 16 federal states and forms the legal basis for gambling in Germany [11].

Playing the lottery is especially popular among older generations; 60 percent of people over the age of 45 take part in the "6 out of 49" lottery at least once a year. Among 18- to 25-year-olds, the proportion of players in Lotto "6 aus 49" is only five percent [11]. Over the years, lotteries in Germany have been lacking in addressing generations under the age of 45 years.

Lotteries face important marketing challenges since the gambling market is highly regulated. For instance, advertising for gambling is subject to numerous restrictions to protect minors and vulnerable consumers. In Germany, the participation in games of chance is generally only permitted from the age of 18. Therefore, targeted advertising for gambling to minors is strictly prohibited. Personalized advertising on most German communication channels is subject to strict conditions based on the General Data Protection Regulation and the younger target groups in general are very hard to address.

Given this complexity, it is not surprising to find little scholarly research on gambling and advertising. This might be the result of the topic's difficulty in finding trustworthy evidence to support academic standards or its sensitivity to governments (as many of the lotteries are state-owned) [12]. Lottery advertising is driven by the ethos of winning; according to [12], the words, signs, myths, and symbols surrounding lottery gambling create the expectation and desire of winning. Especially, young consumers are attracted by concrete and emotive aspects of winning in gambling advertisements. According to [13], the chance of winning is an essential motive to gambling and there are four optional motives which depend on the personal dispositions and preferences of consumers, namely: (1) dream of hitting the jackpot; (2) intellectual challenge; (3) mood change; and (4) social rewards. The dream of hitting the jackpot is the principal motive for participating in games with a small chance of winning. The imagination of winning the jackpot is viewed as pleasant dream. Furthermore, gambling can be an intellectually stimulating hobby [14], although this specific motive tends to be more present in horse betting and poker games. Some gamers experience affective and emotional changes in their mood such as excitement or relaxation. Gambling can offer social rewards such as communion (socializing with other people), competition (competing with others) and ostentation (display conspicuous consumption and the opportunity to gain social recognition).

Digital, mobile, and website advertising are gradually replacing traditional physical media advertising due to its improved tracking and ROI evaluation [3]. According to a critical research review, the research of gambling advertising should take new forms of marketing, such as social media communication into consideration [14]. In fact, social media communication was named the industry trend for lotteries from 2018 to 2022 [2].

2.2   Challenges (not only) for a Climate Lottery Start-Up

In Germany, the market share of the climate lottery among social lotteries is currently less than 1%. ClimaClic is a social purpose lottery with focus on climate projects - a special form of classic lotteries - founded in 2022 by Burda Direct GmbH in conjunction with the non-profit ClimaClic gGmbH. To receive a non-profit status, the financial interest of the climate lottery must not be in the foreground. Thus, the economic interest must be subordinated, according to the Joint Gambling Authority, organizers (in this specific case the ClimaClic gGmbH) may only be non-profit organizations.

With respect to the product range, ClimaClic offers three different tickets which represent separate prize categories. Each prize tier offers the chance to win a weekly jackpot and daily prizes. The amount varies with the selected prize category. This allows ClimaClic as a lottery to address different target groups. Once lottery players have decided on a prize category, they select a funded climate project. ClimaClic supports climate projects which can be divided into three areas: (1) Nature & Landscape; (2) Environment & Resources; and (3) Education & Research. A particularity of ClimaClic is the specific focus on online operation which means that the complete customer onboarding and purchasing process has to be concluded online.

So far, ClimcClic has not been able to achieve high level of awareness. Accordingly, there is an important need to strengthen the brand communication through technologically supported communication campaigns. This is essential, not only to acquire new clients, but also to strengthen the existing customer base in the long term against competitive pressure within the media and entertainment industry where Burda Direct belongs to.

3 Potentials of Personalization and UX-Improvement for a Climate Lottery

Personalization is the process of designing an individualized interaction to enhance customer experience [15]. Unlike customization (which is customer-initiated), personalization is a company-initiated concept [16]. For personalization, insight based on personal and behavioral information is needed to tailor an individualized superior experience [4]. Personalization is conceptualized as the process of customer identification, need identification, customer interaction, and product personalization [17]. This conceptualization has been further developed by adding the measurement of the impact of personalization [18].

Recent literature identified a framework of six questions to be resolved in the personalization process: (1) what is personalized; (2) communication of personalization; (3) kinds of data (4) source of data; (5) type of personalization; and (6) responsible area [19]. Regarding the question on what is personalized, [20] differentiate between the categories of functionality, content, interface, and channel. Content (e.g. text, audio, image, and video) and the rise of channels in influencing and convincing consumers has gained importance, especially considering the ability to tailor any type of content to specific market niches [3]. Thus, the present study focuses on the personalization of content and channels, and this is where the first personalization potential derives from.

Potential 1: Generating unique content. As stated above, ClimaClic offers the opportunity to select between three categories of climate projects where the consumer can choose between several projects to be supported with the purchase of a ticket. In this sense, delivering differentiated content in distinct channels for individual customers can offer potential for personalized marketing.

With respect to the question on how the personalized design is communicated to the customer, three approaches can be differentiated: (1) the self-reference method can establish an interaction with the individual through more specific wording, e.g. [21]; (2) the anthropomorphism method uses a humanization approach for communication including mimics, gestures, voice, or emotional reactions, e.g. [22]; and (3) the system characteristics method where personalized information is presented through intelligent systems, e.g. [23].

When personalizing, the kinds of data can be summarized in three main categories of information: (1) individual-level such as past digital behavior, attitudes and preferences; (2) social-level such as family, friends, classmates and colleagues, and community; and (3) situation-based such as time-based or location-based [19].

The individual-level personalization approach includes data about e.g. purchases, online reviews, sites visited, posts, likes, comments, users' clickstreams or actions in a session. Personalized recommendations that address the decisions made by important social circles are provided to individuals through the recollection of data in online social groups (personalization based on social-level data) [24]. The second potential for the climate lottery is related to the situation-based approach which depends on details regarding the precise places individuals are, the characteristics and time frame of the situations they experience.

Potential 2: Referring to different situations. As the aim of the project is to attract a younger target, the personal information should not only be based on individual and social-level characteristics. In the case of the climate lottery, there could be considered situation-based information such as the specific location or routines of the individuals, in this sense, there can be created a connection to local interests or concerns related to participating climate projects.

The source of data comes from customer and firm sources. One advantage of the creation of an algorithm is being more flexible and independent form third-party sources. Regarding the type of personalization, exposing the right content to the right target requires the management of several systems and constant personalization of content to the needs and use cases of individual scenarios. Since it takes more marketing resources to match the right content to the right client, enterprises intent to advance in the automation of the content management process [3]. Thus, the third potential originates from the idea to match the content to the customer needs through automation.

Potential 3: Delivering correspondent to customer needs. AI and ML approaches allow the automation using design text, images and videos which better suit to the market segments, channels and platforms and additionally accelerating the content delivery and optimization [3]. The principal goal of the project is the production of a functional AI- or ML-based algorithm, which is to be developed and evaluated in close cooperation at Burda Direct GmbH and in the UX laboratories of the HSO (responsible area). The intention of Burda Direct is to further develop aspects of AI or ML-based dialogue marketing automation and its fields of application and the HSO focuses more on the aspects of teaching and research and development in this specific area of interest.

4 A Framework for AI-Implementation

The framework for our research project with the aim of creating of a functional AI- or ML-based algorithm, proposes three stages: (1) basic research; (2) applied research; (3) algorithm development. The first stage comprises the development and testing of AI- or ML-based algorithms. Here, the user experience of the customer should be researched, whether and how machine-generated creative content is recognized as such and accepted as "genuine", "credible" or "reliable". The second stage, an investigation of whether and how an automated communication generates emotional and cognitive reactions in an inhomogeneous target group will be measured. The third stage, initially topic-specific algorithms can be further developed and tested with other content or campaign goals.

To better control the process of AI-implementation, ten work packages (WP) and four milestones were defined. In the first WP a comprehensive desk research is carried out to determine which technologies have already been identified for the implementation of AI or ML-based algorithms for dialogue marketing automation. This includes not only the well-known solutions for personalized content generation, but also competitor and benchmark approaches. The second WP has the objective to evaluate previous campaigns of the ClimaClic brand. In close cooperation with the brand's marketing specialists, new approaches to dialogue marketing automation can be identified and analyzed regarding customer acceptance, implementation risks and market relevance. The focus of the third WP is on the evaluation of success-relevant parameters for automated dialogue marketing campaigns to create the optimal basis for the development and use of AI or ML. The aim of the fourth work package is to investigate the transferability to other applications. With the conclusion of WP 4, the specification of the AI- or ML-based algorithm (Milestone 1) should be reached.

For further application-relevant research of the concepts identified in WP 3 and 4, a suitable research structure must be established in WP 5. This includes a research framework for iterative execution of UX and A/B testing. In WP 6, the personalized content is tested to determine how the automatically generated content is perceived and evaluated by the target group. Based on the results of WP 5 and WP 6, requirements for implementation in a possible AI or ML-based algorithm for dialogue marketing automation are defined and recorded in a specification (WP 7). After WP 7, the second milestone, the prototype generation, should be completed.

The core task of WP 8 is the evaluation of the first prototype by implementing a live campaign. The results can show whether the research activities carried out so far can be confirmed and which further research priorities need to be defined. In this work package, the content playouts of the algorithm from WP 7 will first be tested and evaluated in the UX-laboratory and then in the field. After WP 8, the concept will be evaluated by customers (Milestone 3).

Based on the findings from WP 8, research iterations are to be carried out in WP 9. In the last WP 10, the results are consolidated. Here the success goals are compared with previous campaigns. The algorithm is intended to continuously and independently develop itself on the topic of climate change and optimize it with new trends and insights. After the last WP is completed, the prototype should be evaluated and developed (Milestone 4).

5 Recommendations and Conclusions

AI and ML are becoming necessary in the daily business of nearly all industries. It is important for enterprises to quickly adapt and create skills and knowledge in their respective industries. AI can be effective in situations where the decisions are narrow, and the outcomes are observable and assessable promptly. However, the deployment of AI can be more challenging in situations where decisions are broader and need more time to implement [3]. It is not uncommon that companies face important obstacles in the implementation of new technologies and procedures, therefore, we proposed an IUC on UX-Optimized lottery customer acquisition processes through automated content creation.

With respect to the detected potentials on personalization and UX-improvement for the climate lottery, we propose the following recommendations:

Recommendation for Potential 1: About the generation of unique content, it is necessary to evaluate the perception of customers on different personalization dynamics and approaches to determine its value [24].

In this sense, the framework proposes the UX-testings which can propose valuable information on the distinct personalization dynamics. These insights can help afterwards in the decisions on the allocation of the marketing budget.

Recommendation for Potential 2: The inclusion of different situational factors can help to enhance the customer experience. As mentioned above, situation-based information is mainly related to time and location [19]. Information related to time such as routines or location such as whereabouts (e.g. school, workplace) and features of the place can help to refine the information on the targets to define common patterns for personalization [24].

Recommendation for Potential 3: The delivery of correspondent to customer needs using AI and ML can have a positive impact on the customer experience. In line with [25], we argue that the implementation of AI technology can have an impact of marketing strategies and marketers can gain competitive advantage, deliver enhanced customer experiences, and foster corporate success by embracing and developing new capabilities. Nevertheless, it is necessary to consider the evaluation of the risks for marketers and consumers. Regarding the marketers, AI technology depends on the accuracy of the input information, therefore biased information can provide incorrect output [26]. acceptance of AI-generated content. Consumers might feel insecure or even reject AI in specific contexts. Furthermore, ethical considerations have to be analyzed as there Therefore, as proposed in the framework, it is necessary to apply an extensive prototype testing.

In line with the patterns of cooperation of [5], the success of an IUC can be enhanced through the factors of flexibility, honesty, clarity and awareness.

IU-Cooperation Recommendation 1: Flexibility is especially important in setting the priorities. For the present IUC the selection of the partner has been very important. Sharing the same visions and interests was an important prerequisite to establish the collaboration. Furthermore, the definition of common goals between Burda Direct and HSO helped to accelerate the process and not losing the vision when developing the joint project.

IU-Cooperation Recommendation 2: Honesty refers to open communication. The continuing interchange of information on current developments has been helpful. Furthermore, partners should be honest about their time and commitment to the project.

IU-Cooperation Recommendation 3: Clarity regarding the aims, planning, responsibilities and interests. The planning process is a crucial stage since the timeframe and the responsibilities are defined. This helps the partners to plan and organize themselves (e.g. involve additional personnel).

IU-Cooperation Recommendation 4: Awareness on current economic, legal, political and social developments has to be considered. Especially in the current economic situation, it is important to have scenarios when for example financial restrictions might occur.

This research has two central limitations. First, the postdoc tandem program still is in its initial stages. All recommendations can be given up to certain point and in future work there should be a retrospective of the whole process. Second, the AI implementation is exemplified on the automated content generation, Projects with other focuses will have the need to adapt the objectives, content and time frame to its specific requirements.

Disclosure of Interests.   The authors have no competing interests to declare that are relevant to the content of this article.

References

1. Kords, U.: Digital scheitern: 10 gedankliche Fallstricke für die digitale Transformation des Vertriebs. In: Verband, D.D., e, (eds.) Dialogmarketing Perspektiven 2018/2019: Tagungsband 13. wissenschaftlicher interdisziplinärer Kongress für Dialogmarketing, pp. 93–108. Springer Fachmedien Wiesbaden, Wiesbaden (2019). https://doi.org/10.1007/978-3-658-25583-1_5

2. Businesswire.: Top Emerging Trends in the Global Lottery Market lTechnavio l Business Wire (2018)

3. Stone, M., et al.: Artificial intelligence (AI) in strategic marketing decision-making: a research agenda. Bottom Line 33. Emerald Group Holdings Ltd., pp. 183–200 (2020). https://doi.org/10.1108/BL-03-2020-0022/FULL/XML.

4. Chandra, S., Verma, S., Lim, W.M., Kumar, S., Donthu, N.: Personalization in personalized marketing: Trends and ways forward. Psychol. Market. **39**, 1529–1562 (2022). https://doi.org/10.1002/MAR.21670

5. Rybnicek, R., Königsgruber, R.: What makes industry–university collaboration succeed? A systematic review of the literature. J. Bus. Econ. **89**, 221–250 (2019). https://doi.org/10.1007/S11573-018-0916-6/FIGURES/4

6. Myoken, Y.: The role of geographical proximity in university and industry collaboration: case study of Japanese companies in the UK. Int. J. Technol. Transf. Commercialisation **12**(1/2/3), 43 (2013). https://doi.org/10.1504/IJTTC.2013.064170

7. Barnes, T., Pashby, I., Gibbons, A.: Effective university – industry interaction: a multi-case evaluation of collaborative R&D projects. Eur. Manage. J. **20**(3), 272–285 (2002). https://doi.org/10.1016/S0263-2373(02)00044-0

8. Ankrah, S., Omar AL-Tabbaa,: Universities–industry collaboration: a systematic review. Scand. J. Manag. **31**(3), 387–408 (2015). https://doi.org/10.1016/j.scaman.2015.02.003

9. OECD Science, Technology and Industry Scoreboard 2015. OECD Science, Technology and Industry Scoreboard. OECD (2015). https://doi.org/10.1787/STI_SCOREBOARD-2015-EN

10. Statistiken zum Glücksspiel l Statista (2024). https://de.statista.com/themen/570/glueckssp iel/#topicOverview. Accessed 31 Jan 2024

11. Statistiken zu Lotto l Statista (2024). https://de.statista.com/themen/130/lotto/#topicOver view. Accessed 31 Jan 2024

12. McMullan, J.L., Miller, D.: Wins, winning and winners: the commercial advertising of lottery gambling. J. Gambling Stud. **25**, 73–295 (2009). https://doi.org/10.1007/S10899-009-9120-5/METRICS

13. Binde, P.: Why people gamble: a model with five motivational dimensions. Int. Gambling Stud. **13**, 81–97 (2013). https://doi.org/10.1080/14459795.2012.712150

14. Binde, P.: Gambling advertising: a critical research review per binde responsible gambling trust gambling advertising: a critical research review responsible gambling trust recommended citation. www.responsiblegamblingtrust.org.uk. Accessed 31 Jan 2024

15. Polk, J., McNellis, J., Tassin, C.: Gartner magic quadrant for personalization engines. **13** (2020)

16. Montgomery, A.L., Smith, M.D.: Prospects for personalization on the internet. J. Interact. Mark. **23**(2), 130–137 (2009). https://doi.org/10.1016/j.intmar.2009.02.001

17. Peppers, D., Rogers, M.: The one to one future: building relationship one customer at a time l Enhanced Reader 1997 (2024)

18. AdomaviciusTuzhilin, G.: Incorporating contextual information in recommender systems using a multidimensional approach. ACM Trans. Inf. Syst. **23**, 103–145 (2005)

19. Aksoy, N.C., Kabadayi, E.T., Yilmaz, C., Alan, A.K.: A typology of personalisation practices in marketing in the digital age. J. Mark. Manag. **37**(11–12), 1091–1122 (2021). https://doi.org/10.1080/0267257X.2020.1866647

20. Fan, H., Poole, M.S.: What is personalization? Perspectives on the design and implementation of personalization in information systems. J. Organ. Comput. Electron. Commer. **16**(3–4), 179–202 (2006). https://doi.org/10.1080/10919392.2006.9681199

21. Sinatra, A.M., Robert A.S., Valerie K.S.: The Effects of Self-Reference and Context Personalization on Task Performance during Adaptive Instruction. SAGE PublicationsSage, Los Angeles (2016).https://doi.org/10.1177/1541931213601090

22. Puzakova, M., Kwak, H., Rocereto, J.F.: When humanizing brands goes wrong: the detrimental effect of brand anthropomorphization amid product wrongdoings. J. Mark. **77**(3), 81–100 (2013). https://doi.org/10.1509/JM.11.0510/ASSET/IMAGES/LARGE/10.1509_JM.11.0510-FIG5.JPEG

23. Choi, S.H., Yang, Y.X., Yang, B., Cheung, H.H.: Item-level RFID for enhancement of customer shopping experience in apparel retail. Comput. Ind. **71**, 10–23 (2015). https://doi.org/10.1016/j.compind.2015.03.003

24. Aksoy, C., Nilsah, E.T., Kabadayi, C.Y., Alan, A.K.: Personalisation in marketing: how do people perceive personalisation practices in the business world? J. Electron. Commer. Res. **24**, 2023 (2024)

25. Zhou, W., Zhang, C., Wu, L., Shashidhar, M.: ChatGPT and marketing: analyzing public discourse in early Twitter posts. J. Mark. Anal. **11**, 693–706 (2023). https://doi.org/10.1057/S41270-023-00250-6/FIGURES/4

26. Rivas, P., Zhao, L.: Marketing with ChatGpt: navigating the ethical terrain of GPT-based ChatBot technology. AI **4**(2), 375–384 (2023). https://doi.org/10.3390/ai4020019

Application of Machine Learning in Credit Card Fraud Detection: A Case Study of F Bank

Yuan-Fa Lin[1], Chou-Wen Wang[1(✉)], and Chin-Wen Wu[2]

[1] National Sun Yat-Sen University, Kaohsiung, Taiwan
chouwenwang@gmail.com
[2] Nanhua University, Dalin, Chiayi County, Taiwan

Abstract. Due to the Covid-19 pandemic, people are increasingly engaging in non-face-to-face credit card transactions in their daily lives. However, this trend has also provided opportunities for malicious actors to obtain customer credit card information through various illicit means, leading to a continuous rise in credit card fraud. Traditional fraud detection methods, relying on extensive rules and manual judgment, struggle to effectively prevent the evolving techniques of fraud and often result in significant false positives, requiring substantial time for transaction verification. In recent years, the development of big data and machine learning algorithms has offered an effective solution to this challenge. This study employs three common machine learning algorithms—Logistic Regression, Random Forest, and Extreme Gradient Boosting—for predicting credit card fraud. Utilizing transaction data from Bank F time period from January 2021 to May 2023, including fields such as transaction ID, credit limit, occupation, transaction date, transaction time, transaction amount, etc., the study addresses the issue of imbalanced data in credit card fraud through sampling methods. Different ratios of normal to fraud samples, coupled with varying sampling frequencies, are employed along with ensemble learning techniques to enhance the accuracy and stability of the predictive model. Subsequently, various commonly used machine learning evaluation metrics are applied to identify the best model. The empirical results indicate that the Extreme Gradient Boosting model performs best in detecting credit card fraud. In scenarios with different sampling ratios of normal to fraud samples, the study identifies key features such as changes in the cardholder's transaction behavior concerning transaction region, frequency, and amount. The results of this study provide the bank with references on how to develop more effective strategies for fraud prevention.

Keywords: Credit Card Fraud Detection · Machine Learning · Logistic Regression · Random Forest · Extreme Gradient Boosting

1 Introduction

The COVID-19 pandemic has significantly increased non-face-to-face credit card transactions, leading to a surge in credit card fraud as fraudsters exploit various illicit means to acquire customer data. Traditional fraud detection methods, heavily reliant on rules

F. F.-H. Nah and K. L. Siau (Eds.): HCII 2024, LNCS 14720, pp. 210–222, 2024.
https://doi.org/10.1007/978-3-031-61315-9_15

and manual judgement, have become less effective against evolving fraud techniques and often produce a high number of false positives, resulting in the need for time-consuming transaction verifications. However, the recent advancements in big data and machine learning algorithms offer a powerful solution to this challenge.

In this context, the current study utilizes three popular machine learning algorithms - Logistic Regression, Random Forest, and Extreme Gradient Boosting - to predict credit card fraud. The study is methodically structured into five stages: establishing research motivation and objectives, formulating research methods via literature review, planning research methodology and data sources, conducting empirical model analysis, and drawing conclusions. The data set comprises 176,473 transactions from Bank F's credit card records between January 2021 and May 2023, with a minority marked as fraudulent. The study addresses the issue of data imbalance through under-sampling and ensemble learning techniques, and employs feature engineering to enhance the accuracy of fraud prediction. To tackle the problem of imbalanced data in credit card fraud, the study employs sampling methods with ratios of 10:1 and 20:1 for normal and fraudulent samples, respectively. Through various iterations of these sampling techniques, multiple independent classifiers were trained using ensemble learning methods, with the results of these models then integrated to generate the final prediction.

This approach is designed to improve the predictive accuracy and stability of the models. The study applies various machine learning evaluation metrics, such as the Confusion Matrix, Accuracy, Precision, Recall, F1-Score, and ROC Curve, to identify the most effective model. Empirical findings reveal that the Extreme Gradient Boosting model excels in detecting credit card fraud, outperforming the other models across all metrics. Key features that emerged as significant in different sampling ratios include changes in cardholder transaction behavior, particularly in terms of transaction region, frequency, and amount. These insights are invaluable for banks in developing more effective fraud prevention strategies.

The rapid pace of technological innovation, marked by faster internet speeds, increased computing power, and reduced hardware costs, has greatly accelerated the fields of big data analysis and machine learning. Machine learning models, which learn from training data and improve as they process more data, become increasingly accurate over time. Customized fraud detection models, based on customers' historical transaction data, can be developed through machine learning. However, with the evolution of diverse machine learning models, choosing the most suitable one for credit card fraud detection has become crucial.

In conclusion, the study's findings indicate that the Extreme Gradient Boosting model is superior in detecting credit card fraud, offering key features crucial for fraud detection. These results provide banks with valuable references to enhance their fraud detection systems and reduce fraud risks, thereby helping to minimize the time and cost associated with manual transaction verification.

2 Literature Review

2.1 Application of Machine Learning Models in the Finance

Machine Learning (ML) algorithms learn to predict outcomes from data, automating decision-making with minimal human intervention. They adapt and improve over time by identifying patterns within large datasets. ML has become essential across various fields, such as quantitative finance, computer vision, and natural language processing, thanks to the rise of big data and computational advancements.

ML models are built on training datasets and adjust their parameters through iterative learning to enhance performance based on new data inputs. These algorithms can be trained in several ways, each with its advantages and challenges, broadly categorized into four types:

1. Supervised Learning: Utilizes labeled training data to learn the mapping between inputs and the correct outputs, aiming to predict outcomes for new data. It includes algorithms like Linear Regression, SVM, and Decision Trees, suitable for classification and regression tasks.
2. Unsupervised Learning: Works with unlabeled data to discover hidden structures or patterns, without predefined outcomes. It's used for clustering, dimensionality reduction, and association analysis, employing algorithms like K-Means and PCA.
3. Semi-Supervised Learning: Combines elements of both supervised and unsupervised learning, using a small amount of labeled data alongside a larger set of unlabeled data to improve model performance.
4. Reinforcement Learning: Focuses on training agents to make decisions by interacting with an environment to achieve maximum cumulative rewards. It involves defining agents, states, actions, rewards, and policies to learn the best strategies for given objectives.

In the realm of machine learning applications for credit card fraud detection, several challenges and strategies emerge from recent studies. Weston et al. (2008) utilize real credit card transaction data for peer group analysis, monitoring account holders' behaviors against a peer group with expected similar behaviors. Significant deviations indicate potential fraud, leveraging similarities within peer groups to detect anomalies or suspicious transactions. Prusti and Rath (2019) suggest an innovative approach using ensemble learning, which constructs multiple independent classification algorithms. Each learns and predicts independently before their results are combined into a single prediction, forming a more stable model with superior predictive power for fraud detection.

Baabdullah et al. (2020) apply various machine learning algorithms to handle imbalanced data and detect credit card fraud. Their comparative study reveals that without resampling techniques, focusing on accuracy, sensitivity, and the area under the precision/recall curve (PRC) can enhance fraud detection accuracy and reduce fraudulent incidents.Goyal and Manjhvar (2020) point out the lack of a universal benchmark in evaluating fraud detection systems, leading many studies to use multiple metrics for a comprehensive model assessment. Beyond basic accuracy, precision, recall, F1 scores, confusion matrices, and ROC curves are considered to minimize Type I and Type II errors.Cherif et al. (2023) highlight the issue of imbalanced datasets, where fraudulent transactions are significantly outnumbered by legitimate ones, posing a challenge for prediction models. Most algorithms assume an equal distribution of classes, but in fraud detection, the rarity of fraud cases can diminish algorithm effectiveness. Addressing this issue involves enhancing algorithm adaptability to handle imbalanced datasets or employing sampling methods to balance data distribution.

Collectively, these studies underscore the multifaceted challenges in credit card fraud detection. Imbalanced datasets significantly impact algorithm performance, addressed either by adapting algorithms to manage imbalances or using sampling to equalize data representation. Ensemble learning and peer group analysis offer effective fraud detection methods by combining multiple classifiers' predictions or comparing transaction behaviors within peer groups. However, the absence of a unified evaluation standard necessitates using various metrics for a thorough performance assessment, aiming to enhance accuracy and reduce the occurrence of fraud.

3 Research Methodology

3.1 Data Source

The data for this study were derived from the transaction records of credit card customers at F Bank, spanning from January 1, 2021, to May 31, 2023. The dataset encompasses transactions from 743 accounts, totaling 176,473 credit card transactions. Among these, 175,330 were normal transactions, and 1,143 were identified as fraudulent, with the proportions of fraudulent to normal transactions being 0.65% and 99.35% respectively. The transaction data includes various types of credit card activities within the specified period, such as general purchases, authenticated transactions, tax payments, installment plans, and rewards redemption. Out of the sample, 290 customers had experienced credit card fraud, leaving 453 customers who had not been subjected to fraud. In the entire dataset, fraud transactions accounted for a small fraction of the total transactions, indicating that while the majority of credit card activities are legitimate, fraudulent transactions tend to be concealed within normal activities and are designed to evade detection by the issuing bank.

As shown in Table 1, this study encompasses a total of 35 feature fields, including transaction ID, credit limit, occupation, marital status, education level, transaction date, transaction time, transaction amount, among others. These feature fields provide extensive information about credit card transactions. Preliminary data exploration and processing were conducted using these fields, and feature engineering was applied to generate derived features to assist in enhancing the accuracy of fraud transaction prediction. Derived fields, such as the cumulative transaction amount over the last 360 days, cumulative number of transactions over the last 360 days, maximum transaction amount over the last 360 days, and minimum transaction amount over the last 360 days, offer additional insights into cardholders' transaction behavior patterns. These patterns may help predict potential fraudulent transactions, as fraud may exhibit unusual patterns or changes in behavior in these features. Utilizing these features in training machine learning models can improve the models' effectiveness in detecting fraudulent transactions. Subsequently, this study will employ these data to develop credit card fraud detection models and compare the performance of different machine learning models in fraud detection. By analyzing the data and evaluating the performance of machine learning models, the study aims to identify superior machine learning models for timely approval of legitimate transactions and immediate detection and prevention of fraudulent activities.

The transaction amount ranges for both normal and fraudulent transactions. The majority of normal transactions are concentrated in amounts below 1000 TWD, accounting for 70.53% of the total. However, the interval below 1000 TWD also exhibits a higher risk of fraud, with fraudulent transactions in this range constituting about 27.38%, suggesting a relatively higher proportion of fraud in low-amount transactions.

Data Source: Compiled for this Study. The tendency for fraudulent transactions to opt for lower transaction amounts may be attributed to smaller amounts being less likely to attract the attention of the issuing institution and the cardholder. Another reason might be to minimize the risk of detection, as issuing institutions usually implement control mechanisms for large transactions, such as instant notifications through app push notifications, SMS, or email to the cardholder, or verification calls from personnel. In Taiwan, to enhance the security of online credit card transactions, the Financial Supervisory Commission mandates that issuing banks send transaction confirmation SMS for online transactions exceeding 3000 TWD. Thus, Table 3–3 reveals that transactions below 5000 TWD encompass 86.09% of fraudulent transactions. Nevertheless, the segment above 5000 TWD still covers 13.91% of fraud, likely because fraudsters aim to quickly exhaust the cardholder's credit limit by attempting large-amount transactions to maximize their illicit gains. Therefore, when cardholders conduct unusually large transactions, issuing banks should also pay special attention to whether these transactions are legitimate.

3.2 Data Imbalancing Treatment

In credit card fraud detection, imbalanced data is a key issue because most machine learning algorithms are affected by data imbalance as identified in prior literature reviews. However, this problem can be effectively addressed through sampling methods or ensemble learning. Common sampling methods include data duplication and deletion, which help to form a more balanced dataset. Here are introductions to some common sampling methods:

1. Oversampling: Oversampling increases the number of samples in the minority class by repeatedly sampling to equalize the number of examples across classes. The advantage is maintaining data integrity without losing important information since it balances the dataset by reusing existing data. However, it can be time-consuming and increase computational costs. Moreover, if the proportion of fraud samples is extremely low, it might lead to model overfitting because the model learns predominantly from the repeated minority samples and might struggle to predict accurately in real-world scenarios.
2. Under-sampling: Under-sampling reduces the number of samples in the majority class through random sampling to balance the dataset. For example, in a credit card transaction record, if there are 200,000 normal transactions and only 1,000 fraudulent ones, under-sampling would randomly remove samples from the normal transactions

Table 1. Feature Description

NO	Variables	Explanation	Data Type
1	IDNO	Transaction ID	Category
2	CRLIMIT	Credit Limit	Number
3	POSITION	Occupation	Category
4	MARRIAGE	Marital Status	Category
5	EDUCATION	Education Level	Category
6	TX_DATE	Transaction Date	Number
7	TX_TIME	Transaction Time	Number
8	TX_AMT	Transaction Amount	Number
9	POS_NO	Terminal Machine Number	Category
10	REV_FLAG	Cancellation Mark	Category
11	ADJ_FLAG	Adjustment Mark	Category
12	MCHT_NO	Store Code	Category
13	MCC	Store Category	Category
14	ACQ_BIN	Acquiring Bank Code	Category
15	POS_ENTRY	Terminal Entry Method	Category
16	STIP	Proxy Authorization Mark	Category
17	MANUAL	Authorization Method	Category
18	CUS_CLASS	Credit Rating	Category
19	CUS_AVAIL	Available Balance	Number
20	CUS_LIMIT	Cardholder Limit	Number
21	RESP_ACT	Authorization Result	Category
22	APPR_CODE	Authorization Number	Category
23	POS_COND	Terminal Acquirer Status	Category
24	MERCH_NAME	Store Name	Category
25	COUNTRY	Transaction Country	Category
26	EDC_FUNC	EDC Transaction Code	Category
27	INSTALL_FLAG	Installment Mark	Category
28	BONUS_FLAG	Reward Redemption Mark	Category
29	FALLBACK	Chip to Magnetic Stripe Transaction Mark	Category
30	AC_FLAG	Mobile Payment Code	Category
31	DISPFLG	Fraud Mark	Category
32	FLAG_3D	3D Transaction Mark	Category

(continued)

Table 1. (*continued*)

NO	Variables	Explanation	Data Type
33	AGE	Age	Number
34	SEX_H	Cardholder Gender	Number
35	DATEOPEN1	Credit Card Issuance Date	Number

to bring their numbers down to around 1,000. This method can decrease model training time, which is advantageous for processing large datasets. However, its major drawback is the potential loss of representative data samples.

3. Synthetic Minority Oversampling Technique (SMOTE): Based on minority class data, SMOTE doesn't just copy existing minority class samples but generates synthetic new samples near selected data points to increase data diversity. The algorithm generates similar samples along the path connecting minority class fraud data with its neighbors, thus increasing the volume of data and achieving class balance. The advantage is maintaining data integrity and reducing overfitting risks, but it may also lead to the generation of specific patterns, which could cause overfitting to these patterns and increase computational costs for large samples.

3.3   Brief Review of Machine Learning Models

This study consists of 176,473 data entries, spanning from January 1, 2021, to May 31, 2023. Given the sequential nature of transaction data, this study segmented the data temporally, designating data from January 1, 2021, to December 31, 2022, as the training set, which totals 147,081 entries. Data from January 1, 2023, to May 31, 2023, was used as the test set, comprising 29,392 entries. Within the training data, there were 600 instances of fraudulent transactions and 146,481 normal transactions. In the test data, there were 543 instances of fraudulent transactions and 28,849 normal transactions.

To address the issue of data imbalance, the study employed sampling methods with ratios of 10:1 and 20:1 for normal and fraudulent samples. Through several iterations of different samplings and the application of ensemble learning techniques, multiple independent classifiers were trained. The results of these models were then consolidated to produce the final predictions. We employ three machine learning models for training: Logistic Regression, Random Forest, and XGBoost. Below is a concise concise description of the three models mentioned.

1. Logistic Regression:Logistic regression is a statistical model commonly used for binary classification problems, ideal for scenarios with two possible outcomes, such as default/no default, male/female, etc. It models the probability of an event occurring by transforming the linear combination of input variables (features) into a probability value between 0 and 1 using the logistic (or sigmoid) function. Its strength lies in its simplicity and interpretability, being a go-to model for many classification issues. However, it assumes a linear relationship between features and is sensitive to feature engineering quality.

2. Random Forest:Random forest is an ensemble learning method that builds upon decision tree algorithms, introducing randomness in the construction and prediction processes to enhance performance and generalization. It employs bootstrap sampling for training individual trees and randomly selects features at each node for splitting. The model's predictions are aggregated, usually by majority vote for classification and average or median for regression. Random Forest is powerful, particularly for medium to large datasets, and combats overfitting through its inherent randomness.

3. Extreme Gradient Boosting (XGBoost): XGBoost combines gradient boosting with regularization techniques to improve predictive performance and reduce overfitting, applicable to both regression and classification. Known for high predictive performance, it addresses various problem types and incorporates L1 and L2 regularization. XGBoost can handle missing data, supports parallel computing, offers cross-validation for optimal hyperparameter selection, and allows feature importance evaluation. It typically outperforms other algorithms, especially on large datasets, and provides flexibility with custom loss functions. However, it requires careful hyperparameter tuning and has higher memory demands, which may be challenging in resource-limited environments. XGBoost is primarily used for structured data and is less suited for unstructured data like images and text.

These models are essential tools in machine learning, each with its unique advantages and limitations. They are widely used across various industries for predictive analytics and decision-making processes.

4 Research Methodology

4.1 Model Comparison

In the context of credit card fraud detection, banks aim to identify as many potential fraudulent transactions as possible without generating excessive false positives that inconvenience customers. Therefore, this study utilizes the F1 score as the primary criterion for evaluating model performance, as the F1 score balances precision and recall.

Table 2 shows the predictive results of three machine learning models, including an ensemble learning method, under a sample ratio of normal to fraudulent transactions of 10:1. Overall, the Extreme Gradient Boosting (XGBoost) model demonstrates the best composite performance with the highest F1 and AUC scores, making it the optimal predictive model among the three. The Random Forest model also performs well but slightly less so compared to XGBoost. Logistic Regression, on the other hand, performs poorly in this highly imbalanced scenario, indicating room for model optimization.

In detail, Logistic Regression shows improved performance with an increased number of samplings, yet it scores lower in precision, recall, and F1 across all samplings, with fluctuating AUC values. Random Forest exhibits stability across all samplings, with high precision, recall, F1 scores, and AUC, indicating excellent performance despite the imbalance. XGBoost maintains high accuracy, precision, recall, and F1 scores across all samplings, with consistently high AUC values, suggesting superior comprehensive performance in fraud detection. Individual model results are analyzed as follows:

1. Logistic Regression:

- Accuracy gradually increases with the number of samples, ranging from 83.97% to 93.00%.
- Precision increases with the number of samples, indicating a decrease in the chance of making errors in predicting fraud transactions.
- Recall improves with the number of samples, indicating better capture of fraud transactions.

Table 2. Comparison of Model Results at Different Sampling Numbers

# of Samples	1	3	5	7	9
Logistic Regression					
Accuracy	83.97%	91.63%	93.23%	92.49%	93.00%
Precision	7.09%	20.68%	23.61%	21.26%	21.87%
Recall	61.69%	73.85%	76.76%	78.87%	79.15%
F1 Score	12.72%	31.30%	35.40%	32.71%	33.63%
AUC	73.05%	86.16%	88.47%	90.42%	86.26%
Random Forest					
Accuracy	98.58%	98.50%	98.46%	98.51%	98.49%
Precision	61.07%	59.08%	57.95%	59.10%	58.64%
Recall	69.61%	68.63%	69.61%	69.61%	69.70%
F1 Score	65.06%	63.49%	63.18%	63.90%	63.62%
AUC	97.78%	97.57%	97.62%	97.70%	97.67%
XGBoost					
Accuracy	98.32%	98.42%	98.32%	98.41%	98.44%
Precision	54.04%	55.92%	53.82%	55.73%	56.34%
Recall	76.43%	79.01%	81.69%	79.06%	79.23%
F1 Score	63.31%	65.48%	64.87%	65.36%	65.83%
AUC	98.42%	98.73%	98.50%	98.66%	98.60%

Data Source: Compiled for this study.

- F1 score increases gradually with the number of samples, but it is lower compared to the other two models.

2. Random Forest:

- Performance is stable across different sampling frequencies, with high accuracy (>98%) consistently.
- Precision remains around 60% across different sampling frequencies, indicating stable performance in predicting fraud.
- Recall shows relatively stable performance between 69–70%.
- F1 score maintains a high level across different sampling frequencies, indicating good balance between precision and recall.

3. XGBoost:

- Accuracy remains around 98% across different sampling frequencies, showing very stable performance overall.
- Precision remains above 50% across different sampling frequencies, indicating stable performance in predicting positive cases.
- Recall remains relatively stable around 79% across different sampling frequencies.
- F1 score maintains a high level across different sampling frequencies, indicating good balance between precision and recall.

From Table 2, it is evident that both Random Forest and XGBoost models outperform Logistic Regression in overall performance across different sampling frequencies. They exhibit good performance on highly imbalanced datasets and maintain stable performance. Overall, based on the evaluation metrics, the XGBoost model is the best fraud detection model, providing efficient predictive performance in practical applications.

In this study, we further explores the XGBoost model's top ten important features selected in the 10:1 sampling ratio scenario and provides explanations for their significance:

1. COUNTRY_dchange_180 (Change in transaction country in the last 180 days): Indicates whether there has been a change in the transaction country for the account in the past 180 days. Checking for recent changes in transaction countries may help detect abnormal transaction behavior.
2. TX_AMT (Transaction amount): Indicates the amount of each transaction. Large or very small transactions may carry significant risk, hence the model focuses on this feature.
3. COUNTRY1 (Transaction region is the USA, Canada): Indicates whether the transaction region is in the USA or Canada, suggesting transactions in these regions might be more complex and susceptible to fraud.
4. cum_past360_cnt (Cumulative transaction count in the last 360 days): Indicates the cumulative number of transactions for the account in the past 360 days. This feature helps identify accounts with unusual transaction frequencies.
5. MCC7273 (Merchant category is dating website): Transactions categorized as dating websites may pose risks, as these sites are common targets for fraudulent activities.
6. cum_past30_min_cnt (Cumulative transaction count in the last 30 min): Indicates the cumulative number of transactions for the account in the last 30 min. A sudden increase in transaction frequency may indicate abnormal account behavior.
7. POS_ENTRYphysical (Is it a physical transaction): Physical transactions may pose different fraud risks compared to online or virtual transactions.
8. RESP_ACTD (Is it a declined transaction): This feature indicates whether the transaction is marked as declined. Declined transactions may be triggered by abnormal behavior, making them important indicators for fraud detection.
9. COUNTRY2 (Transaction region is the UK): Similar to COUNTRY1, this feature indicates whether the transaction region is in the UK.
10. MERCH_NAME_dchange_360 (Change in transaction merchant in the last 360 days): Indicates whether there has been a change in the transaction merchant for the account in the past 360 days. Changes may indicate unusual transaction behavior for the account.

Considering these features, the XGBoost model primarily focuses on transaction country, amount, region, frequency, and merchant category. Changes in these transaction behaviors can be considered important information for fraud detection. For issuing banks, any changes in cardholders' transaction behaviors involving these features should warrant increased scrutiny for potential fraud or alternative methods of transaction verification.

4.2 Robustness Check

Apart from establishing models under a 10:1 ratio of fraud samples, Table 3 shows the model prediction results under different sampling frequencies with a fraud sample ratio of 20:1. Overall, when the normal to fraud sample ratio increases to 20:1, all three models still maintain high overall accuracy. Random Forest and XGBoost models remain superior, demonstrating stable and excellent performance in this highly imbalanced dataset. Logistic Regression performs relatively worse in this scenario but still shows some improvement with increasing sampling frequency.

These analyses demonstrate that even with a higher fraud sample ratio of 20:1, Random Forest and XGBoost models maintain superior performance, while Logistic Regression shows relatively poorer performance but still improves with increasing sampling frequency. Additionally, Table 3 presents the results under a 20:1 normal to fraudulent transaction ratio, using ensemble learning methods across different sampling frequencies. The Random Forest and XGBoost models remain preferable, maintaining stability and superior performance in this highly imbalanced dataset. Logistic Regression, while showing improvement with increased sampling, still lags behind the other two models.

In summary, based on the evaluation metrics, XGBoost is determined as the best model for fraud detection, providing efficient predictive performance for practical applications. This suggests that banks should closely monitor transactions with changes in the identified key features to promptly detect and prevent fraudulent activities.

5 Conclusion

This study utilized three commonly used machine learning algorithms, namely Logistic Regression, Random Forest, and XGBoost, for credit card fraud prediction. It conducted preliminary descriptive statistics on the credit card transaction data from F Bank to identify the distribution of features such as sample gender, education level, and transaction time. Moreover, it addressed the issue of imbalanced data in credit card fraud by using undersampling, employing different ratios of normal to fraud samples, and varying sampling frequencies along with ensemble learning methods to enhance model prediction accuracy and stability.

All three machine learning algorithms achieved good accuracy, with XGBoost performing the best among them. It exhibited superior performance in terms of accuracy, precision, recall, and F1 score, while providing key features helpful for fraud detection. Across different sampling ratios of normal to fraud samples, features such as change in transaction country in the last 180 days, transaction amount, transaction region in the USA or Canada, cumulative transaction count in the last 360 days, merchant category as dating websites, cumulative transaction count in the last 30 min, whether it is a physical transaction, and whether it is a declined transaction were selected as important variables. These findings provide reference for fraud detection for issuing banks to improve their fraud detection systems and reduce fraud risks. However, facing the continuously evolving fraud techniques, banks should also continuously monitor changes in cardholder transaction behavior and adjust their control strategies as fraud techniques evolve. This flexibility and continuous adjustment of strategies are crucial for coping with the ever-changing fraud techniques.

This study addressed the issue of imbalanced credit card fraud data while balancing computational efficiency and fully utilizing fraud samples by utilizing undersampling. It established three machine learning models using undersampling with normal to fraud sample ratios of 10:1 and 20:1. However, in practice, the actual ratio of fraud to normal samples may vary. Therefore, future research could explore different sampling ratios and establish different machine learning models to compare results. Additionally, this study only used credit card data from a single bank. If data from multiple banks with different credit cards could be combined, it might be possible to identify more key features and enhance model performance further.

References

Baabdullah, T., Alzahrani, A., Rawat, D.B.: On the comparative study of prediction accuracy for credit card fraud detection with imbalanced classifications. In: 2020 Spring Simulation Conference (SpringSim), pp. 1–12 (2020). https://doi.org/10.22360/SpringSim.2020.CSE.004

Cherif, A., Badhib, A., Ammar, H., Alshehri, S., Kalkatawi, M., Imine, A.: Credit card fraud detection in the era of disruptive technologies: a systematic review. J. King Saud Univ. – Comput. Inf. Sci. **35**(1), 145–174 (2023). https://doi.org/10.1016/j.jksuci.2022.11.008

Duman, E., Ozcelik, M.H.: Detecting credit card fraud by genetic algorithm and scatter search. Expert Syst. Appl. **38**(10), 13057–13063 (2011). https://doi.org/10.1016/j.eswa.2011.04.110

Gadi, M.F.A., do Lago, A.P., Wang, X.: A comparison of classification methods applied on credit card fraud detection. Technical Report (2016)

Goyal, R., Manjhvar, A.K.: Review on credit card fraud detection using data mining classification techniques & machine learning algorithms. SSRN Scholarly Paper 3677692 (2020). https://papers.ssrn.com/abstract=3677692

Kovalenko, O.: Credit Card fraud detection using machine learning | SPD technology. software product development company (2023. https://spd.tech/machine-learning/credit-card-fraud-detection/

Patidar, R., Sharma, L.: Credit card fraud detection using neural network. Int. J. Soft Comput. Eng. (IJSCE) **1**(32–38) (2011). https://citeseerx.ist.psu.edu/document?repid=rep1&type=pdf&doi=0419c275f05841d87ab9a4c9767a4f997b61a50e

Prusti, D., Rath, S.K.: Fraudulent transaction detection in credit card by applying ensemble machine learning techniques. In: 2019 10th International Conference on Computing, Communication and Networking Technologies (ICCCNT), pp. 1–6 (2019). https://doi.org/10.1109/ICCCNT45670.2019.8944867

Receiver operating characteristic. In: Wikipedia (2023). https://en.wikipedia.org/w/index.php?title=Receiver_operating_characteristic&oldid=1184185440

Şahin, Y.G., Duman, E.: Detecting credit card fraud by decision trees and support vector machines (2011). https://openaccess.dogus.edu.tr/xmlui/handle/11376/2366

Weston, D.J., Hand, D.J., Adams, N.M., Whitrow, C., Juszczak, P.: Plastic card fraud detection using peer group analysis. Adv. Data Anal. Classif. **2**(1), 45–62 (2008). https://doi.org/10.1007/s11634-008-0021-8

Explainable AI in Machine Learning Regression: Creating Transparency of a Regression Model

Robbie T. Nakatsu[✉]

Loyola Marymount University, Los Angeles, CA, USA
Robbie.Nakatsu@lmu.edu

Abstract. This paper explores how to develop machine learning regression models that are more explainable and transparent for the end-user. Explainable regression models can be created by rank-ordering the features of the regression model that contribute most to predictive accuracy. In addition, fitting graphs can be generated that show how the addition of each feature in a regression model incrementally improves predictive accuracy. These information graphics are especially useful in understanding the tradeoffs involved in selecting a model that considers both model complexity and model performance. These methods are illustrated with two examples: a multiple regression model using a straightforward application of machine learning regression; and a more complex polynomial regression model that captures higher-order terms and interactions among all variables in the model.

Keywords: Explainable AI · Machine Learning · Regression Modeling · Information Visualization · Fitting Graph · Forward Selection · Stepwise Regression · Polynomial Regression · Human-Computer Interaction (HCI)

1 Introduction

Linear regression is a powerful statistical technique used not only in machine learning, but also across a wide variety of disciplines (e.g., finance, economics, medicine, biology, marketing, to name but a few of the disciplines). It is, arguably, the most popular statistical technique used in business, economics, and the social sciences. Numerous books and research articles have been written on the subject, detailing its methods and techniques (see e.g., [1–3]).

In essence, linear regression involves specifying the relationship between a numeric dependent variable (also referred to as the outcome or target variable) and one or more independent variables (also referred to as features, predictors, or regressors). Assume a linear regression model of the form

$$Y = \beta_0 + \beta_1 X_1 + \beta_2 X_2 + \cdots + \beta_p X_p + \epsilon \tag{1}$$

© The Author(s), under exclusive license to Springer Nature Switzerland AG 2024
F. F.-H. Nah and K. L. Siau (Eds.): HCII 2024, LNCS 14720, pp. 223–236, 2024.
https://doi.org/10.1007/978-3-031-61315-9_16

where Y is the dependent variable, and $X_1, X_2, ..., X_p$ are the independent variables. Ordinary least squares (OLS) regression will find the β coefficients that minimize the sum of squared error (SSE).

$$SSE = \sum_{i=1}^{n} \left(y_i - \widehat{y_i}\right)^2 \qquad (2)$$

where y_i is the actual outcome and $\widehat{y_i}$ is the predicted outcome fitted by using the regression model for i = 1 to n observations. A common metric for evaluating the predictive accuracy of a regression model is **MSE,** which refers to the mean of the squared errors. It is calculated by dividing SSE in Eq. 2 by n:

$$MSE = \frac{1}{n} \sum_{i=1}^{n} \left(y_i - \widehat{y_i}\right)^2 \qquad (3)$$

We can then take the square root of MSE to arrive at the root mean squared error **RMSE**:

$$RMSE = \sqrt{\frac{1}{n} \sum_{i=1}^{n} \left(y_i - \widehat{y_i}\right)^2} \qquad (4)$$

Throughout this paper RMSE will be used to assess the predictive accuracy of a regression model.

Broadly, there are two major categories of linear regression: (1) statistical inference and (2) prediction used in machine learning. Statistical inference studies how a set of independent variables affects a dependent variable, and what the sizes of their effects are. Because it focuses on understanding the relationship between a dependent variable and independent variables, the interpretability of the model is the key objective. On the other hand, prediction—used in machine learning applications—develops a model that predicts a numeric outcome from a set of independent variables. Hence, model interpretability is not as important, if at all—that is, prediction does not require that you have an understanding of the model used to make the prediction. In many cases, the end-user treats the regression model as a black box and is only concerned that the model is making good predictions. As Kuhn and Johnson comment "the predictive models that are the most powerful are usually the least interpretable" [4, p. 50].

However, in recent years, there have been calls by many in the field to make AI systems more transparent [5]. A common complaint levelled against AI user interfaces is that they are black boxes that cannot be probed for deeper understanding. To address this problem, **Explainable AI (XAI)**, sometimes called interpretable AI, refers to AI that can explain its actions in order to better understand system predictions and recommendations [6–8] Provided with such explanations, the end-user would be in a better position to use the AI system's outputs in more thoughtful and critical ways.

Explanations in AI is not a new research topic but has been around for decades. It was a popular topic during the 1970s and 1980s when researchers were looking for ways to make rule-based expert systems more transparent [e.g., 9–11]. But the need for Explainable AI has re-emerged as an urgent research topic, given the enormous impact AI has had on our society, and the proliferation of applications in the world today [5].

In this paper, I explore the topic of Explainable AI in the context of linear regression modeling. I first look at more traditional uses of regression modeling, and then compare and contrast those uses to machine learning regression modeling. I then explore how to create more transparency in linear regression models. I present two examples in this paper to illustrate Explainable AI: first, a multiple regression model using a straightforward application of machine learning regression; and second, a more complex polynomial regression model that includes second-order quadratic terms and interactions among all variables in the model.

2   Regression for Statistical Inference: Understanding Relationships

In traditional regression, the emphasis is on performing statistical inference. Statistical inference looks at how a set of independent variables affects a dependent variable, and what the sizes of their effects are. Furthermore, statistical inference examines whether these effects are statistically significant or not. Let's look at a typical example. I will use the Ames housing dataset [12], which is freely available on Kaggle [13], to perform the regression analysis. The dataset describes the sale of residential properties in Ames, Iowa from 2006 through 2010. It contains 1460 observations and 79 explanatory variables describing various features of the home such as square footage, lot size, number of bedrooms, age of home, condition of the home, type of sale, and so forth.

I ran a multiple linear regression, using the Ames dataset, regressing the dependent variable **SalePrice** (i.e., sales price of the home) on ten independent variables: GrLivArea (square footage), TotalBasementSF (size of a basement), OverallCond (condition of a home), OverallQual (quality of a home), LotArea (lot size), GarageArea (garage size), BedroomAbvGr (number of bedrooms), GarageCars (number of cars that will fit in garage), FullBath (number of full bathrooms), and Age (age of the property). The linear regression model was created using the **OLS** routine found in the Python library **statsmodels**. The OLS routine performs ordinary least squares regression.

Anyone familiar with linear regression modeling will immediately recognize the output that is generated when running OLS regression. See Fig. 1 for this output, which describes the regression model.

In multiple linear regression, you would interpret a model coefficient β_i as *the average effect on Y (the dependent variable) of a one-unit increase in x_i (independent variable) assuming that the values of all the other independent variables are held constant.* For example, Fig. 1 shows that the coefficient (see column coef) for GrLivArea (square footage of home) is 60.49. This means that a one-unit increase in square footage (GrLivArea) increases SalePrice by 60.49, assuming that all the other independent variables are held constant. Similarly, a one unit increase in the number of cars, as specified by GarageCars, will increase SalePrice by 11,530, again assuming that all other variables are held constant.

To check for statistical significance, you would look at the column P > |t|. The column displays the p-values: a low p-value signifies that there is strong evidence to reject the null hypothesis that $\beta_i = 0$. Typically, a p-value $<\,= 0.05$ or a p-value $<\,= 0.01$ is used as a cutoff for rejecting the null hypothesis. As shown in Fig. 1, with the exception of GarageArea and FullBath, we can reject the null hypothesis that $\beta_i = 0$—signifying that there is a statistically significant relationship between the independent variable and the dependent variable: all these p-values are < 0.01.

```
OLS Regression Results
===============================================================================
Dep. Variable:              SalePrice   R-squared:                        0.786
Model:                            OLS   Adj. R-squared:                   0.785
Method:                 Least Squares   F-statistic:                      532.2
Date:                Fri, 17 Nov 2023   Prob (F-statistic):                0.00
Time:                        00:29:28   Log-Likelihood:                 -17419.
No. Observations:                1460   AIC:                           3.486e+04
Df Residuals:                    1449   BIC:                           3.492e+04
Df Model:                          10
Covariance Type:            nonrobust
===============================================================================
                  coef     std err         t      P>|t|      [0.025      0.975]
-------------------------------------------------------------------------------
Intercept     -7.454e+04  8380.356     -8.895     0.000     -9.1e+04   -5.81e+04
GrLivArea        60.4908     3.322     18.211     0.000       53.975      67.006
TotalBsmtSF      25.2393     2.873      8.786     0.000       19.604      30.874
OverallCond    6741.6466   963.406      6.998     0.000     4851.827    8631.466
OverallQual    1.835e+04  1171.242     15.671     0.000      1.61e+04    2.07e+04
LotArea           0.6168     0.103      5.988     0.000        0.415       0.819
GarageArea        7.9900    10.019      0.797     0.425      -11.664      27.644
BedroomAbvGr  -8218.7704  1499.855     -5.480     0.000     -1.12e+04   -5276.651
GarageCars     1.153e+04  2954.069      3.904     0.000     5739.265     1.73e+04
FullBath      -1936.6376  2600.635     -0.745     0.457    -7038.051    3164.775
Age            -451.3224    48.865     -9.236     0.000     -547.175    -355.469
===============================================================================
Omnibus:                      673.461   Durbin-Watson:                    1.980
Prob(Omnibus):                  0.000   Jarque-Bera (JB):            105899.440
Skew:                          -1.062   Prob(JB):                          0.00
Kurtosis:                      44.669   Cond. No.                      1.28e+05
===============================================================================
```

Fig. 1. A linear regression model of SalePrice (dependent variable) on multiple independent variables.

Another part of the output in Fig. 1 is the R^2 statistic (see R-squared and Adj. R-squared), as well as the F-statistic. R^2 is defined as the proportion of variance explained by the least squares regression model. Because it is a proportion, the value must lie between 0 and 1, inclusive, where $R^2 = 0$ means that the regression model does not account for any of the variation, while $R^2 = 1$ means the regression model accounts for all the variation, or perfectly predicts a y value for a given X (set of independent variables). Adjusted R^2 is a modified version of the R^2 formula in that it penalizes the addition of more independent variables. The F-statistic checks whether a group of independent variables is related to the dependent variable. Whereas the t-statistic tests one independent variable at a time, the F-statistic is a multivariate test. In Fig. 1, the results indicate that the F-statistic = 532.2, and Prob (F-statistic) = .00 (p-value for the F-statistic), meaning that we can reject the null hypothesis that the set of independent variables are unrelated to the dependent variable.

In summary, it is clear that regression modeling for statistical inference involves building an interpretable regression model. Interpreting what the size of the coefficients means, and understanding whether the sizes of the coefficients are significant or not are two key functions of the regression model output when used in this way.

3   Regression for Machine Learning: Predicting a Continuous-Valued Numeric Outcome

The emphasis on machine learning regression modeling is on making the best possible prediction on a continuous-valued numeric outcome. For this reason, model interpretability is less important to the analysis. In contrast to model interpretability, the primary benchmark in machine learning regression is predictive accuracy. As noted in the introduction, I will use RMSE (root mean squared error) to assess predictive accuracy of a regression model. To improve predictive accuracy, or reduce RMSE, a regression model will tend to include many more variables, including higher order terms, and interaction effects. Indeed, machine learning can often involve processing huge datasets containing numerous independent variables: It is not uncommon for datasets to contain hundreds, thousands, or even tens of thousands features. Such datasets are referred to as high-dimensional datasets.

With high-dimensional datasets, there naturally arises the problem that several of the features may not contribute to the predictive power of the machine learning model. There is a need to extract the most important features and eliminate the irrelevant ones. Hence, an important processing step in machine learning regression is **feature selection**.

When you are dealing with thousands of features in a single model, the feature selection task can become time-consuming and laborious if you had to manually decide which features to include and which features to drop. Fortunately, the process can be automated using some kind of feature selection algorithm. Ideally, a feature selection algorithm should demonstrate the following properties:

1. The algorithm should run in a reasonable amount of time. With high-dimensional datasets, it can become computationally prohibitive to try out every single combination of the features. More efficient algorithms that perform reasonably well, but do not guarantee the optimum solution, are sought instead. For example, the algorithm could use early stopping criteria when an acceptable level of performance has been achieved.
2. The algorithm should perform cross-validation (or some other validation method) so that it chooses the best subset of features based on predictive accuracy that has been calculated on **out-of-sample** data—that is, data that has not been used to train the machine learning model. A popular method is to perform k-fold cross validation (described in more detail below).

Regarding the second point, a central problem in machine learning is the problem of model overfit. This refers to "finding chance occurrences in data that look like interesting patterns, but which do not generalize to unseen data" [14, p. 111]. By training a dataset using a machine learning algorithm, you may develop a model that fits well to your training data, but does not generalize well to future, unseen cases. The solution is to validate your model on out-of-sample data, also known as the holdout data, because it has been set aside and not used to train the Machine learning algorithm [15]. If you validate your model on the same data that you used to train the machine learning algorithm, you are likely to get overoptimistic results, or an underestimation of RMSE.

k-fold Cross-Validation (k-fold CV) [16, 17] is a well-proven method for model validation that has become standard practice in the data science and machine learning communities. It not only provides an estimate of error on out-of-sample data, but it also results in a more accurate estimate because the error estimate is averaged over k iterations. The method involves the following steps: First, the rows of the dataset are randomly shuffled. Second, the randomized dataset is split into k partitions called folds (5 or 10 folds is most commonly used). Third, the technique iterates k times, once on each of the k folds: For each iteration, one fold is set aside as the test set, and the remaining k-1 folds are designated as the training set. The regression model is then built on the training set and validated on the test set—i.e., RMSE is calculated. After repeating k times this way, the average of the k RMSE calculations is computed. In addition, in a number of recent studies, it has been shown that **repeated k-fold CV** can further improve your estimate of error [18–20]. Repeated k-fold CV means that k-fold CV is repeated a number of times to generate even more accurate estimates of error. In the regression analyses that follow, I performed 5 repetitions of 10-fold CV to obtain more accurate estimates of RMSE.[1]

3.1 Feature Selection Using Forward Selection

When you are trying to decide which features will contribute most to the predictive accuracy of a regression model, a good choice is to use **forward selection**. (An alternative approach, backward elimination, is not addressed in this paper). Forward selection is a **stepwise** approach because it is an iterative algorithm that builds a regression model, incrementally, based on the results of previous iterations [1].

Forward selection begins with a regression model containing no features, referred to as the **null model**. It then adds features, one by one, until an acceptable level of performance is achieved. For each iteration, it finds the best model that has the lowest cross-validated error. For example, it first finds the best 1-feature model. Then it will find the best 2-feature model, the best 3-feature model, up until the best p-feature model—i.e., the full model containing all the features. For each iteration, it always adds an additional feature to the existing best-feature model. That is, once a feature is added, it can never be dropped. (Note that there are variations of forward selection, but this is the standard one that is used in this study). Let's look at an example to see how this algorithm works.

I performed OLS multiple linear regression on the Ames dataset, except this time used a total of 19 features to predict SalePrice. In the first pass, the forward selection algorithm will select the best 1-feature solution. That is, it will go through each of the 19 features, and select the feature that results in the lowest cross-validated RMSE. The RMSE is calculated using repeated k-fold cross validation, so that the RMSE is calculated on out-of-sample data. The procedure is repeated 19 times: By the 19th iteration, the cross-validated RMSE represents the error for the full model containing all 19 features. The results of forward selection are given below:

---

[1] The Ames housing dataset, with only 1460 observations, was small enough to performed repeated k-fold CV. For larger datasets, the computational costs of performing repeated k-fold CV might outweigh the benefits of obtaining more accurate estimates of RMSE.

```
Number of features: 1   RMSE: 48206.35
Number of features: 2   RMSE: 42253.99
Number of features: 3   RMSE: 40164.03
Number of features: 4   RMSE: 38781.35
Number of features: 5   RMSE: 38154.55
Number of features: 6   RMSE: 37502.74
Number of features: 7   RMSE: 37019.52
Number of features: 8   RMSE: 36629.37
Number of features: 9   RMSE: 36313.98
Number of features: 10  RMSE: 36211.97
Number of features: 11  RMSE: 35136.82
Number of features: 12  RMSE: 34776.99
Number of features: 13  RMSE: 34673.35
Number of features: 14  RMSE: 34647.16
Number of features: 15  RMSE: 34658.44
Number of features: 16  RMSE: 34682.19
Number of features: 17  RMSE: 34717.24
Number of features: 18  RMSE: 34782.66
Number of features: 19  RMSE: 34912.85
```

The output shows that the lowest RMSE occurs when 14 features have been added. This means that 5 features (out of 19 total) are not included in the best model. Notice how after 14 features, RMSE increases, leading to poorer performance. This is because after 14 features, the regression model is overfit to the data, and does not generalize as well on out-of-sample data.

3.2 Creating Model Transparency

The 14-feature model, which represents the best-performing regression model, can be used as the final solution to this regression problem. It can be used to predict the sales price of a (future) home in Ames, Iowa that was not included in the original dataset. However, it is a black box model: We do not know which features contributed most to the 14-feature model, nor do we know which features were left out. To better understand which features are most important, I generated an ordered list of the features that were added in the forward selection process. The more important features would be added first, and the less important features would be added last.

The output below displays the order that each feature was added using forward selection. The first feature selected is OverallQual, followed by GrLivArea. The least important feature is GarageArea. (Note: It is not that GarageArea does not matter, but you can see in the list that GarageCars, or number of cars that will fit in the garage, serves as a proxy for GarageArea, so it was selected instead in the forward selection algorithm). The output also displays the 14 features that were added in the best model having the lowest cross-validated RMSE, as well as the 5 features that were left out.

```
The order in which features are selected:
['OverallQual', 'GrLivArea', 'TotalBsmtSF', 'GarageCars',
'Age', 'OverallCond', 'LotArea', 'SaleCondition_Partial',
'BedroomAbvGr', 'KitchenQual_TA', 'KitchenQual_Gd',
'KitchenQual_Fa', 'SaleCondition_Normal',
'SaleCondition_AdjLand', 'CentralAir_Y', 'SaleCondition_Family',
'SaleCondition_Alloca', 'FullBath','GarageArea']

The best solution uses 14 features:
['GrLivArea', 'TotalBsmtSF', 'OverallCond', 'OverallQual',
'LotArea', 'BedroomAbvGr', 'GarageCars', 'Age',
'KitchenQual_Fa', 'KitchenQual_Gd', 'KitchenQual_TA',
'SaleCondition_AdjLand', 'SaleCondition_Normal',
'SaleCondition_Partial']

RMSE for the best solution is: 34647.16

Features not selected:
['FullBath', 'CentralAir_Y', 'SaleCondition_Family',
'GarageArea', 'SaleCondition_Alloca']
```

3.3 Visualizing Model Performance with a Fitting Graph

What else might we do to foster transparency of a regression model? In addition to understanding which features contributed most to the prediction of sales price, it would also be useful to see, graphically, how adding one feature at a time affects the RMSE. The information graphic in Fig. 2 was created for this purpose. It shows how feature selection affects the RMSE as you increase the number of features selected from 1 to 19. As typically happens in a regression model, the most dramatic reductions in RMSE occur at the beginning: the addition of the second feature (GrLivArea) to the first feature (OverallQual) results in a reduction in RMSE from about 48,000 to about 42,000. In general, each additional feature will tend to result in less and less improvement; by the 11th feature, the RMSE has stabilized and does not appear to decrease much after that. The rate of decrease of RMSE decreases so that by the 11th feature, the curve has flattened out showing that RMSE has stabilized and does not appear to decrease much after that. This information graphic, then, shows us how each of the features incrementally contributes to the predictive power of the model.

If we zoom in on features 10 through 19, we can get a clearer picture of what is happening to RMSE. Figure 3 shows that after 14 features, the performance of the model deteriorates slightly—its RMSE increases. This bowl-shaped graph on model performance is what we expect to see in machine learning model performance: In the beginning, with more features added, the RMSE decreases; it then reaches a minimum point, and then starts to increase due to model overfit. The graphs in Fig. 3 and Fig. 4 confirm that our regression model is behaving as expected.

But more than that, the information graphics in Figs. 2 and 3 provide some insight on to how to trade off model complexity (number of features) with model performance

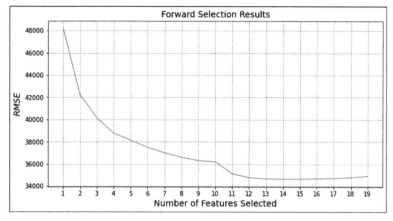

Fig. 2. Forward selection results from 1 through 19 features selected on Ames housing dataset.

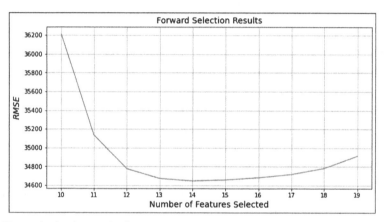

Fig. 3. Forward selection results from 10 through 19 features selected on Ames housing dataset.

(RMSE). When designing a regression model, simplicity may be an important criterion to consider when selecting a model. In this example, the 7-feature model has an RMSE of 37,019 while the 14-feature model has a minimum RMSE of 34,647. That represents an improvement of only 6.4%, achieved at doubling the number of features from 7 to 14. When considering the tradeoff between model simplicity and model performance, the 7-feature model may be chosen instead of the 14-feature model as it results in a suitable performance level that is achieved with 50% fewer features. The fitting graph enables us to visualize this trade off more clearly.

4   A More Complex Example

One of the benefits of the machine learning approach to regression modeling is that you can include a lot of different combinations of the independent variables when building the regression model: all possible interaction effects, and higher-order polynomial terms can be added to the regression model. Forward selection can then be used to determine which variables and combinations of variables result in the best predictive model. In this next example, I have used the same Ames dataset, but this time have created all terms of degree = 2 or less.

4.1 Building the Polynomial Regression Model

Conveniently, the scikit-learn library in Python has a function called **PolynomialFeatures**, which will automatically generate a new feature matrix consisting of all polynomial combinations of features with a degree less than or equal to the specified degree. For example, assume four features A, B, C, and D. PolynomialFeatures, specified with degree = 2, will generate the following 14 features, which includes all possible squared terms and two-way interaction terms:

```
       Original terms:    A, B, C, D
        Squared terms:    A², B², C², D²
    Interaction terms:    A*B, A*C, A*D, B*C, B*D, C*D
```

With 19 features to choose from in the Ames housing dataset, PolynomialFeatures (with degree = 2) will generate a total of 210 features. By eliminating the duplicate columns, and the intercept column, 185 features are left to build the regression model.

4.2 Visualizing Performance with a Fitting Graph

Figure 4 shows a fitting graph that displays how RMSE varies, using forward selection, from 1 through 185 features. See the blue line in Fig. 4, which shows RMSE calculated on out-of-sample data (labeled holdout data in Fig. 4) using k-fold cross-validation. From Fig. 4, you can see that the fitting graph has the familiar bowl shape: The RMSE first decreases, then reaches a minimum when number of features is 65; and after that increases, representing model overfit. Finding the right number of features to select in forward selection involves finding the point on the fitting graph where the minimum occurs. You will also notice in Fig. 4 that I have included, for comparative purposes, a second orange line, which represents RMSE on the training data. As expected, the RMSE on the training data will be lower than the RMSE calculated on holdout data and will be monotonically decreasing as you increase the number of features to 185. In general, you should not use RMSE that has been calculated on training data, as it will produce overoptimistic results, or underestimated RMSE, as the orange line in Fig. 4 illustrates.

When considering model simplicity as a criterion for model selection, the fitting graph (blue line) in Fig. 4 suggests that you may not want to select the 65-feature model because model performance does not improve that much between 20 and 65 features. The 20-feature model will generate results similar to that of the 65-feature model. Roughly, from 20 to 130 features, the holdout RMSE does not change that much. From 130 onwards, with more and more features added to the model, performance starts to degrade more significantly, as model overfit becomes more pronounced. After 170 features, performance degrades very sharply: Model overfit is most dramatic in the upper extreme (right) end of the holdout curve.

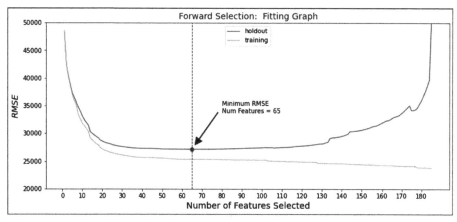

Fig. 4. Fitting Graph of the polynomial regression model (degree = 2) on the Ames housing dataset. The total number of features is 185. The top blue line represents the error on the out-of-sample holdout data, while the bottom line represents the error on the training data. (Color figure online)

What are the 20 features that contributed most to predictive accuracy using forward selection? Here are the twenty most important features, as indicated by the order in which they were added during forward selection:

```
 1. OverallQual
 2. GrLivArea CentralAir_Y
 3. TotalBsmtSF
 4. OverallQual^2
 5. GarageCars
 6. OverallQual GarageCars
 7. CentralAir_Y
 8. LotArea SaleCondition_Normal
 9. TotalBsmtSF LotArea
10. OverallQual LotArea
11. Age CentralAir_Y
12. OverallCond
13. GarageCars SaleCondition_Partial
14. BedroomAbvGr CentralAir_Y
15. GrLivArea LotArea
16. GrLivArea GarageCars
17. BedroomAbvGr
18. OverallQual KitchenQual_Gd
19. GrLivArea KitchenQual_TA
20. GrLivArea
```

The 20-feature degree-2 regression model, as you can see above, uses several interaction terms, but only one squared term (OverallQual^2). This regression model results in an RMSE of 28,194. This is a significant improvement over the 14-feature degree-1 regression model, described in Sect. 3.1, which had an RMSE of 34,647: an 18.6% reduction in error. The 65-feature degree-2 regression model, which represents the minimum RMSE, as seen in Fig. 4, resulted in an RMSE of 27,150. This is only marginally better than the 20-feature degree-2 model, but requires a considerably more complex model, or more than three times as many features (20 vs. 65).

5 Discussion

In the preceding discussion I illustrated and demonstrated two applications of regression modeling. First, the more traditional approach performs statistical inference. Its primary objective is to understand the relationships between the independent variables and the dependent variable. Second, the machine learning approach is focused on prediction, or finding the regression model that makes the most accurate predictions. A key insight is that while model explainability is a fundamental and critical aspect of building a regression model in statistical inference, it is oftentimes overlooked as an objective in machine learning. In machine learning, it is more important to devise a regression model that makes the best predictions: In fact, many of the best regression models are complex and hard to interpret, because they may involve combinations of the variables, including higher-order terms and interaction effects. I illustrated such a model in Sect. 4.

The two different approaches to regression modeling represent two different paradigms of model building. Table 1 summarizes how the two approaches differ in terms of model interpretability, sample size, number of features, additivity (i.e., the presence of interaction effects), and uncertainty (i.e., testing whether the relationships are statistically significant or not).

Although far less important in machine learning than in statistical inference, explainability has re-emerged as an important issue in the development of machine learning algorithms. The ability to understand a regression model has become important for many end-users who may be tasked to understand when a regression model does not work as expected, or when the model fails. A related issue to explainability is trust: How can we trust an AI system if we do not understand how it works? As Varshney [21] notes, it is not enough for AI to demonstrate competence in performing its task; the system should also be capable of explaining its actions so that is more comprehensible to all stakeholders across the organization, whether the executives sponsoring the AI project, the designers who must build the AI models, the developers who are tasked to implement the system and deploy it into the organization, or the end-users who may use the AI system to help make them make better and quicker decisions. Finally, it is important for the system to be aligned to broader organizational and societal goals. For example, this might include safeguards to ensure that an AI system is acting in an ethical and socially responsible way. This paper describes some of the steps that a regression modeler can take to make the model more transparent for everyone in the organization, and this, ultimately, can go a long way toward fostering more trust in the system.

Table 1. Comparing Statistical Inference to Machine Learning on Regression Modeling

	Statistical Inference	Machine Learning
Objective	Understand the relationships between a dependent variable and a set of independent variables	Find a regression model that most accurately predicts a continuous-valued numeric outcome
Model Interpretability	Important	Not as important
Sample Size	Tends to be small	Can be large
Number of features	Tends to be small	Can be large
Additivity	Additivity is the primary way that predictors affect the outcome	Non-additivity (interaction effects) are predicted to be strong
Uncertainty	Hypothesis testing is performed to understand whether model effects are significant	No hypothesis testing performed

References

1. James, G., Witten, D., Hastie, T., Tibshirani, R.: An introduction to statistical learning. Springer, New York (2013)
2. Allison, P.D.: Multiple regression: A primer. Pine Forge Press, Thousand Oaks, CA (1999)
3. Montgomery, D.C., Peck, E.A., Vining, G.G.: Introduction to linear regression analysis. John Wiley & Sons, Hoboken, NJ (2021)
4. Kuhn, M., Johnson, K.: Applied predictive modeling. Springer, New York (2013)
5. Confalonieri, R., Coba, L., Wagner, B., Besold, T.R.: A historical perspective of explainable artificial intelligence. Wiley Interdisciplinary Rev. Data Mining Knowl. Dis. **11**(1), e1391 (2021)
6. Burkart, N., Huber, M.F.: A survey on the explainability of supervised machine learning. J. Artifi. Intell. Res. **70**, 245–317 (2021)
7. Angelov, P.P., Soares, E.A., Jiang, R., Arnold, N.I., Atkinson, P.M.: Explainable artificial intelligence: an analytical review. Wiley Interdisciplinary Rev, Data Mining Knowl. Dis. **11**(5), e1424 (2021)
8. Tjoa, E., Guan, C.: A survey on explainable artificial intelligence (xai): toward medical xai. IEEE Trans. Neural Netw. Learn. Syst. **32**(11), 4793–4813 (2020)
9. Buchanan, B.G., Shortliffe, E.H.: Rule-based expert systems: the MYCIN experiments of the Stanford heuristic programming project. Addison-Wesley, (1984)
10. Swartout, W.R.: XPLAIN: A system for creating and explaining expert consulting programs. Artif. Intell. **21**(3), 285–325 (1983)
11. Chandrasekaran, B., Tanner, M.C., Josephson, J.R.: Explaining control strategies in problem solving. IEEE Intell. Syst. **4**(1), 9–15 (1989)
12. De Cock, D.: Ames, Iowa: Alternative to the Boston housing data as an end of semester regression project. J. Statist. Educ. **19**(3) (2011)
13. https://www.kaggle.com/datasets/marcopale/housing
14. Provost, F., Fawcett, T.: Data science for business: what you need to know about data mining and data-analytic thinking. Sebastopol, CA: O'Reilly Media (2013)

15. Nakatsu, R.T.: Information visualizations used to avoid the problem of overfitting in supervised machine learning. In: Nah, F.H., Tan, C.H. (eds.) HCIBGO 2017. LNCS, vol. 10294, pp. 373–385. Springer International Publishing (2017). https://doi.org/10.1007/978-3-319-58484-3_29
16. Stone, M.: Cross-validatory choice and assessment of statistical predictions. J. Roy. Stat. Soc.: Ser. B (Methodol.) **36**(2), 111–133 (1974)
17. Geisser, S.: The predictive sample reuse method with applications. J. Am. Stat. Assoc. **70**(350), 320–328 (1975)
18. Molinaro, A.M., Simon, R., Pfeiffer, R.M.: Prediction error estimation: a comparison of resampling methods. Bioinformatics **21**(15), 3301–3307 (2005)
19. Nakatsu, R.T.: An evaluation of four resampling methods used in machine learning classification. IEEE Intell. Syst. **36**(3), 51–57 (2021)
20. Nakatsu, R.T.: Validation of machine learning ridge regression models using Monte Carlo, bootstrap, and variations in cross-validation. J. Intell. Syst. **32**(1), 1–18 (2023)
21. Varshney, K.R.: Trustworthy machine learning. Independently Published, Chappaqua, NY (2022)

Designing for AI Transparency in Public Services: A User-Centred Study of Citizens' Preferences

Stefan Schmager[1], Samrat Gupta[1,2(✉)], Ilias Pappas[1,3],
and Polyxeni Vassilakopoulou[1]

[1] Department of Information Systems, University of Agder, Kristiansand, Norway
samrat.gupta@uia.no
[2] Information Systems Area, Indian Institute of Management Ahmedabad, Ahmedabad, India
[3] Department of Computer Science, Norwegian University of Science and Technology,
Trondheim, Norway

Abstract. Enhancing transparency in AI enabled public services has the potential to improve their adoption and service delivery. Hence, it is important to identify effective design strategies for AI transparency in public services. To this end, we conduct this empirical qualitative study providing insights for responsible deployment of AI in practice by public organizations. We design an interactive prototype for a Norwegian public welfare service organization which aims to use AI to support sick leaves related services. Qualitative analysis of citizens' data collected through survey, think-aloud interactions with the prototype, and open-ended questions revealed three key themes related to: articulating information in written form, representing information in graphical form, and establishing the appropriate level of information detail for improving AI transparency in public service delivery. This study advances research pertaining to design of public service portals and has implications for AI implementation in the public sector.

Keywords: Transparency · Human-Centered AI · Action Design Research · Design Preferences

1 Introduction

Artificial intelligence (AI) has advanced rapidly since its inception in the mid-1940s [1, 2]. In the public sector context, there are several examples of successful AI deployments. For instance, AI helps tax authorities in classifying individuals and corporate taxpayers to customize services and prevent fraud, predicting the arrival, departure and delays in public transport for commuters, pinpointing locations for police department in which they should concentrate their efforts in to deter crime, and immigration authorities to identify potential red flags and recommend immigration applications that should be approved [3–6]. There has also been some criticism of the use of AI applications in public services. For instance, concerns regarding prejudice and discrimination were raised when the Austrian labor administration established a system to classify job searchers according

F. F.-H. Nah and K. L. Siau (Eds.): HCII 2024, LNCS 14720, pp. 237–253, 2024.
https://doi.org/10.1007/978-3-031-61315-9_17

to their chance of landing a position [7]. AI is envisaged to change society in both favorable and unfavorable ways, whether it be through bettering healthcare outcomes or streamlining supply chains [1, 8]. Globally, public service organizations are competing to be the first to capitalize on the AI hype and, in the process, are attempting to establish themselves as trustworthy, transparent and open AI stakeholders [1, 9–11].

Transparency is one of the important cultural values, particularly for Nordic countries [12, 13]. Transparency in AI caters to the right and ability to obtain information about the use of AI in organizations [1]. Transparency can take many forms, but among its many elements is the availability of data regarding internal operations and performance while improving the ability of outside entities to monitor decisions or actions occurring within that organization [14]. AI transparency in public service organizations is deemed highly important by numerous actors engaged with elucidating AI processes [1, 15]. Lack of transparency in the public services is often linked with dysfunctional government, poor decision-making, lack of accountability of public officials, social media induced polarization and spread of misinformation [10, 16–18]. It is generally believed that transparency is vital for maintaining a balance of power between the public service authorities and citizens thereby increasing the likelihood of exposing wrongdoings or abuse of power [19].

A common mechanism to promote transparency is providing access to data through Application Programming Interfaces (APIs), dedicated apps and portals [20]. Strictly regulated operations on such data are enabled through an interface for analyzing, visualizing, and exploring by using data-driven algorithms and big data technologies [10]. However, achieving full AI transparency in public services and its practical realization is challenging due to several reasons [19]. First, creating transparency from public services data is hampered by numerous socio-technical constraints, so simply opening the data is insufficient [21, 22]. Second, public services data portals may reinforce the opinions of their designers through the presentation of selected and aggregated data, thereby stifling the range of perspectives of a diverse society [10, 23]. Given the challenges, research advancements can be helpful for establishing transparency in rendering public services. To this end, we seek to address the following research question:

What are the effective design strategies for AI transparency in public services?

In doing so, we consider the context of a Norwegian public services organization that plays a key role in Norwegian government's public administration in managing different types of citizens' benefits. This study specifically focusses on one of the organization's several AI initiatives which is to develop a model to predict the length of sick leaves. Such a model can serve as an additional resource for case handlers assisting them in concentrating their efforts towards the goal of providing effective services intended for "user-adapted follow-up" [3]. Such a context is especially befitting to our study because Nordic countries are known for their innovative approaches to ensuring the transparent functioning of government, and they are highly regarded within the European Union for their high standards of transparency [1, 11]. This study contributes to the evolving discourse on human AI interaction and citizens' needs for AI transparency in the context of public services [8, 24, 25].

This article is structured as follows. Section 2 covers the related literature and background of this study while elaborating more on the role and importance of AI transparency in rendering public services. The next section discusses the methodology employed to conduct this research thereby explaining the details pertaining to action design research and data collection. Section 4 presents an overview of the design of AI prototype used for this study. Subsequently, we discuss the findings of this study. A discussion of implications and limitations of this study is carried out in Sect. 6. Finally, Sect. 7 concludes this study.

2 Background

Transparency in public services is holistically defined as "the availability of information about an organization or actor that allows external actors to monitor the internal workings or performance of that organization" [14]. As such, ensuring transparency in rendering public services through AI has the potential to improve quality, enable personalization, ensure informed decision-making and more efficient use of resources in public service delivery [9, 26, 27]. Although, the awareness about ethical importance of algorithms and AI has been growing, the general public is unaware of even the most fundamental aspects of AI [28]. There are initiatives in place to make AI more understandable, such as a national online AI course of Finland which inspired a global course [1]. However, a lot needs to be done to overcome the misconceptions and reservations about the operation of AI. In the past, citizens' concerns have led to termination of several public service AI initiatives after their inception [29]. One such example pertains to System Risk Indication (SyRI) which was a Dutch initiative for profiling citizens to detect fraud with social services [30]. The Court found that the SyRI Act is in violation of the European Convention on Human Rights (ECHR) thus calling for ceasing the further use of SyRI [31]. The plaintiffs contended that because of a lack of transparency, the citizens are unaware of the risk model or algorithm that was employed and therefore unable to defend themselves against decisions made using SyRI [30]. Another such example relates to a lawsuit filed by a number of Brazilian legal organizations demanding a halt on the use of face recognition technology for public safety[1]. They claimed that the technology violated citizens' fundamental rights to free speech, assembly, and association, and would worsen issues such as inequality and structural racism. Therefore, it is crucial for citizens to know that AI is used in a transparent and accountable manner upholding all established rules for public service delivery [9, 32]. National strategic policies for AI strengthen notions of openness and trust while fostering sociotechnical understanding and resulting in a deeper comprehension of AI transparency [1].

2.1 Role of National Culture in AI Transparency

Though several cultures share the same values of transparency, trust, openness, accountability, and fairness, they do not give those values the same priority [33]. These variations may cause AI policies from various nations to enshrine the same values in different ways [1]. Moreover, with inherent economic, moral, and socio-cultural differences between

---

[1] https://www.aljazeera.com/economy/2023/7/13/facial-recognition-surveillance-in-sao-paulo-could-worsen-racism.

countries and regions, perceptions on how human-AI interactions should operate do not follow a common mindset [34]. For example, preference disparities in interface design were found when older adult East Asians and Caucasians were compared to two distinct interface designs [35]. As such, various governments (primarily in Nordic countries) have started to create national strategic policies for AI. It is essential to understand the socio-cultural shifts and how they should be maintained in a nation's strategic AI policy [36]. However, researchers and practitioners who can inform respective governments about the national outlook on AI are themselves grappling with its understanding due to scarcity of appropriate research studies in this field.

The concept of transparency has been frequently debated in the field of AI and Human Computer Interaction (HCI) [14, 37]. Proponents of transparency stress that transparency fosters understanding, a tighter bond between citizens, and a better understanding of public services [12]. Thus, a number of authors have contended that the public's lack of faith in public services is mainly because they are not provided with real information regarding the operations and efficacy of public services [38]. On the flip side, critics of transparency argue that showing the outcomes of public services may not actually boost citizens' trust [39]. They contend that transparency could possibly "delegitimize" the administration [14]. They assert that, since transparency requires users who can process it to function and most of the material released by public service portals is too difficult for even professionals to understand, transparency in AI just increases ambiguity and lack of trust [40, 41].

Despite contrasting viewpoints about AI transparency in public services, governments across the world are jockeying to make their public services more transparent. However, they come across different types of barriers in the design of open data applications and portals [42]. These barriers can be classified as data quality barriers, technical barriers, economic barriers, organizational barriers, ethical barriers, human barriers, political barriers, and usage barriers [10].

2.2 Norwegian Guidelines on AI Transparency

Transparency is highlighted as a key value in AI development in the Norwegian policy guide, which describes the country's national strategies for AI [34]. This guide also makes serious efforts to promote and practice transparency in its discussions of AI [43]. As compared to other national policy documents which provide truncated technical overviews, the first few pages of Norwegian policy guide explain AI technologies in detail thus providing any stakeholder fundamental AI literacy [1]. Furthermore, the Norwegian guide recognizes that "lack of transparency" is an issue when citizens attempt to access AI-based public services [1]. The policy states that in order to minimize burdens on citizens, improve access to services, and eventually enhance perceptions of transparency, "the Government has established a 'once only' principle to ensure that citizens and businesses do not have to provide the same information to multiple public bodies" [44].

This study responds to recent calls for research citizens' preferences and AI adoption in public services [26, 45–47].

3   Methodology

3.1   Research Context

This study is conducted in collaboration with a large Norwegian public welfare service organization responsible for different kinds of social benefits. The organization has established a specialized team to explore the potential of AI and machine learning technology in improving the efficiency, robustness, and quality of public services. One direction for the use of AI systems the team explores is to reduce administrative burdens for citizens and employees by improving the use of time and resources. The organization has developed a predictive model for estimating the probable duration of people being absent from their work due to sickness. The model is envisioned to be used as a decision support system for case handlers to decide whether a legally mandated meeting to discuss additional support for the person on sick leave will be necessary. These meetings can be resource-heavy and involve many actors, but Norwegian law allows for them to be omitted if deemed not to be necessary.

3.2   Action Design Research

The study employs the Action Design Research (ADR) methodology, which is a collaborative research approach that allows researchers to work closely with organizations to solve real-world problems [48]. The ADR methodology involves multiple iterative "build-intervene-evaluate" (BIE) cycles, where both researchers and stakeholders jointly shape the research artefact. We held bi-weekly collaboration sessions with the organizational partner to discuss design choices, provide feedback, and plan for the next iterations of the prototype. The emerging artefact then served as a tool to explore and refine the practical application of AI in public service, ensuring that the resulting systems are not only technologically advanced but also ethically aligned and human-centric [49]. Consistent with ADR, we reflected on the problem understanding, theories, and emerging designs, evaluating potential knowledge contributions to the field.

3.3   Data Collection and Analysis

We recruited 28 individuals from the age range of 18–65 to mirror the demographic distribution of Norwegians on sick leave, as reported by the central statistics office in Norway (2022). Because the study focused on understanding specific user perspectives in a given context through rich qualitative data, the insights of 28 participants were found to be sufficient in providing diverse and in-depth empirical data [50–52]. The data collection included three consecutive parts. First, participants provided general demographic information and self-assessed their prior AI knowledge and technology use frequency. Next, a moderated task-based user study was conducted, where participants interacted with the interactive prototype and made decisions based on a scenario. The scenario led to a decision point at which they were asked whether they would consent to the use of the AI-supported sick leave duration prediction system. Participants were asked to think aloud while interacting with the prototype, which helped to capture thoughts and feelings better and provided insights into reasoning [53] Finally, participants rated their agreement with a set of statements related to the scenario and answered open-ended questions about their decision-making processes. The study followed ethical standards and legal regulations and was approved by the Norwegian Center for Research Data (NSD).

The collected data was then categorized into three types: scale ratings, verbal feedback, and interactions with the interactive prototype. Initially, each data type was examined independently before being integrated with the other two. In the first step, we analyzed participants' responses to predefined questions. We compiled the scale ratings into a spreadsheet, which provided a comprehensive view of the data, and revealed diversity, trends, and allowed us to track any shifts in opinions throughout the study. We then undertook an exhaustive analysis of the qualitative data, including verbal feedback from the think-aloud recordings and responses to open-ended questions to different explanation types. This involved multiple reviews of the study recordings and a detailed examination of interview transcripts. During this process, we highlighted notable incidents, quotes, and expressions from participants, and engaged in in-depth discussions within our research team to collaboratively interpret these insights. Finally, we analyzed how participants interacted with the prototype, drawing from session recordings and notes. We focused on identifying which information elements were accessed, how participants engaged with these elements and the time spent on them. By integrating this analysis with the insights from the earlier stages, we aimed to enhance our overall examination and identify emerging concepts and recurrent themes, cross-referencing observations and findings across the different data categories.

4   Prototype Development

To evaluate the different types of explanations and information in a naturalistic context, we developed an ensemble artefact, embodied in an interactive prototype [54]. The prototype mimics a public service portal, depicting a specific interaction sequence of users being asked whether they would consent to the optional use of an AI-based prediction system. The development of our prototype was informed by the Google PAIR framework, which is a comprehensive collection of guidelines formalizing Google's insights from industry and academia[2]. Google's PAIR framework is the most comprehensive and balanced when comparing different Human-Centered AI frameworks [55]. Auernhammer, 2020 emphasizes that it is important to embed technologies within a user-centric interface that aligns with users' workflows and cognitive models, beyond just crafting a robust algorithm or machine-learning model [56].

4.1   Information Elements

The prototype replicates a public service portal, presenting a scenario where users are asked about their willingness to consent to the use of an AI-based prediction system. The scenario starts with a standard portal notification that directs users to the introduction page of the AI prediction tool. Consistent with the PAIR framework's first chapter (User

---

[2] https://pair.withgoogle.com/guidebook/chapters.

Needs + Defining Success), it's crucial to identify areas where AI's capabilities align with user requirements. Ehsan & Riedl, 2020 emphasize that for AI systems to be considered trustworthy and ethically sound, explanations must be clear enough to enable users to make informed judgments based on these understandings [57]. In the specific public service context, it's also vital to consider and convey the rights and responsibilities of all parties involved [58]. For our case, this includes explaining the government's mandate on mandatory dialogue meetings, the possibility of waiving these meetings, and the citizen's right to dissent to the use of the AI system. This information is provided in a textual format (Fig. 1). Furthermore, the information includes the collective benefit derived from consenting to use AI in the particular context. This entails addressing considerations of social ethics rather than imposing moral obligations on citizens.

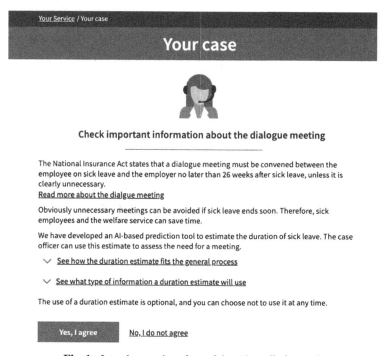

Fig. 1. Introductory interface of the AI prediction tool

To enhance transparency regarding the process, data usage, and the model's functionality, participants were allowed to explore different information elements and descriptions of the AI prediction tool. One such element is a process diagram, designed to illustrate the AI's role within the overall process and emphasize the case handler's critical involvement (Fig. 2). As per Chapter 3 of the PAIR guidelines (Mental Models), such visual explanations are vital in setting realistic expectations about the role of an AI system, its capabilities, limitations, and the value it offers (Google PAIR, 2022). The diagram shows the AI system within the context of the full process flow and delineates areas where AI supplements human decision-making. It also depicts the procedure in the absence of the AI-enhanced prediction tool, in case a citizen would not consent to its use. The chart aims to provide a comprehensive view of AI's role in the process, thereby fostering transparency and a deeper understanding, and enabling citizens to also better understand their rights and role in the process.

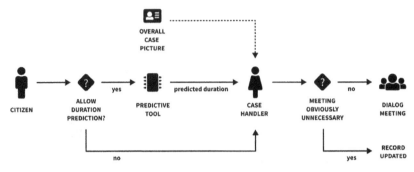

Fig. 2. Process visualization

The prototype also includes a data table to provide information about the data used by the AI model and the rationale (Fig. 3). This element is based on the PAIR recommendations in Chapter 2, which emphasizes the importance of explaining the usage and benefits of specific data to users. The data table consists of an introductory paragraph clarifying that no new data will be collected, but existing data reused. The table itself consists of two columns: one for the type of data and the other for why it's necessary. This transparent approach to data aims to enhance participants' understanding of the system and provide insights into the personal data used.

INFORMATION ELEMENT	NEED
Start date of sick leave	Calculation of duration of the sick leave.
Change in degree in the sickness	Identify degree changes in sickness to detect improvements or deterioration
Diagnosis and co-diagnosis, and change in diagnose / co-diagnose	Different types of diagnoses / co-diagnoses are associated with short / long sick leave durations.
Gender and age of sick leaver	Individal age and gender are associated with sick leave duration
Sick leave history	Relationship between previous sickness absence and current sickness absence.
Unique ID of attesting doctor	Variations in sick leave practice affects sick leave.

Fig. 3. Information about the data used by the AI prediction tool

Lastly, we integrated an interactive chart showing Shapley Values for feature importance visualization (Fig. 4) [59]. This chart highlights the importance of various variables across different age groups, providing a visual representation of the model's logic. This enhances exploratory transparency and helps citizens adjust their trust in the system

[58]. This integration aligns with Chapter 4 of the PAIR framework (Explainability + Trust), which states that while providing detailed explanations of the model's logic may not always be possible, the reasoning behind predictions can sometimes be made accessible. The interactive feature importance chart aims to fill this gap by presenting users with a visually engaging way to comprehend the model's inner workings.

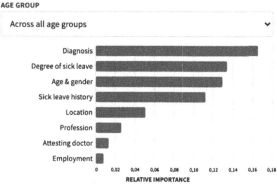

Fig. 4. Feature importance chart

At the end of the interaction sequence, citizens reach a consent decision point where they must decide whether to agree to the use of AI or to opt-out (Fig. 5). This emphasizes the voluntary nature of their agreement and the option to not provide consent. Consistent with Chapter 5 (Feedback + Control) of the PAIR framework, it's essential to convey citizens' control over their data usage and the decisions they make.

Fig. 5. Consent decision point

The chapter also stresses the importance of feedback mechanisms, which can help refine the AI model's performance and user experience. While opting out can be considered implicit feedback, this stage also presents an opportunity to request more detailed feedback from citizens. Within the prototype, we encouraged users to express their opinions and objections (Fig. 6). This not only facilitates the direct participation of citizens but also upholds their right to object to personal data usage without negative consequences or justification requirements. This approach ensures transparency and upholds the principle that the right to object is an inherent right of citizens within the process.

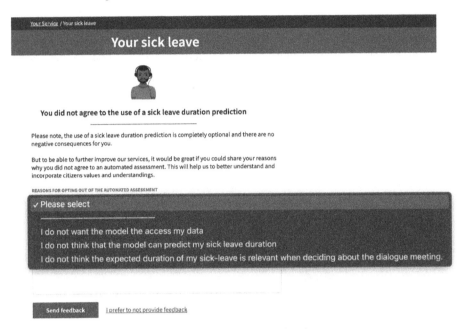

Fig. 6. Feedback form in case of opting out

5 Findings

The participants of the study were generally positive about the use of AI in public service, as indicated by their responses when asked about their level of comfort with the use of AI. Before interacting with the prototype, around 53% of the participants reported feeling comfortable with the general use of AI in public services, while 30% were unsure or declined to say, and 17% felt uneasy about the idea. These initial responses suggest a mix of cautiousness and openness towards AI in the public sector, with some participants expressing concerns about the potential risks and uncertainties associated with AI use. However, after interacting with the prototype, the participants' outlook became more positive overall. 65% stated feeling at ease, 23% remained uncertain or preferred not to answer, and only 12% maintained reservations. This slight shift in the level of comfort suggests that the interactive experience provided by the prototype helped address some of the concerns and uncertainties that participants had about AI in public services. To further unpack how these interactions might have influenced participants' stances on the use of AI in public service, we also analyzed the verbalized comments from the think-aloud protocol as well as the actual interactions with the prototype. This analysis revealed three key themes: a) Articulating Information in Written Form, b) Representing Information in Graphical Form, and c) Establishing the Appropriate Level of Information Detail.

5.1 Articulating Information in Written Form

The first theme emphasizes the importance of copywriting in AI interfaces. Participants carefully evaluated the textual explanation used in the system. While the feedback on the provided explanations was generally positive, several participants expressed comments suggesting that the explanations as well as other interface text were still unclear. There were multiple remarks about challenges in comprehending the textual content, with some participants suggesting that the language used was too technical or complex. Participants stated: *"There were some words I didn't understand"* and *"I think that here it needs to be made much easier. Sit down really. It's not designed for a simple user"*. Some participants connected their comments about the language to the organization. One participant expressed: *"It's their terminology that I think is... is a bit wrong... or I think a lot of people will stop here and really need help."* Another participant remarked: *"That's kind of a typical [name of organization] description. [...] They can write it a little easier, but I understand what's there."* This underscores the critical role of clear and understandable communication in AI-driven public services and highlights the nuanced role of language in shaping user experiences and trust in AI systems.

5.2 Representing Information in Graphical Form

The second theme identified in our data analysis is centered around the visual elements used in the prototype, specifically focusing on the process visualization (Fig. 2) and the feature importance charts (Fig. 4). Our analysis delved into which graphical representations captured the participants' attention and their subsequent feedback on these visual elements. Upon examining the engagement with this visualization, we found that different graphical representations resonated differently with users, influenced significantly by their backgrounds. For instance, the process visualization was particularly well-received by participants who had familiarity with such representations in their professional contexts. These participants stated: *"I really love this"* and *"I love to see how the process is with and without the tool"*. At the same time, this type of visualization was difficult to comprehend and puzzled users who had no prior familiarity with such representations. Some users expressed their general favor for visual explanations but still had difficulties understanding the visualization: *"I found it exciting that it was explained with pictures or icons, but I also found it a bit confusing"*. Others stated that the provided visualization did not match their personal Mental Model of the process: *"I don't quite feel my understanding of the situation is similar to that on the process diagram then."* However, some participants stated that they did not understand the process visualization at all: *"Hard to figure out the chart"* and *"I don't understand anything about this"*.

The second graphical information in the prototype was an interactive feature importance chart (Fig. 4). Despite aiming to provide more transparency into the inner workings of the sick-leave prediction model, the feature importance chart was found to be too difficult to comprehend by participants, describing it as *"too academic"*, or it was unclear what was the underlying reason for including it: *"Yes, I understand what it shows. But I don't understand why it shows it really."*. The findings hint towards the necessity of considering the diversity of potential prior experience with certain forms of visualizations, as well as incorporating diverse graphical representations to accommodate different citizen groups' needs. The right combination of visual approaches tailored to diverse users is key to building transparent, intuitive interfaces that help citizens of all backgrounds interact successfully with AI systems.

5.3   Establishing the Appropriate Level of Information Detail

The third theme identified in our analysis explores the optimal level of detail for the information presented to citizens, particularly concerning the data table (Fig. 3) as well as textual information. The data table provided a comprehensive overview of the data utilized by the sick leave prediction model, including explanations for the necessity of each data type. Several participants perceived the level of detail as overly extensive, describing it as "*too much to read*" or "*Way too much text*". However, a few participants appreciated the availability of detailed information, delving into it to thoroughly grasp the scope and reason for the data repurposing. One participant noted, "*I think it is useful that it says what the purpose of collecting information is, and what kind of information is collected.*" However, despite not fully reading the details, many participants acknowledged the value of having this information available. As one participant expressed: "*I thought it was really nice. Even though I haven't read it very thoroughly right now, I thought it was really good to know exactly why they're using it and that information and what it's being used for. [...] And then I get some confidence in understanding that that's why and that they're not just gathering any information and kind of that it's to give rise to something.*"

The level of detail in the provided information was also brought up in other places. One other instance was the explanation of the governmental mandate for dialogue meetings. One participant remarked that: "*I would rather have said that it is required by law, not necessarily the National Insurance Act. In other words, for someone who has nothing to do with the National Insurance Act, it is not so important which law it is.*" But the same participant also appreciated the availability of more information if desired, stating that: "*[...] if you don't know what it entails then it's nice that you can read a little more about what it entails.*" The overarching insight from this theme is that citizens appreciate the presence of detailed information, even if they do not engage with it extensively. The availability of thorough explanations contributed to a sense of ease among participants, illustrating that the mere provision of detailed explanations can foster trust in AI systems. Leveraging progressive disclosure, which presents essential information first and then elaborates on more detailed aspects, can help maintain focus on the primary objectives. This approach is about empowering citizens and fostering confidence through a commitment to openness and accountability.

6   Discussion and Contributions

As evident from our findings, in order to enhance transparency and drive citizens' engagement with AI enabled public services, there is a need to restructure information elements and realign content according to preferences of citizens. To generate trust, citizens must play active role in using public services and provide actionable feedback about their perceptions of the use of AI in these services. AI capabilities in public services can advance the creation of public value in the form of trust, flexibility, adaptability, and responsiveness [26]. One of the most significant obstacles that public services must overcome is driving engagement with their stakeholders amidst growing speculations about responsible use of AI [10]. In order to ensure that AI policy initiatives are accurate and pertinent, it is imperative that public offices increase the level of awareness and involvement with citizens. Thus, evaluating the preferences of the public can promote continued cooperation between all stakeholders, accelerating the flow of information and the ensuing chances for further development of public AI based interfaces.

6.1 Contribution to Research

The findings of this study underscore the link between the government and the citizens thereby broadening the scope of current research on adoption of AI in the public sector. The three themes identified in this study emphasize the critical role of clear and comprehensible communication in AI-enabled public services. Educating the public about the features of AI systems and the role that humans play in decision-making significantly improves public perceptions of AI, which is crucial for its adoption. Our research goes beyond earlier conceptual and review-based studies [1, 10, 15, 19, 60] by being able to assess individual preferences and perceptions about AI transparency. The information citizens receive strengthens their faith in the public service portals, and having a human in the loop gives them confidence that they can clarify their unique situation in an open, accountable and trustworthy manner.

6.2 Contribution to Practice

This study sheds light on how the interactive engagement with the AI prototype enhances its positive perception, suggesting that structured information and hands-on engagement with AI can positively shape public opinion about transparency and trust in public services. The right combination of comprehensible communication, visual approaches, and requisite information availability is the key to building transparent, intuitive interfaces that help citizens of all backgrounds interact successfully with AI enabled public services. Public entities may find it easier to provide their services efficiently if citizen's trust is enhanced and they derive a value from usage of AI in public services. Through the identification of characteristics that contribute to a positive attitude toward AI, we offer practical guidance to practitioners about the transparent design of AI systems within a public service setting.

6.3 Limitations and Future Research

Citizens' attitudes toward AI are influenced by their socio-cultural identity and, in particular, their general level of trust in the government, which is notably high in Norway (OECD 2022). Consequently, it is critical to carry out more targeted research with larger samples as well as studies in other sociocultural contexts. Secondly, in this study, we focus on citizens' preferences of transparency in AI based public services. Further research may delve into the interplay of key characteristics such as transparency, trust and openness of AI based public services. Trust is systemic in nature, embedded in the broader system of public and private entities connected to AI [1, 61]. As such, there is a fascinating opportunity for research partners from around the world for investigating and creating a framework for human-centered in public services.

7 Conclusion

This study provided valuable insights into how citizens perceive and interact with AI systems introduced in public services. The findings demonstrate that directly engaging users helps improve generally positive views by addressing concerns through informed experience. The three key themes that emerged from our analysis showed citizens have high standards for transparent communication. The implications of this research can help guide development of future citizen-facing AI to maximize benefits and allay worries through responsible, user-centered practices. Hence, it aims to contribute with practical insights in the extant body of research on normative guidelines for the responsible design and development of AI with the goal of human benefit [9, 26, 55, 62, 63]. Our research contributes to the broader understanding of public sentiment towards AI in public services and offers practical implications for the design of AI based systems in the public sector.

References

1. Robinson, S.C.: Trust, transparency, and openness: how inclusion of cultural values shapes Nordic national public policy strategies for artificial intelligence (AI). Technol. Soc. **63**, 101421 (2020)
2. Ulnicane, I., Knight, W., Leach, T., Stahl, B.C., Wanjiku, W.G.: Governance of Artificial Intelligence: Emerging International Trends and Policy Frames: The Global Politics of Artificial Intelligence. Taylor & Francis, Milton Park, Oxfordshire (2022)
3. Schmager, S., Grøder, C.H., Parmiggiani, E., Pappas, I., Vassilakopoulou, P.: What do citizens think of AI adoption in public services? Exploratory research on citizen attitudes through a social contract lens. In: Proceedings of the 56th Hawaii International Conference on System Sciences (2023)
4. Kuziemski, M., Misuraca, G.: AI governance in the public sector: three tales from the frontiers of automated decision-making in democratic settings. Telecommun. Policy **44**(6), 101976 (2020)
5. Höchtl, J., Parycek, P., Schöllhammer, R.: Big data in the policy cycle: policy decision making in the digital era. J. Organ. Comput. Electron. Commer. **26**(1–2), 147–169 (2016)
6. Molnar, P.: Technology on the margins: AI and global migration management from a human rights perspective. Camb. Int. Law J. **8**(2), 305–330 (2019)
7. Lopez, P.: Bias does not equal bias: a socio-technical typology of bias in data-based algorithmic systems. Internet Policy Rev. **10**(4), 1–29 (2021)
8. Zhang, P., Nah, F.F.H., Benbasat, I.: Human-computer interaction research in management information systems. J. Manag. Inf. Syst. **22**(3), 9–14 (2005)
9. Tiwari, A.A., Gupta, S., Zamani, E.D., Mittal, N., Agarwal, R.: An overarching conceptual framework for ICT-enabled responsive governance. Inf. Syst. Front. (2023). https://doi.org/10.1007/s10796-023-10415-4

10. Matheus, R., Janssen, M., Janowski, T.: Design principles for creating digital transparency in government. Gov. Inf. Q. **38**(1), 101550 (2021)
11. Persson, P., Zhang, Y., Asatiani, A., Lindman, J., Rudmark, D.: Toward citizen-centered digital government: design principles guided legacy system renewal in a Swedish municipality (2024)
12. Kassen, M.: Understanding transparency of government from a Nordic perspective: open government and open data movement as a multidimensional collaborative phenomenon in Sweden. J. Glob. Inf. Technol. Manag. **20**(4), 236–275 (2017)
13. Vidaver-Cohen, D., Brønn, P.S.: Reputation, responsibility, and stakeholder support in Scandinavian firms: a comparative analysis. J. Bus. Ethics **127**, 49–64 (2015)
14. Grimmelikhuijsen, S., Porumbescu, G., Hong, B., Im, T.: The effect of transparency on trust in government: a cross-national comparative experiment. Public Adm. Rev. **73**(4), 575–586 (2013)
15. Felzmann, H., Villaronga, E.F., Lutz, C., Tamò-Larrieux, A.: Transparency you can trust: transparency requirements for artificial intelligence between legal norms and contextual concerns. Big Data Soc. **6**(1), 2053951719860542 (2019)
16. Gupta, S., Jain, G., Tiwari, A.A.: Polarised social media discourse during COVID-19 pandemic: evidence from YouTube. Behav. Inf. Technol. **42**(2), 227–248 (2023)
17. Kaur, K., Gupta, S.: Towards dissemination, detection and combating misinformation on social media: a literature review. J. Bus. Ind. Mark. **38**(8), 1656–1674 (2023)
18. Qureshi, I., Bhatt, B., Gupta, S., Tiwari, A.A.: Introduction to the role of information and communication technologies in polarization. In: Qureshi, I., Bhatt, B., Gupta, S., Tiwari, A.A. (eds.) Causes and Symptoms of Socio-Cultural Polarization, pp. 1–23. Springer, Singapore (2022). https://doi.org/10.1007/978-981-16-5268-4_1
19. Fung, A.: Infotopia: unleashing the democratic power of transparency. Polit. Soc. **41**(2), 183–212 (2013)
20. Luna-Reyes, L.F., Bertot, J.C., Mellouli, S.: Open government, open data and digital government. Gov. Inf. Q. **31**(1), 4–5 (2014)
21. Conradie, P., Choenni, S.: On the barriers for local government releasing open data. Gov. Inf. Q. **31**, S10–S17 (2014)
22. Janssen, M., Charalabidis, Y., Zuiderwijk, A.: Benefits, adoption barriers and myths of open data and open government. Inf. Syst. Manag. **29**(4), 258–268 (2012)
23. Kitchin, R., Lauriault, T.P., McArdle, G.: Knowing and governing cities through urban indicators, city benchmarking and real-time dashboards. Reg. Stud. Reg. Sci. **2**(1), 6–28 (2015)
24. Saldanha, D.M.F., Dias, C.N., Guillaumon, S.: Transparency and accountability in digital public services: learning from the Brazilian cases. Gov. Inf. Q. **39**(2), 101680 (2022)
25. Vössing, M., Kühl, N., Lind, M., Satzger, G.: Designing transparency for effective human-AI collaboration. Inf. Syst. Front. **24**(3), 877–895 (2022)
26. Zamani, E.D., Smyth, C., Gupta, S., Dennehy, D.: Artificial intelligence and big data analytics for supply chain resilience: a systematic literature review. Ann. Oper. Res. **327**(2), 605–632 (2023)
27. Vassilakopoulou, P., Haug, A., Salvesen, L.M., Pappas, I.O.: Developing human/AI interactions for chat-based customer services: lessons learned from the Norwegian government. Eur. J. Inf. Syst. **32**(1), 10–22 (2023)
28. Kieslich, K., Keller, B., Starke, C.: Artificial intelligence ethics by design. Evaluating public perception on the importance of ethical design principles of artificial intelligence. Big Data & Soc. **9**(1), 20539517221092956 (2022)
29. van Veenstra, A.F., Grommé, F., Djafari, S.: The use of public sector data analytics in the Netherlands. Trans. Gov. People Process Policy **15**(4), 396–419 (2020)

30. Bekker, S.: Fundamental rights in digital welfare states: The case of SyRI in the Netherlands. In: Spijkers, O., Werner, W.G., Wessel, R.A. (eds.) Netherlands Yearbook of International Law 2019. NYIL, vol. 50, pp. 289–307. T.M.C. Asser Press, The Hague (2021). https://doi.org/10.1007/978-94-6265-403-7_24

31. Van Bekkum, M., Borgesius, F.Z.: Digital welfare fraud detection and the Dutch SyRI judgment. Eur. J. Soc. Secur. **23**(4), 323–340 (2021)

32. Henman, P.: Improving public services using artificial intelligence: possibilities, pitfalls, governance. Asia Pac. J Public Adm. **42**(4), 209–221 (2020)

33. Schwartz, S.: A theory of cultural value orientations: explication and applications. Comp. Sociol. **5**(2–3), 137–182 (2006)

34. van Berkel, N., Papachristos, E., Giachanou, A., Hosio, S., Skov, M.B.: A systematic assessment of national artificial intelligence policies: perspectives from the Nordics and beyond. In: Proceedings of the 11th Nordic conference on Human-Computer Interaction: Shaping Experiences, Shaping Society, pp. 1–12 (2020)

35. Haddad, S., McGrenere, J., Jacova, C.: Interface design for older adults with varying cultural attitudes toward uncertainty. In: Proceedings of the SIGCHI Conference on Human Factors in Computing Systems, pp. 1913–1922 (2014)

36. Kulesz, O.: Culture, platforms and machines: the impact of artificial intelligence on the diversity of cultural expressions. Intergovernmental Committee Prot. Promot. Diversity Cult. Expr. **12** (2018)

37. Abdul, A., Vermeulen, J., Wang, D., Lim, B.Y., Kankanhalli, M.: Trends and trajectories for explainable, accountable and intelligible systems: an HCI research agenda. In: Proceedings of the 2018 CHI Conference on Human Factors in Computing Systems (CHI '18). Association for Computing Machinery, New York, NY, USA, Article 582 (2018)

38. Cook, T.E., Gronke, P.: The skeptical American: revisiting the meanings of trust in government and confidence in institutions. J. Politics **67**(3), 784–803 (2005)

39. Bannister, F., Connolly, R.: The trouble with transparency: a critical review of openness in e-government. Policy Internet **3**(1), 1–30 (2011)

40. Heald, D.: Varieties of transparency. In: Heald, D. (ed.) Transparency: the key to better governance? British Academy (2006). https://doi.org/10.5871/bacad/9780197263839.003.0002

41. Im, T., Cho, W., Porumbescu, G., Park, J.: Internet, trust in government, and citizen compliance. J. Public Adm. Res. Theory **24**(3), 741–763 (2014)

42. Chen, C.P., Zhang, C.Y.: Data-intensive applications, challenges, techniques and technologies: a survey on Big Data. Inf. Sci. **275**, 314–347 (2014)

43. Dexe, J., Franke, U.: Nordic lights? National AI policies for doing well by doing good. J. Cyber Policy **5**(3), 332–349 (2020)

44. Stani, E., Barthélemy, F., Raes, K., Pittomvils, M., Rodriguez, M.A.: How data vocabulary standards enhance the exchange of information exposed through APIs: the case of public service descriptions. In: Proceedings of the 13th International Conference on Theory and Practice of Electronic Governance, pp. 807–810 (2020)

45. Asatiani, A., Malo, P., Nagbøl, P.R., Penttinen, E., Rinta-Kahila, T., Salovaara, A.: Sociotechnical envelopment of artificial intelligence: an approach to organizational deployment of inscrutable artificial intelligence systems. J. Assoc. Inf. Syst. (JAIS) **22**(2), 325–252 (2021)

46. Wirtz, B.W., Langer, P.F., Fenner, C.: Artificial intelligence in the public sector-a research agenda. Int. J. Public Adm. **44**(13), 1103–1128 (2021)

47. Mikalef, P., et al.: Enabling AI capabilities in government agencies: a study of determinants for European municipalities. Gov. Inf. Q. **39**(4), 101596 (2022)

48. Sein, M.K., Henfridsson, O., Purao, S., Rossi, M., Lindgren, R.: Action design research. MIS Q., 37-56 (2011).https://doi.org/10.2307/23043488

49. Schmager, S., Pappas, I., Vassilakopoulou, P.: Defining human-centered AI: a comprehensive review of HCAI literature. In: Proceedings of the 2023 Mediterranean Conference on Information Systems (2023)
50. Creswell, J.W., Poth, C.N.: Qualitative Inquiry and Research Design: Choosing Among Five Approaches. Sage publications. Thousand Oaks, California (2016)
51. Marshall, B., Cardon, P., Poddar, A., Fontenot, R.: Does sample size matter in qualitative research? A review of qualitative interviews in IS research. J. Comput. Inf. Syst. **54**(1), 11–22 (2013)
52. Faulkner, L.: Beyond the five-user assumption: benefits of increased sample sizes in usability testing. Behav. Res. Methods Instrum. Comput. **35**, 379–383 (2003)
53. Ericsson, K.A., Simon, H.A.: Verbal reports as data. Psychol. Rev. **87**(3), 215 (1980)
54. Miah, S.J., Gammack, J.G.: Ensemble artifact design for context sensitive decision support. Australas. J. Inf. Syst. **18**(2), 5–20 (2014)
55. Shneiderman, B.: Human-centered artificial intelligence: reliable, safe & trustworthy. Int. J. Hum. Comput. Interact. **36**(6), 495–504 (2020)
56. Auernhammer, J.: Human-centered AI: the role of Human-centered design research. In: Boess, S., Cheung, M. Cain, R. (eds.), The Development of AI Synergy - DRS International Conference (2020)
57. Ehsan, U., Riedl, M.O.: Human-centered explainable AI: towards a reflective sociotechnical approach. In: Stephanidis, C., Kurosu, M., Degen, H., Reinerman-Jones, L. (eds.) HCII 2020. LNCS, vol. 12424, pp. 449–466. Springer, Cham (2020). https://doi.org/10.1007/978-3-030-60117-1_33
58. Schmager, S.: From commercial agreements to the social contract: human-centered AI guidelines for public services. In: The 14th Mediterranean Conference on Information Systems (MCIS), Catanzaro, Italy (2022)
59. Lundberg, S.M., Lee, S.I.: A unified approach to interpreting model predictions. Adv. Neural Inf. Process. Syst. **30** (2017)
60. Vassilakopoulou, P.: Sociotechnical approach for accountability by design in AI systems. In: 2020 European Conference on Information Systems, ECIS (2020)
61. Thiebes, S., Lins, S., Sunyaev, A.: Trustworthy artificial intelligence. Electron. Mark. **31**, 447–464 (2021)
62. Vassilakopoulou, P., Parmiggiani, E, Shollo, A., Grisot, M.: Responsible AI: concepts, critical perspectives and an information systems research agenda. Scand. J. Inf. Syst. **34**(2) (2022)
63. Akbarighatar, P., Pappas, I., Vassilakopoulou, P.: A sociotechnical perspective for responsible AI maturity models: findings from a mixed-method literature review. Int. J. Inf. Manage. Data Insights **3**(2), 100193 (2023)

Sustainable E-commerce Marketplace: Reshaping Consumer Purchasing Behavior Through Generative AI (Artificial Intelligence)

Jung Joo Sohn[1]([✉]) [iD], Nickolas Guo[1] [iD], and Youri Chung[2]

[1] Purdue University, West Lafayette, IN, USA
{jjsohn,guo682}@purdue.edu
[2] Ewha Womans University, Seoul, Korea
yurichung@ewhain.net

Abstract. In a world marked by rapid technological advancements and a burgeoning consumer product market, addressing the critical aspects of sustainability, affordability, and user experience has become imperative. This paper details the development of *Fetch*, an innovative mobile application that seamlessly integrates sustainable e-commerce, AI-driven personalized recommendations, and efficient warranty management. By combining secondary research, primary research, development, and usability testing, the study delves into consumer behavior, attitudes toward pre-owned products, and perspectives on sustainability. The current mobile-centric focus lays the foundation for future expansions, including web browsers and other device applications.

Keywords: Artificial Intelligence (AI) · Human-Centered AI · Sustainability · Mobile products/services including mobile TV/video · Service Design · E-commerce · Sustainability · Generative AI · Warranty Management

1 Introduction

With the escalating trend of consumerism and its impact on the rapidly deteriorating predicament of our environment, the need to become more self-conscious of our carbon footprint and purchasing behavior has never been more crucial [1]. The world generates 2.01 billion tons of municipal solid waste annually, with at least 33% of that not managed in an environmentally safe manner [2]. The waste that is produced has significant negative impacts that all contribute to issues such as air pollution, water contamination, transmittable diseases, and infections [3]. In response to this environmental challenge, there is a growing recognition of the importance of sustainable practices, not only on an individual level but also on a societal and global scale. This means making informed purchasing choices that consider the long-term impact on the environment and prioritize the well-being of future generations. As consumers, one of the significant avenues to make a positive impact is through the conscious decision to reuse and recycle consumer products.

F. F.-H. Nah and K. L. Siau (Eds.): HCII 2024, LNCS 14720, pp. 254–269, 2024.
https://doi.org/10.1007/978-3-031-61315-9_18

While second-hand marketplaces like eBay, Facebook Marketplace, and Craigslist provide a commendable platform for buying and selling used items, there exists untapped potential for improving the user experience emphasizing convenience and efficiency to allow for a more widespread adoption of sustainable practices [4]. This is essential as we found that although people believe that sustainability is important, it is difficult to decide to sacrifice the convenience of simply discarding products rather than reselling and reusing second-hand products. Moreover, there is room for improvement in user engagement, education, and incentivization to foster a cultural shift towards embracing the benefits of second-hand commerce. By addressing these limitations and leveraging advanced technologies, we can elevate the effectiveness of such platforms and inspire greater participation in sustainable consumer practices.

This research paper details the development of a mobile application called Fetch that helps facilitate the exchange of second-hand consumer products through an AI-driven e-commerce marketplace. The marketplace is complemented by a robust warranty management system that actively tracks product warranties while also providing a seamless experience in transferring product information to the marketplace and listing it for sale.

2 Research Methodology

Our design process is based heavily on a research-driven approach where the research guides us to identify unique features, differentiation from competitors, and shape the overall design of our solution. We initiated a comprehensive research methodology encompassing both secondary and primary research phases. In the initial stage, we conducted rigorous secondary research to gain valuable insights into the existing landscape of e-commerce marketplaces and warranty management applications. This endeavor aimed to provide us with a comprehensive understanding of the services offered by competing platforms and to identify any pain points in the user experience that warranted improvement.

Building on the foundation laid by secondary research, our approach extended to primary research methods, which included surveys, interviews, and usability testing. These primary research methods were instrumental in obtaining firsthand perspectives, opinions, and preferences regarding the second-hand marketplace. Through surveys and interviews, we gathered qualitative and quantitative data from a diverse sample of participants, allowing us to delve deeper into individual views on sustainability, product life cycles, and consumer behaviors.

3 Secondary Research

Direct competitors that provide similar services as Fetch were selected to conduct competitive analysis for cross examination. This phase aims to achieve three primary objectives: gaining a deeper understanding of existing e-commerce platforms facilitating second-hand product transactions, evaluating mobile applications dedicated to product warranty management, and assessing text-to-image generative AI software.

	eBay	Facebook Marketplace	Craigslist
Description	eBay is a global online marketplace where individuals and businesses can buy and sell a wide range of new and used products. Users can list items for sale through auction-style or fixed-price listings.	Facebook Marketplace is an integrated platform within the Facebook social network, allowing users to buy and sell items locally. It leverages Facebook's extensive user base for local transactions and typically involves face-to-face interactions.	Craigslist is a classified advertisements website where users can find services, jobs, housing, and a marketplace for buying and selling second-hand items locally. It follows a simple and minimalistic design.
User Base	135M annual users	3.96B monthly users	250.5M visits/month
Interface			
Key Features	• Create fixed-price listings or auction-style listings. • Establish trust through feedback and ratings • International shipping options	• List items for sale to nearby buyers • Communicate through Facebook Messenger • View seller profiles to gauge reliability • Join buying/selling groups for specific categories	• Post free ads for various categories • Option to remain relatively anonymous during transactions
Differentiation	• Allows sellers to list items for auction • Has global reach through its vast international user base	• Integration with Facebook profiles facilitate trust through social connections. • Emphasis on local transactions makes it suitable for face-to-face exchanges	• Post free classified ads, attracting a wide range of sellers. • emphasis on local transactions
Negative Reviews (web sources)	"There are too many hidden costs. Old it was getting a bit better, but there are still a lot of bugs. Also, the "watchlist item" reminders can't be individually turned off which is super frustrating and annoying. I get random notifications daily! I've had to just turn off all notifications instead."	"You may not find a large buyer pool for your product. Because there are so many buyers and product categories, your item may get lost in the shuffle. It may take longer to sell your items since there isn't a highly specific, targeted audience."	"Before I downloaded the app, I was using Craigslist on the web. When you went into the cars and trucks section, your thumbnail or gallery would show the mileage on most vehicles. When you download the app the mileage is no longer visible unless you go into the ad. This happens both on the web and the app."
Positive Review (web sources)	"I have used eBay for a few years now purchasing a wide variety of items both new and used. All but one was a great experience. Items exactly as described, and quick shipping, only one was a scam and the item was shipped all around the world for about six weeks, according to the tracking, but in the end, eBay made it right with a refund without resolution."	"The nice thing about Facebook Marketplace is that you can list an item for sale in a buy/sell Facebook Group AND in Facebook Marketplace - Facebook makes that very easy to do. You just choose Facebook Marketplace and the groups you want to list the item in, and your item is then for sale in multiple groups as well as the marketplace."	"Overall app is great. Brings a lot of filtering options to CL. No cost, Ad-free, and dark mode. Could be more intuitive and feature but hey, you get what you pay for. Unlike many other ads, it's not a downloaded version of the website adds capability, which is great. Wish there was something like this for FB Marketplace."

Fig. 1. Comparative Overview of eBay, Facebook Marketplace, and Craigslist

Features	eBay	Facebook Marketplace	Craigslist
Auction-style Listings	O	X	X
Fixed-price Listings	O	O	X
Localized Transactions	Varies	O	O
In-App Messaging	O	O	X
Social Integration	O	O	X
User Ratings/Feedback	O	O	X
Buyer/Seller Protection	O	X	X
Multiple Payment Options	O	O	O
Shipping Options	online	local	Varies (usually local)
User Profiles	O	O	X
Premium Features	O	X	X
Ad Promotions	O	O	X
Mobile App Availability	O	O	O
International Transactions	O	Limited	Varies (by location)

Fig. 2. Feature Comparison of Online Marketplace

3.1 Second-Hand Marketplace

Figure 1 plays a crucial role in our research, as it compiles data gathered through online searches regarding three major second-hand e-commerce platforms: eBay, Facebook Marketplace, and Craigslist. eBay, with 135 million active users, offers versatility through auction-style and fixed-price listings, fostering competition among buyers and appealing to diverse seller preferences [5]. Facebook Marketplace, with a massive 3.96 billion active user base in 3Q of 2023, harnesses social integration to facilitate local transactions within the Facebook network, enhancing trust and community engagement [6]. Meanwhile, Craigslist's straightforward classified advertisements system prioritizes simplicity and directness in local transactions, making it a popular choice for specific niches [7]. These insights underscore the unique strengths of each platform and inform our approach to developing the Fetch app concept, aiming to enhance the user experience in sustainable e-commerce by integrating the best features of these platforms.

Looking further into specific features offered by each platform, Fig. 2 breaks down all the relevant features into a chart for cross-examination. Each feature is color-coded to denote its utility, with green indicating high usefulness, yellow denoting moderate usefulness, and blue highlighting unique aspects. Taking the popular features among users into consideration is essential in guiding our thought process behind developing the concept of the Fetch app.

With a comprehensive understanding of the feature sets presented by these three platforms, our objective is to identify advantageous functionalities from each platform into the Fetch application. We recognize the significance of social integration, inspired by Facebook Marketplace, and have seamlessly integrated this aspect to build trust and foster community engagement among our users. Furthermore, we are incorporating in-app messaging, user ratings, and buyer/seller protection to ensure a secure and transparent buying experience, drawing from the strengths of platforms like eBay and Facebook Marketplace.

3.2 Warranty Management Mobile Applications

Fig. 3 — Comparative Overview Warranty Management IOS Applications

	About	Pricing	Features	Insights
Asurion	• Seller: Asurion Mobile Applications Inc. • 4.4 / 5 star reviews • 257 ratings • Compatibility: iPhone, iPad • Last updated: Jan 2, 2024	• Free Plan • Asurion Home+ Entertainment: $14.99/month • Asurion Home+: $24.99/month • Asurion Appliance+: $29.99/month	• File claims • Track repairs and replacement • Chat with experts • Troubleshoot issues • Understand benefits • Tech tips	• Well designed user-friendly interface • Helps with repairs but does not actively track warranties • Lots of functions and features
Warranty Keeper	• Seller: Yaron M • 4.3 / 5 star reviews • 13 ratings • Compatibility: iPhone, iPod touch, Mac • Last updated: 8 Sept 2023	• Free	• Manage active warranties • Cloud backup • Notifies when warranty is close to expiration • Input personalized categories/item warranty	• Straightforward app concept: sign up, add items, upload purchase proof • Does not have a lot of functions • UI design is relatively plain
Expired	• Seller: Expired Solutions • 3.3 / 5 star reviews • Compatibility: iPhone, iPod touch, Mac • No longer available	• Free	• Manage active warranties • reminders for expiry dates and receipts • Storage of documents and receipts • Scan warranty cards • Generate and export warranty status reports	• Offers a clean, user-friendly interface • Users have requested more detailed categorization for items • Some users report occasional sync issues with cloud backups
Register	• Seller: Affect Marketing LLC • 1.3 / 5 star reviews • 23 ratings • Compatibility: iPhone, iPod touch, Mac • Last updated: Jun 7, 2022	• Free	• Manage active warranties • Check warranty and service plan information • Tech tips & tricks • Learn about loyalty programs and exclusive deals by brands	• Can only add items to listed brands (no custom categories) • Very confusing to use • The app does not perform as stated
iWarranty	• Seller: Innovatech Solutions Ltd • 2.7 / 5 star reviews • 23 ratings • Compatibility: iPhone, iPod touch, Mac • No longer available	• Free	• Comprehensive warranty tracking • Integration with major electronics brands • Custom tags and filters • Notification system • Customer service chatbot	• Relatively more attractive UI design • Oftentimes times out during item input

Fig. 4 — Feature Comparison of Warranty Management iOS Applications

	Features	Asurion	Warranty Keeper	Expired	Register	iWarranty
	Actively tracks warranties	X	O	O	O	O
	File claims	O	X	X	X	O
	Track repairs and replacement	O	X	X	X	O
	Customer Support	O	X	X	X	O
	Troubleshoot issues	O	X	X	X	X
	Tech tips	O	X	X	O	X
	Custom categories	X	O	O	X	O
Add item	Brand	X	X	X	X	O
	Name	X	O	O		O
	receipt	X	O	O		X
	category	X	O	O		O
	Purchase date	X	O	O		
	Expiry date	X	O	O		O
	Warranty period	X	months	Months, years, lifetime		
	Extended warranty	X	X	O	N/A	
	Multiple warranties	X	X	O		N/A
	image	X	O	O		
	notes	X	O	O		
	Model number	X	X	O		
	Serial number	X	X	O		
	price	X	X	O		
	Merchant info	X	X	O		
	Expiration notifications	X	O	O	X	X
	Search	X	O	O	O	X
	iCloud Backup	X	O	Upgraded version	X	X
	passcode	X	X	O	X	X
	Customer support	O	X	X	X	O

Fig. 3. Comparative Overview Warranty Management IOS Applications

Fig. 4. Feature Comparison of Warranty Management iOS Applications

Since another major feature of the Fetch app is the warranty management system, we conduct a competitive analysis of mobile apps that assist users in managing their warranty details. Figure 3 displays the research results and key findings of testing five different mobile applications surrounding warranty management. Each app was downloaded and cross-examined based on its functionality, usability, and user interface (UI) design. The core concept underlying these mobile applications is straightforward yet highly valuable: users input their product warranty information into the app, and the application takes on the role of a digital warranty tracker. This means that users can rely on the app to keep track of their expiration dates, coverage details, terms and conditions, and contact information for customer support or service center. These testing results are based on data from August 2023 and only pertain to iPhone Operating System (iOS) applications, as we had limited access to Android devices for testing. Any changes or updates made after this date may not be reflected in our findings (Fig. 4).

Upon further inspection and a prolonged period of usability testing, some apps yielded unfavorable results as most did not perform as stated. Feelings of confusion and dissatisfaction were prevalent throughout most of the testing. Some applications did not function as they got stuck in a loading screen while others had poorly designed input systems that were difficult to use, leaving users confused as to how the app works. We believe that the concept behind warranty management is very attractive, but current solutions in the market are not desirable due to the lack of performance and functionality.

This market opportunity will be one of the focal points of Fetch as we aim to improve the user experience by offering a seamless integration and user-friendly product registration process.

3.3 Generative-AI Image Software

	DALL-E 3	Mid Journey	Adobe Firefly
Developer	OpenAI	Midjourney, Inc.	Adobe, Inc.
Date released	January 2021	July 2022	March 2023
Platform	ChatGPT, Microsoft Image Creator	Discord	Stand-alone website, Photoshop, Illustrator, Adobe Express, and Adobe Stock
Key Features	• Enhanced detail and nuance in image generation • Integrated brainstorming with ChatGPT • Ethical constraints	• Versatile image styles (realistic to abstract) • Handles complex textual prompts • Four subscription tiers with varied perks	• Text-to-Image generation • Text Effects Generative Fill • Text to Vector Graphic conversion • Generative Recolor for images
Differentiation	Exceptional attention to detail and ethical considerations in image generation	Community-centric operates through Discord and offers a variety of subscription models.	Seamless integration with Adobe's existing creative applications like Photoshop and Illustrator.
Pricing (USD)	$20 / month	Basic: $10/month Standard: $30/month Pro: $60/month Meg: $120/month	Free
Researcher insights	• Less control and configuration • Conversation-style prompt generation	• The interface is very difficult and complex to use and master	• Presets and effects make it more versatile and controlled. • Generative fill is very useful for editing images

Dalle-3	Midjourney	Adobe Firefly
Modern style, simple, clean, living room, modern furniture, attractive design accessories. harmonious, night, city view, BangandOlufsen, minimal decoration, tastefully arranged, comfortable, realistic, high quality product, beige limestone background, 8K, high detail, photo realistic, cinematic lighting, by dieter rams, —ar 9 16 	Modern style, simple, clean, living room, modern furniture, attractive design accessories. harmonious, night, city view, BangandOlufsen, minimal decoration, tastefully arranged, comfortable, realistic, high quality product, beige limestone background, 8K, high detail, photo realistic, cinematic lighting, by dieter rams, —ar 9 16 	Modern style, simple, clean, living room, modern furniture, attractive design accessories. harmonious, night, city view, BangandOlufsen, minimal decoration, tastefully arranged, comfortable, realistic, high quality product, beige limestone background, 8K, high detail, photo realistic, cinematic lighting, by dieter rams, —ar 9 16 

Fig. 5. Comparative overview of Generative-AI Image Software

Fig. 6. Image Quality Evaluation Across AI-Generated

Given that the central feature of Fetch hinges on the integration of AI-generated images, it was imperative for us to explore existing AI software. As illustrated in Fig. 5, we evaluated three prominent Generative AI software options: DALL-E 3, Midjourney, and Adobe Firefly. Through this examination, we sought to identify their strengths, limitations, and overall suitability for our envisioned user experience. Our objective is to ascertain whether AI-generated images could effectively facilitate product suggestions, aiding users in visualizing how various items would complement their living spaces.

To test our hypothesis, we input the same prompt into each of the three AI software to compare the image results as shown in Fig. 6. Looking at the results, all three softwares produced very high-quality images that are suitable for the use of product suggestions. In further examination, Dalle-3 seems to have a particular stylized result similar to digital art while Midjourney and Adobe Firefly produced very photorealistic results [8]. Ultimately, we determined to utilize Midjourney to create AI-generated images for this project. These images from Midjourney simulate the final, completed version of Fetch and represent what users can expect from Fetch when using the product suggestion feature.

To further test the capabilities of generative AI, we used Midjourney as a proof of concept during our usability testing phase. We conducted in-person interviews with 17 participants, asking their feedback and input on areas of strength as well as improvement through a prototype of the Fetch app. The usability testing allowed us to validate the practicality and potential impact of utilizing AI-generated images in product recommendations. The results of the usability tests will be further discussed later in the paper.

4 Primary Research

To amass comprehensive insights and investigate various aspects related to our research, our research strategy encompassed both qualitative and quantitative methodologies. We seek to retrieve data through the deployment of online surveys, in-person interviews, and rigorous usability testing involving a prototype form of Fetch.

4.1 First Survey

The first survey gathered quantitative data and general information about the population as well as people's attitudes towards the second-handed market, sustainability, and warranty management. The following graphs illustrate a few of the many questions that were asked during the survey.

Fig. 7. Consumer Frequency of Purchasing Second-Hand Products

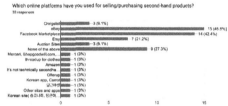

Fig. 8. Usage of Online Platforms for Second-Hand Transactions

Fig. 9. Distribution of Purchase Frequency for Second-Hand Goods

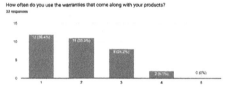

Fig. 10. Preferred Online Platforms for Second-Hand Transactions

The research yielded compelling insights into consumer behavior and attitudes. It's noteworthy that while a significant portion of the population recognizes the environmental benefits of second-hand products in promoting sustainability (Fig. 9), there remains a prevalent preference for purchasing brand-new items and eventually disposing of them in conventional ways once their usefulness wanes (Fig. 7). This paradox highlights the existing gap between awareness of sustainable practices and the actual choices consumers make (Fig. 8).

Furthermore, our investigations delved into participants' utilization of product warranties and their methods for tracking and managing warranty information. Surprisingly, a substantial majority of respondents revealed that they do not fully leverage the advantages and protections afforded by warranty policies offered by companies (Fig. 10). This finding suggests that there is considerable untapped potential for cost savings and improved consumer experiences through the implementation of a more efficient system that simplifies the application and management of warranties.

4.2 Second Survey

The second survey delved into the factors that influence participants' shopping and purchasing decisions, focusing on two pivotal questions highlighted in the graph above. Notably, the majority of respondents expressed a belief in the positive influence of contextual backgrounds accompanying products (Fig. 11). This resounding agreement

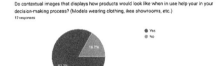

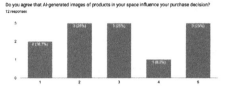

Fig. 11. Influence of Contextual Product Images on Consumer Decision-Making

Fig. 12. Consumer Attitudes Toward AI-Generated Product Images in Purchase Decisions

aligns with our initial hypothesis, which posited that an improved product display and visualization system would empower users to make more informed purchasing decisions. In essence, this survey result affirms that users value visual cues that provide context and utility to products, thereby reinforcing our commitment to incorporating such features into the Fetch platform (Fig. 12).

Following these inquiries, we directed participants towards the subject of AI-generated images, specifically exploring whether such images would influence their purchasing decisions positively or negatively. Respondents who endorsed the idea of AI-generated images were inclined to believe that AI could elevate the level of personalization in terms of product visualization, enhancing the shopping experience. In contrast, those who disagreed with this notion cited concerns over the current capabilities of AI in producing high-quality images or were unconvinced that AI-generated visuals would significantly impact their purchasing choices.

5 Proposed Solution

Fetch is a paradigm shift in how we engage with second-hand consumer products, underpinned by the seamless integration of sustainable e-commerce, AI-driven recommendations, and a robust warranty management system. Designed to empower users and reshape their interaction with used products, Fetch stands at the forefront of the sustainable shopping revolution. Fetch is born from the vision of fostering a culture of conscious consumerism and environmentally responsible choices. In a world grappling with mounting environmental challenges, from global warming to escalating waste production, we recognize that individual actions can collectively drive substantial change. Fetch aims to be the catalyst for such change by making sustainable shopping not just a choice but a lifestyle.

At its core, Fetch transforms the way we buy, sell, and interact with second-hand consumer products. Figure 13 displays the five main features of Fetch: onboarding, generative-AI product suggestions, product registration, warranty management, and profile. Through the power of generative AI, Fetch offers an immersive shopping experience that allows users to visualize how used items seamlessly integrate into their personal spaces. Whether it's furniture, electronic goods, or used vehicles, Fetch fortifies confidence in user's purchases by allowing them to see these items in their own environment. It's also a trusted companion in managing product warranties. With Fetch, you don't have to worry about losing track of your warranties or struggling to navigate the complexities

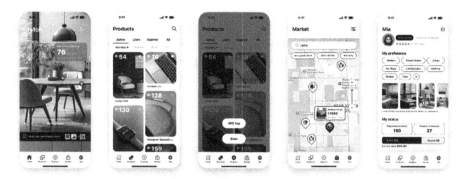

Fig. 13. Interface Overview of Fetch

of warranty claims. Our integrated warranty management system actively tracks your product warranties, reminds you of expiration dates, and facilitates smooth interactions with customer service centers for efficient repairs.

5.1 Onboarding

Fetch offers a unique and personalized onboarding experience that sets the stage for an enriching user journey. Our onboarding process is designed to better understand users' preferences and the types of products that resonate with them, ultimately facilitating more accurate and tailored product recommendations.

Fig. 14. User Preference Onboarding Screens for Personalized Product Suggestions

Upon first launching the Fetch app, users are invited to embark on a journey of self-expression and style exploration. They are prompted to select their preferred themes, styles, moods, and favorite products from a diverse array of options (Fig. 14). Whether it's a minimalist aesthetic, a vibrant and eclectic style, or a preference for vintage treasures, Fetch's onboarding process empowers users to define their unique tastes.

The selections made during onboarding provide valuable insights into users' aesthetic preferences and help Fetch create a personalized shopping experience. As users explore and engage with the app, our AI-driven recommendation engine utilizes this initial input to curate product suggestions that align with their chosen themes and styles. This tailored approach enhances the relevance of product recommendations, ensuring that users discover items that genuinely resonate with their individual preferences.

5.2 Generative AI Product Suggestions

Fig. 15. Navigating AI-Driven Product Suggestions

Fetch's generative AI feature transcends the conventional, catalog-style approach to listing products for sale. Instead of scrolling through static product listings, users are immersed in an interactive and personalized shopping experience. Figure 15 illustrates the tutorials to assist users in using the AI features.

Fig. 16. Post-Tutorial AI Product Suggestion Interface

Imagine a user is looking to purchase a new furniture, Fetch utilizes AI to generate a life-like image of your living room and populates it with multiple furniture options. The generative AI leverages data on your existing furniture, color schemes, and spatial dimensions to provide a truly personalized selection (Fig. 16). Users can visualize how different furniture styles, colors, and sizes harmonize with their current décor, allowing for an informed and confident decision-making process.

This innovative feature not only enhances the user experience but also contributes significantly to sustainability. By visualizing how a second-hand item complements its existing setup, users are encouraged to opt for pre-owned products, promoting the reuse and upcycling of goods. This, in turn, contributes to waste reduction and a more eco-conscious approach to consumption.

Fetch's generative AI feature is a testament to the potential of AI in reshaping the e-commerce landscape. It empowers users to make choices that resonate with their unique preferences and living environments, transforming the way they shop for second-hand items. This sophisticated yet user-friendly technology underscores Fetch's commitment to sustainability, affordability, and a user-centric approach in the world of e-commerce. It represents a bold step toward a more personalized, visually immersive, and environmentally responsible future in consumer purchasing.

5.3 Product Registration

Once the user has chosen to purchase a product through Fetch, the benefits of its integrated warranty management feature come to the forefront. Fetch enables users to seamlessly input their product's warranty information into the app, initiating an efficient and worry-free tracking process. Users have three options to quickly insert their product's warranty information: scanning the physical warranty card using a phone camera, QR codes supported by the manufacturer, and NFC tags that are also supported by the manufacturer.

If the purchased product comes with a physical warranty card with all essential information, the registration process is most effectively optimized using the scan feature with text-recognition capabilities that will automatically input all warranty information into the Fetch app. Similarly, if the warranty details are located on the product's packaging, it would be recommended to use the scan feature to quickly automatically input warranty information. Users also have the option to manually input additional information that is not provided for extra security.

Fig. 17. Product Registration via NFC Tag   **Fig. 18.** Product Registration via QR Code

Utilizing QR codes and NFC tags is a concept idea that has the potential to revolutionize the warranty process. Essentially, we want to replace physical warranty cards with QR codes and/or NFC tags that house all the warranty information related to the purchased product. Figure 17 shows the registration process of using NFC tags while Fig. 18 shows the registration process using QR Codes. Each QR code/NFC tag can be individually coded with specific product information and be distributed within the product's packaging through the manufacturer. Users can then scan the QR code/NFC tag and have all the warranty information accessible on their mobile devices. The advantage of using this technology allows users to efficiently gain instant access to their warranty eligibility/coverage and file repair/replacement claims. However, we also recognize the increased costs and labor associated with introducing and altering the warranty process flow. Therefore, further research and testing will be necessary to fully verify this hypothesis.

5.4 Warranty Management System

Fig. 19. Product Warranty Management Interface

One of the most significant pain points in warranty management is the process of filing repair or replacement claims when a product needs attention. Traditionally, this entails hunting down physical warranty cards, navigating complex company websites, and enduring prolonged response times. Fetch revolutionizes this experience by simplifying the entire process. After users have successfully registered their product, Fetch actively monitors the warranty period and sends timely notifications as the expiration date approaches. With the app's user-friendly interface, users can view and manage all their products and filter by active, liked, and expired warranties (Fig. 19). Users can also swiftly initiate the process of filing a claim, sparing them the frustration of traditional methods and expediting the resolution of any issues (Fig. 20). More details and information is available when clicked into a product usch as an overal report displaying its remaining period, value, and number of buyers interested. Users can also update any warranty information post-registration.

In cases where the seller has already inputted the warranty information into Fetch, the process of transferring warranty information is as straightforward as a few clicks (Fig. 21). This ensures that the new owner inherits the warranty coverage seamlessly, eliminating the administrative hassles that often accompany such transfers.

Fetch's warranty management feature embodies our commitment to providing users with a hassle-free and intuitive experience. By streamlining the often cumbersome warranty-related tasks, Fetch not only simplifies the process but also empowers users

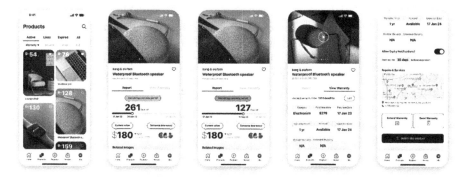

Fig. 20. Details View of Product Warranty Information

Fig. 21. Warranty Transfer

to make the most of their product warranties. This functionality represents a significant step forward in enhancing user convenience and ensuring that they derive maximum value from their purchases, reinforcing Fetch's mission to redefine sustainable and user-friendly e-commerce.

5.5 Used Product E-commerce Marketplace

Fig. 22. Interactive Map-Based Used Product Marketplace

Fetch's marketplace is designed to bring together buyers and sellers in a convenient and visually engaging manner. Upon entering the marketplace, users are greeted by an interactive map that displays a comprehensive view of all available products within their local area (Fig. 22). This map serves as a visual directory, allowing users to explore a diverse range of items for purchase in their vicinity. Users have the option to filter the visible products by quality/condition, price, location, and product category. When a user identifies a product of interest on the map, they can effortlessly navigate to the product's dedicated page. Here, they encounter a wealth of essential information that empowers them to make informed decisions. In addition to product details and images, users gain access to vital information, including seller information and ratings. Fetch ensures transparency by displaying the seller's profile, enabling prospective buyers to gauge reliability and trustworthiness.

Fig. 23. Product Listing and Communication Interface

Fetch fosters direct communication between buyers and sellers through integrated messaging features (Fig. 23). This allows users to seek further information, negotiate terms, or clarify any doubts about the product directly with the seller. Whether it's arranging a meet-up in person to inspect the product or opting for a convenient online transaction with delivery, Fetch accommodates the diverse needs and preferences of its users. Fetch prioritizes safety and convenience in all user interactions. Users can confidently engage with one another, knowing that they have access to essential information and a secure messaging platform. The option to meet in person or choose online transactions with delivery provides flexibility while ensuring a seamless and secure shopping experience.

This online marketplace not only enhances user convenience but also promotes sustainability by encouraging the local exchange of pre-owned items. It aligns with Fetch's mission to reduce waste and foster a culture of conscious consumerism (Fig. 24).

5.6 Profile

The Profile section serves as a hub of personalization, enabling users to fine-tune their Fetch experience according to their evolving tastes and needs. Users have the flexibility to revisit and modify their chosen themes, styles, moods, and favorite product categories at any time. Whether their preferences shift towards a new design trend, a different mood, or an updated product interest, Fetch adapts to these changes with ease.

This dynamic approach to personalization ensures that users consistently receive tailored product recommendations that align with their current preferences and lifestyle. As

Fig. 24. User Profile and Preferences Dashboard

users explore and refine their profile, Fetch's AI-driven recommendation engine leverages this updated information to curate a selection of second-hand consumer products that resonate with their evolving tastes.

6 User Feedback

To assess the effectiveness and user-friendliness of Fetch's innovative features, we conducted a comprehensive usability testing phase, including individual interviews with 17 participants ranging in age from 19 to 36. Their insights helped validate the practicality and potential impact of our innovative features. The primary objective of this research was to gather valuable insights into user perspectives, preferences, and feedback on the usability of Fetch's key functionalities.

Most Valuable Feature	Issues/Concerns	Strengths	Weakness	Suggestions/Improvement
AI functionalities (9)	Authentication & fraud concerns	Advanced technology integration	Potential for misleading AI interpretations	Enhance AI accuracy & realism
Warranty management (5)	User interface complexity	User-friendly design	Limited market reach & brand cooperation	Simplify warranty claim process
Marketplace (3)	Safety & harassment issues	Innovative warranty features	User privacy concerns	Broaden market and product categories
Environmental impact (2)	Market saturation & differentiation worries	Environmental benefits	Need for more diverse & localized content	Implement user verification for trust
Product recommendations (2)	Warranty transfer legality	Personalized shopping experience		Add customizable filters and privacy protections

Fig. 25. Usability User Feedback Summary on App Features

Figure 25 displays a summary of the raw data collected from the interviews during the usability testing. One of the central findings from our usability testing was the overwhelming appreciation for Fetch's AI functionalities, with nine participants highlighting the AI-driven product suggestions as the most valuable feature of the application. Users were particularly impressed by the ability to visualize how second-hand items would fit into their living spaces, emphasizing the importance of this feature in their decision-making process.

On the other hand, several issues and concerns emerged from user feedback. Authentication and fraud concerns were raised, highlighting the need for robust user verification processes to ensure that users do not get scammed for purchasing counterfeit items. Safety and harassment issues were mentioned, emphasizing the need for measures to protect users from unwanted interactions. Some participants who have prior experience using Facebook Marketplace expressed their negative experience meeting up with strangers to purchase products and noted that implementing a security system will be very helpful in increasing trust and reliability.

Interviewees also revealed several strengths that they believe make Fetch a prominent player in the second-hand e-commerce market. Participants commended the advanced technology integration, recognizing Fetch's potential to redefine how they engage with second-hand consumer products. The user-friendly design of the application was highly praised, underscoring the importance of a visually appealing and intuitive interface. The environmental benefits of promoting second-hand shopping were highlighted as a significant strength, aligning with users' desires to make more sustainable choices.

While Fetch demonstrated numerous strengths, it was not without its weaknesses. Participants expressed concerns about the potential for misleading AI interpretations, emphasizing the need for AI-generated images to accurately reflect product quality and appearance. Limited market reach and brand cooperation were identified as areas for improvement, indicating that expanding partnerships and reaching a broader audience would be beneficial.

User feedback provided valuable insights into potential areas for enhancement. Participants suggested several improvements, including enhancing AI accuracy and realism to ensure that generated images align with user expectations. Simplifying the warranty claim process was a common recommendation, aiming to streamline the experience and reduce friction for users. Broadening the market and product categories available on the platform was advised to attract a more extensive user base and offer a more comprehensive selection. Implementing user verification mechanisms to build trust within the platform was suggested, emphasizing the importance of security. Users also recommended the addition of customizable filters and privacy protections to offer a more tailored experience and address privacy concerns effectively.

7 Conclusion

In a world marked by environmental concerns and growing consumerism, Fetch represents a pioneering step toward reshaping consumer purchasing behavior. This paper has detailed the development of Fetch, an innovative mobile application that seamlessly integrates sustainable e-commerce, AI-driven personalized recommendations, and efficient warranty management. Throughout our journey, from secondary and primary research to competitive analysis and usability testing, we've explored the intricate landscape of second-hand commerce, discerned users' desires and apprehensions, and harnessed cutting-edge technologies to pave the way for a profound transformation in consumer practices. Through sustainable shopping, informed choices, and streamlined warranty management, Fetch endeavors to ameliorate the environmental footprint and contribute to a brighter future.

Looking ahead, our long-term vision for Fetch involves developing our own in-house AI system tailored specifically for product suggestions, aiming to further personalize the user experience. While we acknowledge the capabilities of existing AI software, we aspire to further refine and personalize the user experience through a new generative AI system. This AI would be intricately designed to understand user preferences, analyze existing products, and generate compelling, context-aware images that resonate with individual users. We also recognize the potential of QR codes and NFC tags in streamlining warranty management and will conduct further research and testing to explore this innovative concept.

It's important to note that Fetch, as presented here, remains a concept with a carefully designed interface but no functional capabilities. To bring this concept to life, extensive programming, testing, and business development efforts lie ahead. Fetch is poised to empower users to make sustainable and informed choices, turning sustainable shopping into a lifestyle. As we embark on this journey, we envision Fetch becoming a powerful tool that not only transforms the way we buy and sell pre-owned items but also contributes to a more sustainable and environmentally conscious future.

Disclosure of Interests. Images included in this research paper have been sourced from online platforms exclusively for academic and prototyping testing. Existing commercial products were used for proof of concept. There is no commercial or personal gain associated with the use of these images.

References

1. Assadourian, E.: Transforming cultures: from consumerism to sustainability. J. Macromark. **30**(2), 186–191 (2010). https://doi.org/10.1177/0276146710361932
2. What a waste 2.0. https://datatopics.worldbank.org/what-a-waste/. Accessed 21 Sep 2023
3. Solid waste management. https://www.unep.org/explore-topics/resource-efficiency/what-we-do/cities/solid-waste-management. Accessed 16 Nov 2023
4. Borusiak, B., Szymkowiak, A., Lopez-Lluch, D.B., Sanchez-Bravo, P.: The role of environmental concern in explaining attitude towards second-hand shopping. Entrepreneurial Bus. Econ. Rev. **9**(2), 71–83 (2021). https://doi.org/10.15678/EBER.2021.090205
5. eBay Fast Facts. https://investors.ebayinc.com/fast-facts/default.aspx. Accessed 4 Jan 2024
6. Meta reports third quarter 2023 results. https://investor.fb.com/investor-news/press-release-details/2023/Meta-Reports-Third-Quarter-2023-Results/default.aspx. Accessed 4 Jan 2024
7. Kroft, K., Pope, D.G.: Does online search crowd out traditional search and improve matching efficiency? Evidence from craigslist. J. Labor Econ. **32**(2), 259–303 (2014). https://doi.org/10.1086/673374
8. Brynjolfsson, E., Li, D., Raymond, L.: Generative AI at. Work (2023). https://doi.org/10.3386/w31161

Equipping Participation Formats with Generative AI: A Case Study Predicting the Future of a Metropolitan City in the Year 2040

Constantin von Brackel-Schmidt[1]([✉]), Emir Kučević[1], Stephan Leible[1] [iD],
Dejan Simic[1] [iD], Gian-Luca Gücük[1] [iD], and Felix N. Schmidt[2] [iD]

[1] Universität Hamburg, Hamburg, Germany
{constantin.schmidt,emir.kucevic,stephan.leible}@uni-hamburg.de,
{dejan.simic,gian-luca.guecuek}@studium.uni-hamburg.de
[2] Clinical Imaging and Analytics, University Medical Center Hamburg-Eppendorf, Hamburg,
Germany
fel.schmidt@uke.de

Abstract. Urban planning is a complex field that requires the integration of diverse skills, knowledge, and perspectives. Engaging the public in this process promises to achieve high acceptance for resulting endeavors but poses significant challenges, particularly due to the varying levels of technical and artistic skills among participants. Moreover, discussions and collaborative efforts in urban planning often lack practical visualization tools, making it difficult to imagine and debate future urban scenarios. Generative Artificial Intelligence (GenAI) has emerged as a promising technology capable of bridging these gaps. GenAI can generate natural language output and visual content from textual descriptions, offering a novel way to facilitate more inclusive and productive participation in urban planning processes. In this paper, we explore the application of GenAI in a participatory urban planning event utilizing the novel Prompt-a-thon format. We conducted a study involving 64 participants who utilized GenAI in small groups to envision, design, and discuss the future urban landscape of the metropolitan city of Hamburg in Germany. Through this process, we examined the human-AI collaboration, assessing the technology's ability to accommodate varying skills and improve visualization in planning discussions. Our findings indicate that GenAI contributes to leveling diverse skill sets of participants, enabling more people to contribute insights into urban planning discussions. The technology also proved effective in overcoming visualization barriers, thus facilitating more engaging and fruitful discussions about future urban scenarios. These results underscore the potential of GenAI as a valuable tool in the evolving landscape of urban planning and public participation.

Keywords: Urban Planning · Generative Artificial Intelligence · GenAI · Public Participation · Human-AI Collaboration

© The Author(s), under exclusive license to Springer Nature Switzerland AG 2024
F. F.-H. Nah and K. L. Siau (Eds.): HCII 2024, LNCS 14720, pp. 270–285, 2024.
https://doi.org/10.1007/978-3-031-61315-9_19

1 Introduction

Historically, urban planning was predominantly the domain of specialists, perceived as a "technocratic procedure" [1, p. 84]. However, this approach encountered challenges due to the growing demand for increased public engagement in planning processes [2]. In contemporary times, cities are compelled to rapidly adapt and become more agile [3] to effectively address their citizens' evolving needs [4, 5]. It is essential for urban development to prioritize citizen interests and requirements, which necessitates their inclusion in the urban planning process [6]. This can be achieved, for example, through various public participation formats. Many urban development projects now embrace the concept of public participation, recognizing its multiple benefits. Involving citizens can provide unique perspectives and valuable insights, potentially leading to innovative solutions that experts alone might not envisage [7, 8]. Additionally, this involvement can increase the acceptance of proposed plans [7], foster a shared community identity [9], and be integral to a *"socially inclusive innovation process"* [10, p. 1].

While public participation often yields positive outcomes, it is crucial to acknowledge and address its potential drawbacks. The engagement of diverse stakeholders can incur increased costs, both in terms of labor and financial resources, as well as lengthen the implementation process [7]. Furthermore, participation may be restricted when specific skills are required to contribute effectively to the urban planning process [11–13]. For example, the disparity in expertise among participants, such as in artistic or technical skills, can hinder participation. This includes using digital tools, creating relevant texts, or producing appropriate visual representations for ideas, prototypes, and visions. Such skill disparities can result in a biased evaluation of ideas, favoring those based on the skill level of the contributors rather than the intrinsic merit of the ideas themselves, thereby limiting equitable participation opportunities. Furthermore, the lack of a visual representation is a challenge when discussing something that does not yet exist, such as ideas and creating common ground.

Artificial Intelligence (AI) plays a pivotal role in addressing various modern challenges, including climate change, public health, mobility, and spreading misinformation. The introduction of ChatGPT has significantly heightened interest in AI, particularly in the realm of Generative AI (GenAI). Large language models like GPT-3 can produce outputs that closely resemble human responses based on prompts typically presented in natural language through interfaces [14] such as ChatGPT. The rapid emergence of GenAI as a focal point in both practice and research [15, 16] presents opportunities not only for individual users but also for organizations [17]. GenAI has been successfully implemented in various contexts within a short time span, including aiding in idea generation [18], education [19], and problem-solving [20]. Owing to its accessibility and natural language interface, GenAI has the potential to mitigate the previously mentioned skill gaps among participants, especially considering its widespread availability [21].

However, the application of AI in participatory urban planning has not been extensively explored in academic research [22]. Given these factors, this paper seeks to connect these developments and explore the potential of GenAI in addressing the mentioned shortcomings of urban planning processes and enriching and enhancing them by aiding the creative process and minimizing the skill gap of participants. Our research question (RQ) is:

Can Generative AI contribute to the enrichment of participatory urban planning processes in the development of future city scenarios?

To investigate this RQ, we organized a 'Prompt-a-thon' (PaT), a human-AI collaboration event focusing on the use of GenAI for complex problem-solving [23]. This event involved 64 interdisciplinary participants tasked with creating visions for the future of a metropolitan city, Hamburg, Germany, for the year 2040. We aim to contribute to urban planning research, with a particular emphasis on the integration of GenAI in facilitating citizen participation in envisioning the future of cities. We elucidate the key learnings and explore the potential applications of GenAI for future endeavors. Moreover, our research offers insights into the concept of GenAI chaining, demonstrated through our specific approach.

This paper is structured as follows: The subsequent Sect. 2 delves into the theoretical background encompassing urban planning and GenAI. Following this, we describe the research method employed in Sect. 3, focusing on the PaT and our use case-driven approach. Subsequently, we present our findings in Sect. 4 and thoroughly discuss the results and limitations of our paper in Sect. 5. Finally, the paper concludes with a comprehensive summary of our key findings in Sect. 6.

2   Theoretical Background

This section provides an overview of the theoretical underpinnings relevant to our paper. It explores key elements of urban planning and related work in Subsect. 2.1 and examines the use, potential, and challenges of GenAI in Subsect. 2.2.

2.1   Urban Planning

Urbanization, a longstanding phenomenon [24], has reached a stage where, according to the United Nations [25], over half the global population resides in urban areas. The increasing urban populace and evolving needs of residents necessitate innovative urban planning approaches [5]. Urban planning is a multifaceted discipline aimed at developing cities in line with specific objectives through targeted planning activities. Its definitions vary: Næss [26] views it as the *"planning of land use, built-up areas, and infrastructure in urban and metropolitan regions"* [26, p. 1229], while Fischler [24] offers a broader perspective, describing it as the *"collective management of urban development, the use of purposeful deliberation to give shape to human settlements. It is the mobilization of community will and the design of strategies to create, improve, or preserve the environment in which we live. This environment is at once physical (natural and built) and cultural (social, economic, and political)"* [24, p. 108]. He emphasizes urban planning as both a profession and a collective endeavor.

Public participation plays a crucial role in urban planning. Participatory approaches can take various forms, ranging from active citizen engagement in generating ideas [5] to more passive involvement through surveys [27], public hearings [28], or solely information provision [2]. The diversity of these approaches underlines the multifaceted nature of public participation, a concept lacking a universal definition and often used interchangeably with terms like citizen participation [29]. However, there are efforts to categorize this domain. The International Association for Public Participation's IAP2 framework, for instance, outlines five levels of public involvement: (1) Information dissemination to the public, (2) Consultation for deeper engagement, (3) Direct community involvement, (4) Establishing partnerships and collaborations, and (5) Empowering the public to make decisions [30].

For each involvement level, specific methods can be employed. These can include on level (1) fact sheets for basic information dissemination, (2) surveys for consultation, (3) workshops for community involvement, (4) participatory decision-making processes, and (5) citizen juries [10]. The integration of digital tools in these techniques leads to the concept of digital (or e-) participation [31], a realm where information and communications technology (ICT) is used to facilitate participation. The COVID-19 pandemic has notably accelerated the adoption of e-participation as a method [32].

Inclusivity is a crucial aspect of e-participation [33], and digital tools have been instrumental in reducing barriers to participation [34]. Consequently, many traditional participation methods have been adapted for online use. Examples include online surveys, digital workshops, digital decision-making processes, and the utilization of specific digital tools like image creation software [35, 36]. The scope of participation processes has recently broadened to encompass AI applications [37, 38]. Given this expansion, it becomes both socially and scientifically pertinent and promising to explore the potential of GenAI in this evolving landscape.

2.2 Generative Artificial Intelligence Usage, Potentials, and Challenges

Since the advent of OpenAI's ChatGPT, GenAI has garnered significant attention across various social strata, industries, and political spheres. Its rising popularity is likely attributed to its capability to produce 'human-like' outputs from natural language input-prompts [39], a feat it achieves without the need for fine-tuning or task-specific training [14]. GenAI is being explored in a wide array of fields and tasks, including brainstorming for idea generation [18], architectural design [40], knowledge management [41], educational applications [19], and problem-solving [17, 20].

The potential of GenAI is particularly noteworthy: it demonstrates capabilities in domains traditionally thought to be exclusively human, such as creativity [42–44]. Furthermore, the capabilities of large language models are continually evolving. For instance, OpenAI's GPT models have seen significant advancements since their inception. They now support multimodality, application programming interfaces for connecting to other services and web browsing, and are integrated into Microsoft Office 365. OpenAI's models are not alone in this landscape; other large language models like Google's Gemini (formerly Bard), Aleph Alpha, and Claude are also making strides.

Despite these developments, critical issues associated with GenAI are highlighted in the literature. In certain instances, these models can produce biased outputs [45], inaccurate [46], or even potentially usable for illegal activities [47]. Furthermore, ongoing ethical debates surround issues like intellectual property rights, accountability, the explainability of generated content, and resource consumption, particularly the energy associated with operating such systems [48].

Considering the potential misuse of tools like ChatGPT, certain environments have opted to ban them [49]. However, not all sectors demand such stringent measures. Instead, some advocate for the establishment of guidelines to regulate GenAI [50] and promote adequate GenAI education [51]. Education in AI usage is pivotal in harnessing the perceived intelligence of AI systems to augment human capabilities. GenAI, when effectively utilized, can aid humans in diverse tasks. These interactions, which can be viewed as a symbiosis for value co-creation [52, 53], illustrate the collaborative relationship between humans and AI. These dynamics can vary depending on the context.

Siemon [54] categorizes human-AI teams into four types: (1) coordinators, (2) creators, (3) perfectionists, and (4) doers. Each category excels in different areas: coordinators excel in team management, creators in generating solution ideas, perfectionists in completing tasks meticulously, and doers in efficient task execution. Such collaborative and complementary settings of co-creation herald the emergence of hybrid intelligence, a concept where AI and humans synergistically combine their strengths and augment their capabilities. This collaboration aims to achieve a level of intelligence or competence neither could attain independently [55]. A format exploring these potential benefits of human-AI collaboration that we utilized in our case study is the PaT [23].

3   Research Method

This research aimed to explore the application of GenAI with its potential and challenges in the complex and socially significant domain of urban planning. Specifically, we sought to investigate whether GenAI could enrich the urban planning process, particularly in developing future city scenarios. Given the limited existing knowledge about the use of AI in this field [22] and considering the novelty and potential of GenAI, we adopted a phenomenological approach [56]. As Yin [57] recommends, we conducted a case study to investigate a contemporary phenomenon in a naturalistic scenario, focusing on qualitative evidence. To gather our research data, we conducted the novel PaT format [23].

The PaT, inspired by a hackathon, is a collaborative format where participants work together over a designated period to devise solutions for specific challenges or problems. Uniquely, this approach integrates GenAI into the solution development process. This format aligns well with the demands of urban planning, as it can attract a diverse group of participants, fostering networking and idea exchange. Additionally, the PaT has proven effective in generating research data [23, 58].

For this implementation, we aimed to assemble a participant group with diverse backgrounds, expertise, and demographics. We imposed no restrictions on these or other factors, ensuring open access for all interested individuals. To promote the event, we employed various communication channels, including word of mouth, personal invitations, newsletters, and social media, to maximize outreach and participation. Pre-event, ChatGPT Plus, and DALL-E accounts were set up for each group based on the number of registrations, enabling the extraction of prompt and image logs for analysis.

The challenge presented to participants was to envision and visually represent the future of Hamburg by the year 2040. The objective was to use DALL-E 2 to visualize their urban visions through compelling images. We have given participants the freedom to use other GenAI interfaces as well. At the event's outset, attendees received a comprehensive introduction to GenAI and the event's agenda and objective. This included a briefing on a working process we propose, starting with using ChatGPT to refine or expand upon initial ideas about potential future scenarios for Hamburg. It was highlighted that, beyond conventional usage, ChatGPT could also process uploaded files for enhanced interaction. Participants were then encouraged to employ ChatGPT to create prompts tailored for the image-generating GenAI interface, DALL-E 2. Following the introduction, the participants formed 18 groups of three to four persons in size, each provided with a tablet containing the instructions, objective, and a login to ChatGPT and DALL-E 2. They were given the freedom to choose the focus of their visions, with suggestions including various city areas like the town hall, harbor, city park, or themes such as health, mobility, work, culture, and education.

In the subsequent prompting session, each group had its own space to discuss and work with the GenAI interfaces. We utilized ChatGPT Plus (GPT-4) with the version dated August 28, 2023 and DALL-E 2. Post-event, a retrospective session was held to facilitate discussion on the results, key learnings, and standout outcomes. Through a crowd-voting process, the most compelling vision image was selected. The event saw participation from 64 individuals with diverse professional backgrounds, including managing directors, AI engineers, entrepreneurs, product designers, consultants, flame artists, and journalists. The duration of the event was 2.5 h.

4 Results

This section outlines the results obtained from the event. We begin by detailing the utilization of ChatGPT as per the participants' experiences. This is followed by exploring the use of image-producing GenAI interfaces, focusing on DALL-E 2. Finally, a comprehensive analysis of the event's collective outcomes is presented.

4.1 Usage of ChatGPT

The event's initial phase introduced the prospective role of ChatGPT in enhancing or refining ideas about future scenarios in Hamburg and its application in generating prompts for image-generating GenAI interfaces. Following this introduction, participants actively engaged with ChatGPT. Their experiences highlighted several key insights. Firstly, some participants noted and gave feedback to us that interactions with ChatGPT often took the form of iterative dialogues, where a more concise approach yielded better results. They observed that synthesizing outputs from these prompts required human intervention and interpretation. This shows how important the quality of the prompts is in connection with the intended goal and purpose, which plays a role in the context of prompt engineering [59].

Additionally, we noted that engaging with GenAI interfaces facilitated the creation of new connections and insights beyond the initial task, such as uncovering additional information relevant to the given challenge, which enriches the participants' views by providing diverse perspectives. Another observation was the advanced maturity of ChatGPT outputs relating to the topic, making it an effective tool for brainstorming, as Memmert and Tavanapour [18] confirm in their study. Specifically, for envisioning future scenarios, ChatGPT proved to be effective. However, it was noted that when transitioning from ChatGPT to DALL-E 2, it was crucial to use concrete keywords and precisely define the length of the prompts for optimal results, as DALL-E 2 limits the input of characters.

4.2 Usage of DALL-E 2, Midjourney, and Stable Diffusion

DALL-E 2 was suggested as a GenAI interface for generating images visually representing the participants' visions of the future city scenario. However, participants also employed other interfaces like Midjourney and Stable Diffusion based on their preferences. Their experiences with these image-producing GenAI interfaces, particularly DALL-E 2, yielded diverse observations during the event. In terms of DALL-E's performance, participants reported mixed results. Some found that providing detailed descriptions led the outcomes to deviate from their intended visions. Conversely, others observed that supplying concrete ideas resulted in more accurate representations. Achieving the desired level of detail can require multiple adjustments. These may be due to differences in the quality of the prompts, the idea itself, or coincidence.

A notable issue with DALL-E 2 was its tendency to focus on specific details while neglecting others, leading to a partial representation of the input. Moreover, not all entered parameters were consistently reflected in the generated images. Additionally, the participants mentioned limitations in DALL-E's prompt length and its ability to automatically create variations and permit selective adjustments. However, there were instances where DALL-E 2 unexpectedly included elements in the images that were explicitly omitted in the prompts. In contrast, some participants appreciated DALL-E's capability to produce images closely aligned with their specifications, albeit without extending beyond the given concepts.

Regarding GenAI interfaces in general, several points were raised. The quality of the images was sometimes deemed unsatisfactory. The effectiveness of prompts in English was noted, but the process of translating ChatGPT results into suitable prompts for image generation was considered challenging. This difficulty could be attributed to a lack of familiarity with the GenAI interfaces. The conversion from text to image was perceived as complex, particularly in achieving results that harmoniously blend text and visual elements. The output of text in images was regularly erroneous in all GenAI interfaces utilized. Also, significant differences were observed among various image-producing GenAI interfaces, underscoring the importance of selecting the appropriate underlying large language model for specific needs and applications. Certain participants exercised their discretion by utilizing alternative GenAI interfaces, expressing dissatisfaction with the outcomes of initial trials with DALL-E 2. Specifically, they turned to Stable Diffusion and Midjourney. Upon employing these alternatives, discernible disparities were noted in the presentation of the generated images, thereby meeting the preferences of these particular participants.

4.3 Collective Analysis of the Prompt-A-Thon Participation Event

Our findings reveal a dynamic utilization of GenAI with different interfaces and large language models to envision the future of Hamburg in 2040. Participants from varied backgrounds demonstrated the ability to rapidly visualize their concepts for the city's future without needing specific artistic or technical expertise. They often brought two different genKI interfaces into a procedural chain by designing a suitable prompt via ChatGPT and entering it into DALL-E 2 for image generation, also known as GenAI chaining. The prompts by ChatGPT were usually of good quality, but the maximum length of the prompt for DALL-E 2 was not always adhered to, so manual post-processing was necessary.

In terms of the participants' behavior, they predominantly chose to reimagine familiar cityscapes rather than invent new areas. For instance, one group was tasked with envisioning a well-known waterfront square, now featuring floating green spaces connected by bicycle bridges. Another envisioned the harbor as a beacon of technological advancement, harmonizing nature and innovation alongside a prominent city landmark. Furthermore, a vision was articulated where Hamburg emerges as a hub for AI, boasting autonomous vehicles on intelligent roads and holographic displays enhancing public safety and information dissemination. In this future, buildings are adorned with gardens, embodying a fusion of innovation, quality of life, mobility, and augmented reality. The content of the generated images was often associated with content reminiscent of science fiction visuals, for example, known from pictures and movies, and disregarded the laws of physics as we know them today.

From a comprehensive viewpoint, the participant groups selected scenes that not only highlighted key urban areas but also placed them within a broader context of social relevance. The chosen locales included iconic landmarks and sites of interest, such as the Port of Hamburg, the radio telecommunication tower, the central train station, the Elbphilharmonie concert hall, and the town hall. The guiding themes the participants chose reflected socially pertinent contexts and included education, culture, science, mobility, and quality of life. Upon examining the specific terminology employed in the prompt logs, it is possible to categorize the expressions into several key themes. The following themes appeared to represent the participants' vision of essential elements for the city's future:

- sustainability,
- integration with nature,
- autonomous mobility,
    - (including drones, cars, aircraft, and ferries)
- AI-driven technology,
    - (including dynamic bridges and holography)
- renewable energy sources,
    - (including solar panels, wind turbines, and hydroelectric turbines)
- innovation, and
- architecture.

However, when comparing the input prompts with the resulting images, it was observed that the GenAI image generators often struggled to translate the specific phrases and concepts into corresponding visuals accurately. For example, they failed to incorporate recognizable landmarks as requested. It can be assumed that these were not present in the underlying training data or that there were problems interpreting the participants' prompts. Similarly, elements such as renewable energy installations, skyscrapers, drones, and holographic projections were frequently omitted despite being specified in the prompts. Intriguingly, there were instances where the underlying large language model appeared to autonomously decide on the emphasis of the imagery, introducing drones and aircraft into scenes without explicit instructions.

In terms of the PaT format duration, several participants expressed concerns regarding the time constraints. They found it challenging to accurately translate their visions into images, noting difficulties in making the large language models comprehend their ideas, which often resulted in mismatches between their intentions and the output. Conversely, others pointed out the benefits of an iterative process, emphasizing that familiarity and practice with such GenAI technologies and interfaces could enhance outcomes, particularly when transferring concepts and prompt proposals from ChatGPT to DALL-E 2. Aside from observations related to GenAI, participants also appreciated the event for its networking opportunities, remarking on the enjoyable and collaborative atmosphere it fostered.

Fig. 1. The winning image highlights the harbor and the coexistence of nature and technology.

Fig. 2. A generated image indicating more of a dystopian vision even though the belonging prompt did not include such an aspect.

The process of voting for the best image at the end of the event facilitated a lively discussion about GenAI, large language models, and the technology's potential. The immediately generated visualizations enabled discussions directly and on certain levels, such as the details of the realization of the mentioned terminology. Figures 1 and 2 depict images generated during the event. Figure 1 showcases the winning image, while Fig. 2 illustrates a deviation from the creators' likely intent of portraying a positive scenario, instead revealing the AI's rendition of a somewhat dystopian vision.

5 Discussion and Limitations

The outcomes of our PaT conduction and the subsequent utilization of GenAI underscore the potential of this technology in facilitating public engagement in urban planning, specifically in envisioning the future of a major city by the year 2040 through the generation of visualizations of the future. Our initiative attracted diverse participants from various backgrounds, highlighting the inclusivity of the PaT format and GenAI as a tool for public participation. This approach engaged the community and garnered media attention [60], further amplifying public discourse on urban futures, which aligns with our goal of fostering a more inclusive form of public participation [30].

Our findings in terms of the RQ in this paper reveal that participants could quickly generate a wide range of future visions, serving as a catalyst for vibrant discussions on urban futures and technological advancements. The prompts employed during the event reflected key themes prioritized by participants, including the coexistence of nature and technology, sustainable mobility, and the integration of renewable energy sources. This not only provides a basis for fruitful discussions but also a documentation of the participants' thoughts that can be considered in urban planning processes. Specific technologies such as drones, holographic displays, and autonomous vehicles were frequently mentioned, indicating a strong interest or at least expectation in seeing these innovations shape the future urban landscape. Interestingly, there was a preference for reimagining existing urban spaces over creating entirely new ones, suggesting a desire to preserve the essence of current locales while integrating future technologies. This inclination can be linked to the intricacy of the endeavor; the creation of something novel demands a heightened cognitive load and creativity, thereby rendering it more formidable than the enhancement of pre-existing locales or spaces. Notably, the prevailing consensus regarding the future was largely optimistic, highlighting a shared conviction in the capacity of technological advancement to tackle contemporary challenges and mold a favorable urban landscape ahead.

The use of GenAI by participants was favorably received, demonstrating its efficacy in swiftly generating visions of the future. This suggests that GenAI systems, operated via natural language, could serve as an intermediary layer, potentially reducing barriers related to skills or abilities beyond basic language proficiency, such as artistic or specialized technical skills [61]. Nevertheless, feedback indicated that prior experience with GenAI could be beneficial, implying that while GenAI may lower the entry barrier to basic participation for those lacking in specific skills—like drawing or sketching—it might not eliminate but shift the need for skill development. Familiarity with GenAI, similar to other skills, could lead to enhanced outcomes.

However, the GenAI interfaces employed during our event were not solely reliant on natural language inputs, which introduced challenges in their utilization. These challenges are akin to those exemplified in developing geomedia competencies in the study of Hennig and Vogler [61], suggesting that familiarity with the tool's operation remains a hurdle. Future advancements in GenAI could simplify user interaction, thereby improving participation equity. This evolution would shift the focus from the intricacies of tool usage to the quality of outcomes based on input characteristics.

Since our PaT implementation, technologies and GenAI interfaces such as GPT-4 and ChatGPT have been upgraded to offer multimodality, simplifying the user experience.

Users no longer need to manually navigate the transition from text to image generation but can directly instruct the AI to create visuals for the context. This advancement has the potential to alleviate participants' apprehensions regarding the perceived constraints of DALL-E 2. It generates images strictly based on predefined system parameters or customized user specifications without exceeding them. For instance, extending the maximum prompt length could enhance its functionality as a dynamic tool, thereby addressing users' individual (and complex) concerns more effectively.

Our findings also revealed a significant limitation: The generated images often did not align with the input prompts, either by omitting certain elements or failing to capture the intended essence, as noted in the study by Bendel [62]. This discrepancy highlights the ongoing challenge of accurately translating conceptual (natural language-based) visions into visual representations. Based on our results, the variability in participant experience and the nascent developmental stage of GenAI are some reasons for these issues. Indeed, participants observed qualitative differences among the large language models and GenAI interfaces, a sentiment echoed in related research [40, 63].

For future PaT implementations for participatory purposes, it could be beneficial to offer participants a broader selection of large language models and GenAI interfaces. This approach would enable them to choose the most suitable option for their specific needs or to employ a setup of multiple large language models, and GenAI interfaces through an abstraction layer, allowing for a single prompt to yield diverse visual outcomes from several AIs. This chaining of GenAIs, as suggested in our case study, was well-received and widely utilized by participants, indicating its potential value.

However, this chaining also unveiled areas for improvement. For instance, Chat-GPT's lack of awareness regarding the specific requirements of other GenAI interfaces, such as prompt length limitations in DALL-E 2, meant participants had to navigate these constraints independently. This particular issue is solved today due to DALL-E 3 and its direct multimodal integration into ChatGPT. However, it underscores the need to better understand the unique requirements of large language models and GenAI interfaces to specific applications in future iterations. Moreover, our findings revealed that the option to upload PDF files as additional knowledge sources, introduced at the event's start, was underutilized. This suggests that the participants might have been overwhelmed by the range of possibilities within the allotted time, pointing to the necessity of allowing more time for exploration and utilization of the available functions. The participants also confirmed this. Time management and resource availability should be carefully considered in future events to ensure participants can fully engage with and benefit from the technology's capabilities.

Overall, we equipped participants with all necessary tools and hardware, merging onsite activities with a comprehensive introduction to these tools and facilitating e-participation in groups. Our results demonstrated the applicability and enriching facets of GenAI in participatory urban planning processes. They highlight a collaborative human-AI framework, echoing the benefits of brainstorming sessions noted in the literature [18], where participants jointly generated visions of the future. With GenAI being a relatively nascent technology, our case study pioneered the exploration of cutting-edge technologies in urban planning. It can be used as an inspiration for cities and urban planners to engage the public in similar events. The collaborative use of GenAI, combined with guided instructions, suggests a promising strategy to lower participation barriers and enrich the discussions in urban planning.

Looking ahead, the evolution of GenAI holds significant promise. Integrating such events with a competitive element to select a winning image introduces gamification into public participation [64], potentially increasing engagement in such events. In our case study, these playful elements had a fun and engaging rather than a competitive effect. Ultimately, our results add to the growing body of knowledge on the application of GenAI, especially in the area of image generation, which still has little research. Drawing on Brabham's [7] observation that *"the medium of the Web enables us to harness collective intellect among a population in ways that face-to-face planning meetings cannot"* [7, p. 242], we propose GenAI as a novel participant in these meetings, embodying collective intelligence in a human-AI collaboration setup as these models are trained on extensive datasets derived from global human knowledge [14, 54].

Our study is subject to certain limitations. Notably, the event was conducted independently, without formal collaboration with city officials or urban planners. Future iterations could benefit from such partnerships, especially in projects with tangible outcomes, such as redesigning pedestrian zones or parks. Furthermore, our study's findings are based on a single event; replication with other participants and in other cities could provide additional support and validation for our conclusions. A more detailed documentation of participants' backgrounds, particularly concerning their prior knowledge, would enhance the understanding of the impact of diverse experiences on the event outcomes. Additionally, while providing suggested topics, themes, and GenAI interfaces offered a structured starting point, it might have inadvertently influenced participants' directions and outcomes.

Although our event facilitated the chaining of GenAI interfaces by participants, this aspect was rudimentary and manually executed. Future events could benefit from a more sophisticated and integrated approach to GenAI chaining. Lastly, our findings open avenues for more comprehensive analyses, bridging the gap between participatory practices and creative endeavors. Further research could deepen the theoretical understanding and practical applications of GenAI in participatory settings, enriching the dialogue between technology and urban development.

6   Conclusion and Outlook

GenAI is being explored across various fields and contexts to discern its practical applications and limitations. Our study sheds light on the potential use of GenAI within urban planning, particularly in formats that engage public participation. Through analyzing human-AI collaborations, we expand upon previously identified usage parameters and delve into participants' learning experiences as they navigate potential future scenarios for a metropolitan city in the year 2040. Our findings underscore the practicality of GenAI in such contexts, highlighting its significant potential.

Moreover, we offer preliminary insights into the concept of GenAI chaining, which involves the integration of different types of GenAIs, such as text generators, for example, ChatGPT, and image generators, for example, DALL-E 2, Stable Diffusion, and Midjourney. This exploration points to the need for further research to assess the impact of GenAI comprehensively and to evaluate how the ongoing advancement of AI capabilities, such as the transition from non-conversational models like DALL-E 2 to conversational interfaces such as ChatGPT or the integration of multimodality in large language models like GPT-4, affects the quality and applicability of this technology in urban planning. Future studies should be executed in diverse urban planning contexts and can build on our findings, ideally in collaboration with city administrations and urban planners. Such partnerships will be critical in harnessing the full potential of GenAI, enabling more informed and innovative approaches to urban development.

Disclosure of Interests.. The authors declared no potential conflicts of interest with respect to the research, authorship, and/or publication of this article.

References

1. Pissourios, I.A.: Top-down and bottom-up urban and regional planning: towards a framework for the use of planning standards. Eur. Spat. Res. Policy **21**(1), 83–99 (2014). https://doi.org/10.2478/esrp-2014-0007
2. Li, W., Feng, T., Timmermans, H.J., Li, Z., Zhang, M., Li, B.: Analysis of citizens' motivation and participation intention in urban planning. Cities **106**, 102921 (2020). https://doi.org/10.1016/j.cities.2020.102921
3. World Economic Forum. Inspiring Future Cities & Urban Services: Shaping the Future of Urban Development & Services Initiative (2016). http://www3.weforum.org/docs/WEF_Urban-Services.pdf. Accessed 28 Feb 2024
4. Borchers, M., Tavanapour, N., Bittner, E.: Toward intelligent platforms to support citizen participation in urban planning. In: Proceedings of the Pacific Asia Conference on Information Systems (PACIS2022) (2022)
5. Kamacı, E.: A Novel discussion on urban planning practice: citizen participation. Int. J. Archit. Planning **2**(1) 1–19 (2014)
6. Marsal-Llacuna, M.-L., Leung, Y.T., Ren, G.-J.: Smarter urban planning: match land use with citizen needs and financial constraints. In: Computational Science and Its Applications - ICCSA 2011. LNCS, 6783, pp. 93–108 (2011). https://doi.org/10.1007/978-3-642-21887-3_8
7. Brabham, D.C.: Crowdsourcing the public participation process for planning projects. Plann. Theory **8**(3), 242–262 (2009). https://doi.org/10.1177/1473095209104824
8. Ludzay, M., Leible, S.: A bottom-up e-participation process: empowering citizens to innovate the public administration and its sphere of influence. In: Proceedings of the International Conference on Wirtschaftsinformatik (WI2022) (2022)
9. Biondi, L., Demartini, P., Marchegiani, L., Marchiori, M., Piber, M.: Understanding orchestrated participatory cultural initiatives: mapping the dynamics of governance and participation. Cities **96**, 102459 (2020). https://doi.org/10.1016/j.cities.2019.102459
10. Foroughi, M., de Andrade, B., Roders, A.P., Wang, T.: Public participation and consensus-building in urban planning from the lens of heritage planning: a systematic literature review. Cities **135**, 104235 (2023). https://doi.org/10.1016/j.cities.2023.104235
11. Tappert, S., Mehan, A., Tuominen, P., Varga, Z.: Citizen participation, digital agency, and urban development. Urban Plann. **9** (2024). https://doi.org/10.17645/up.7810
12. Gryl, I., Jekel, T.: Re-centring geoinformation in secondary education: toward a spatial citizenship approach. Cartographica: Int. J. Geogr. Inf. Geovisualization **47**(1), 18–28 (2012). https://doi.org/10.3138/carto.47.1.18

13. Wolff, A., Gooch, D., Cavero, J., Rashid, U., Kortuem, G.: Removing barriers for citizen participation to urban innovation. In: de Lange, M., de Waal, M. (eds.) The Hackable City: Digital Media and Collaborative City-Making in the Network Society, pp. 153–168. Springer Singapore, Singapore (2019). https://doi.org/10.1007/978-981-13-2694-3_8

14. Brown, T., et al.: Language models are few-shot learners. In: Proceedings of the 34th International Conference on Neural Information Processing Systems, pp. 1877–1901 (2020)

15. Dwivedi, Y.K., et al.: "So what if ChatGPT wrote it?" Multidisciplinary perspectives on opportunities, challenges and implications of generative conversational AI for research, practice and policy. Int. J. Inf. Manage. 71, 102642 (2023). https://doi.org/10.1016/j.ijinfomgt.2023.102642

16. Leible, S., Gücük, G.-L., Simic, D., von Brakel-Schmidt, C., Lewandowski, T.: Zwischen Forschung und Praxis: Fähigkeiten und Limitationen generativer KI sowie ihre wachsende Bedeutung in der Zukunft. HMD Praxis der Wirtschaftsinformatik 61(2), 344–370 (2024). https://doi.org/10.1365/s40702-024-01050-x

17. Dell'Acqua, F., et al.: Navigating the jagged technological frontier: field experimental evidence of the effects of AI on knowledge worker productivity and quality. SSRN Electron. J. (2023). https://doi.org/10.2139/ssrn.4573321

18. Memmert, L., Tavanapour, N.: Towards Human-AI-Collaboration in Brainstorming: empirical insights into the perception of working with a generative AI. In: Proceedings of the European Conference on Information Systems (ECIS2023) (2023)

19. Hsu, Y.-C., Ching, Y.-H.: Generative artificial intelligence in education. Part Dyn. Frontier. TechTrends 67, 603–607 (2023). https://doi.org/10.1007/s11528-023-00863-9

20. Wu, T., Terry, M., Cai, C.J.: AI Chains: transparent and controllable human-AI interaction by chaining large language model prompts. In: Proceedings of the 2022 CHI Conference on Human Factors in Computing Systems, pp.1–22 (2022). https://doi.org/10.1145/3491102.3517582

21. Jiang, E., et al.: PromptMaker: prompt-based prototyping with large language models. In: Extended Abstracts of the 2022 CHI Conference on Human Factors in Computing Systems. (2022). https://doi.org/10.1145/3491101.3503564

22. Son, T.H., Weedon, Z., Yigitcanlar, T., Sanchez, T., Corchado, J.M., Mehmood, R.: Algorithmic urban planning for smart and sustainable development: systematic review of the literature. Sustain. Cities Soc. 94, 104562 (2023). https://doi.org/10.1016/j.scs.2023.104562

23. Kučević, E., von Brakel-Schmidt, C., Lewandowski, T., Leible, S., Memmert, L., Böhmann, T.: The prompt-a-thon: designing a format for value co-creation with generative AI for research and practice. In: Proceedings of the Hawaii International Conference on System Sciences (HICSS2024) (2024)

24. Fischler, R.: Fifty theses on urban planning and urban planners. J. Planning Educ. Res. 32(1), 107–114 (2012). https://doi.org/10.1177/0739456X11420441

25. United Nations. World Urbanization Prospects: The 2018 Revision (2019). https://population.un.org/wup/Publications/. Accessed 28 Feb 2024

26. Næss, P.: Critical realism, urban planning and urban research. Eur. Planning Stud. 23(6), 1228–1244 (2015). https://doi.org/10.1080/09654313.2014.994091

27. Glass, J.J.: Citizen participation in planning: the relationship between objectives and techniques. J. Am. Planning Assoc. 45(2), 180–189 (1979). https://doi.org/10.1080/01944367908976956

28. Callahan, K.: Citizen participation: models and methods. Int. J. Public Ad. 30(11), 1179–1196 (2007). https://doi.org/10.1080/01900690701225366

29. Schroeter, R., Scheel, O., Renn, O., Schweizer, P.-J.: Testing the value of public participation in Germany: theory, operationalization and a case study on the evaluation of participation. Energy Res. Soc. Sci. 13, 116–125 (2016). https://doi.org/10.1016/j.erss.2015.12.013

30. International Association of Public Participation. IAP2 Spectrum of Public Participation (2018). https://cdn.ymaws.com/www.iap2.org/resource/resmgr/pillars/Spectrum_8.5x11_Print.pdf. Accessed 28 Feb 2024

31. Sæbø, Ø., Rose, J., Skiftenes Flak, L.: The shape of eParticipation: characterizing an emerging research area. Gov. Inf. Quart. **25**(3), 400–428 (2008). https://doi.org/10.1016/j.giq.2007.04.007

32. Pantić, M., et al.: Challenges and opportunities for public participation in urban and regional planning during the covid-19 pandemic—lessons learned for the future. Land **10**(12), 1379 (2021). https://doi.org/10.3390/land10121379

33. Leible, S., Ludzay, M., Götz, S., Kaufmann, T., Meyer-Lüters, K., Tran, M.N.: ICT Application types and equality of e-participation - a systematic literature review. In: Proceedings of the Pacific Asia Conference on Information Systems (PACIS2022) (2022)

34. Sissel Hovik, G., Giannoumis, A.: Linkages between citizen participation, digital technology, and urban development. In: Sissel Hovik, G., Giannoumis, A., Reichborn-Kjennerud, K., Ruano, J.M., McShane, I., Legard, S. (eds.) Citizen Participation in the Information Society: Comparing Participatory Channels in Urban Development, pp. 1–23. Springer International Publishing, Cham (2022). https://doi.org/10.1007/978-3-030-99940-7_1

35. Al-Kodmany, K.: Visualization tools and methods for participatory planning and design. J. Urban Technol. **8**(2), 1–37 (2001). https://doi.org/10.1080/106307301316904772

36. Törnroth, S., Day, J., Fürst, M.F., Mander, S.: Participatory utopian sketching: a methodological framework for collaborative citizen (re)imagination of urban spatial futures. Futures **139**, 102938 (2022). https://doi.org/10.1016/j.futures.2022.102938

37. Arana-Catania, M., et al.: Citizen participation and machine learning for a better democracy. Digital Gov. Res. Pract. **2**(3), 1–22 (2021). https://doi.org/10.1145/3452118

38. Borchers, M., Tavanapour, N., Bittner, E.: Exploring AI supported citizen argumentation on urban participation platforms. In: Proceedings of the Hawaii International Conference on System Sciences (HICSS2023), pp. 1643–1652 (2023)

39. Floridi, L., Chiriatti, M.: GPT-3: its nature, scope, limits, and consequences. Minds Mach. **30**, 681–694 (2020). https://doi.org/10.1007/s11023-020-09548-1

40. Ploennigs, J., Berger, M.: AI art in architecture. AI in Civil Eng. **2** (2023). https://doi.org/10.1007/s43503-023-00018-y

41. Alavi, M., Leidner, D.E., Mousavi, R.: Knowledge management perspective of generative artificial intelligence. J. Assoc. Inf. Syst. **25**(1), 1–12 (2024). https://doi.org/10.17705/1jais.00859

42. Chen, L., Sun, L., Han, J.: A comparison study of human and machine-generated creativity. J. Comput. Inf. Sci. Eng. **23**(5) (2023). https://doi.org/10.1115/1.4062232

43. Gero, K.I., Liu, V., Chilton, L.B.: Sparks: inspiration for science writing using language models. In: Proceedings of the 2022 ACM Designing Interactive Systems Conference (DIS '22), pp. 1002–1019 (2022). https://doi.org/10.1145/3532106.3533533

44. Kirkpatrick, K.: Can AI demonstrate creativity? Commun. ACM **66**(2), 21–23 (2023). https://doi.org/10.1145/3575665

45. Chan, A.: GPT-3 and INSTRUCTGPT: technological dystopianism, utopianism, and "Contextual" perspectives in AI ethics and industry. AI Ethics **3**, 53–64 (2023). https://doi.org/10.1007/s43681-022-00148-6

46. Emsley, R.: CHATGPT: these are not hallucinations - they're fabrications and falsifications. Schizophrenia **9**, 52 (2023). https://doi.org/10.1038/s41537-023-00379-4

47. Al-Hawawreh, M., Aljuhani, A., Jararweh, Y.: CHATGPT for cybersecurity: practical applications, challenges, and future directions. Cluster Comput. **26**(6), 3421–3436 (2023). https://doi.org/10.1007/s10586-023-04124-5

48. Ray, P.P.: ChatGPT: a comprehensive review on background, applications, key challenges, bias, ethics, limitations and future scope. Internet Things Cyber-Phys. Syst. **3**, 121–154 (2023). https://doi.org/10.1016/j.iotcps.2023.04.003

49. Lim, W.M., Gunasekara, A., Pallant, J.L., Pallant, J.I., Pechenkina, E.: Generative AI and the future of education: Ragnarök or reformation? a paradoxical perspective from management educators. Int. J. Manag. Educ. **21**(2), 100790 (2023). https://doi.org/10.1016/j.ijme.2023.100790

50. Hacker, P., Engel, A., Mauer, M.: Regulating CHATGPT and other large generative AI models. In: Proceedings of the 2023 ACM Conference on Fairness, accountability, and transparency, pp. 1112–1123 (2023) https://doi.org/10.1145/3593013.3594067

51. Kishore, S., Yvonne Hong, V., Nguyen, A., Qutab, S.: Should ChatGPT be Banned at Schools? Organizing Visions for Generative Artificial Intelligence (AI) in Education. In: Proceedings of the International Conference on Information Systems (ICIS2023) (2023)

52. Fu, Z., Zhou, Y.: Research on human–AI co-creation based on reflective design practice. CCF Trans. Pervasive Comput. Interact. **2**, 33–41 (2020). https://doi.org/10.1007/s42486-020-00028-0

53. Wu, Z., Ji, D., Yu, K., Zeng, X., Wu, D., Shidujaman, M.: AI creativity and the Human-AI co-creation model. In: Human-Computer Interaction. Theory, Methods and Tools. Lecture Notes in Computer Science, vol. 12762, pp. 171–190 (2021). https://doi.org/10.1007/978-3-030-78462-1_13

54. Siemon, D.: Elaborating team roles for artificial intelligence-based teammates in Human-AI collaboration. Group Decis. Negotiation **31**, 871–912 (2022). https://doi.org/10.1007/s10726-022-09792-z

55. Dellermann, D., Ebel, P., Söllner, M., Leimeister, J.M.: Hybrid intelligence. Bus. Inf. Syst. Eng. **61**(5), 637–643 (2019). https://doi.org/10.1007/s12599-019-00595-2

56. Schwarz, G., Stensaker, I.: Time to take off the theoretical straightjacket and (Re-)Introduce phenomenon-driven research. J. Appl. Behav. Sci. **50**(4), 478–501 (2014). https://doi.org/10.1177/0021886314549919

57. Yin, R.K.: Case Study Research: Design and Methods, (4th ed.). Applied Social Research Methods Series, 5. Sage, Los Angeles, CA (2009)

58. von Brackel-Schmidt, C., et al.: A User-centric taxonomy for conversational generative language models. In: Proceedings of the International Conference on Information Systems (ICIS2023) (2023)

59. White, J., et al.: A Prompt Pattern Catalog to Enhance Prompt Engineering with ChatGPT (2023). https://doi.org/10.48550/arXiv.2302.11382

60. Hasse, M.: Neues Experiment: Wie ChatGPT & Co. Hamburgs Zukunft sehen. Hamburger Abendblatt (2023)

61. Hennig, S., Vogler, R.: Geomedia Skills–a Required Prerequisite for Public Participation in Urban Planning? In: Proceedings of the REAL CORP 2013 (2013)

62. Bendel, O.: Image synthesis from an ethical perspective. AI Soc. (2023). https://doi.org/10.1007/s00146-023-01780-4

63. Göring, S., Ramachandra Rao, R.R., Merten, R., Raake, A.: Appeal and quality assessment for AI-generated images. In: Proceedings of the 15th International Conference on Quality of Multimedia Experience (QoMEX). IEEE, pp. 115–118 (2023). https://doi.org/10.1109/QoMEX58391.2023.10178486

64. Hamari, J., Hassan, L., Dias, A.: Gamification, quantified-self or social networking? Matching users' goals with motivational technology. User Model. User-Adapted Interact. **28**, 35–74 (2018). https://doi.org/10.1007/s11257-018-9200-2

Workplace, Wellbeing and Productivity

Requirements of People with Disabilities and Caregivers for Robotics: A Case Study

Anke Fischer-Janzen[1(✉)] , Markus Gapp[1] , Marcus Götten[2] ,
Katrin-Misel Ponomarjova[1] , Jennifer J. Blöchle[1] , Thomas M. Wendt[1] ,
Kristof Van Laerhoven[3] , and Thomas Bartscherer[2]

[1] Faculty of Economics, Work-Life Robotics Institute, Offenburg University of
Applied Sciences, Max-Planck-Str. 1, 77656 Offenburg, Germany
Anke.Fischer-Janzen@hs-offenburg.de
[2] Umwelt-Campus Birkenfeld, Trier University of Applied Sciences, Campusallee,
55768 Hoppstädten-Weiersbach, Germany
[3] Ubiquitous Computing, Department of Electrical Engineering and Computer
Science, University of Siegen, Hölderlinstr. 3, 57076 Siegen, Germany
https://wlri.hs-offenburg.de/wlri

Abstract. Robotics offers new solutions for digital customer interaction. Social robots can be used in applications such as customer support, guiding people to a location on company premises, or entertainment and education. An emerging area of research is the application in community facilities for people with disabilities. Such facilities face a shortage of skilled workers that could be addressed by robotics. In this work, the application of social and collaborative robots in care facilities and workshops for the disabled is presented by providing a requirements analysis. The use of the humanoid robot Pepper in assisted living was tested and subsequently evaluated in interviews with caregivers who initiated and observed the interaction between the group and the robot. Additionally, robotic applications in assisted work were assessed, resulting in a divergence from the industrial use of robots. A comparative overview with recent literature is presented. The connection between the community home and the workshop raised the question of whether the use of different robots in both places could lead to conflicts.

Keywords: Assistive Robotics · Assisted Living · User Experience · Social Robots

1 Introduction

An ongoing trend is the use of robotics in social and medical environments. Early versions of social robots are represented in the early 2000s with toys such as Furby (Tiger Electronics) or Aibo (Sony) that change their behavior depending on the user's behavior [2]. Today, social robots are motivated by new challenges such as the prevention of loneliness and social isolation as seen in the Covid-19

F. F.-H. Nah and K. L. Siau (Eds.): HCII 2024, LNCS 14720, pp. 289–301, 2024.
https://doi.org/10.1007/978-3-031-61315-9_20

pandemic [8,13], assistance in activities of daily living (ADL) through cognitive and physical stimulation [13], and human-friendly navigation to avoid collisions in logistic tasks, but also in facility tours to present the location to customers [18]. In healthcare, they are used to assist healthcare workers by reducing time-consuming tasks such as picking up and delivering bed sheets and other supplies, allowing caregivers to focus on patient needs and more critical tasks [10]. As a result, the market and research interest in social robots is growing [1,15].

The application of social robots varies greatly depending on the audience and purpose. According to the IFR (International Federation of Robotics), they are categorized as a subgroup of service robotics, as opposed to industrial robotics [5]. Service robots are designed to perform human-centered tasks outside of industrial automation applications [6]. A division into social and non-social robots and domestic or professional use further defines the application of the robot [12]. For this reason, design focuses not only on functionality, but also on form and material to increase acceptance [13]. This can be seen in the diverse field of social robotics, from humanoid robots such as Pepper and NAO (Aldebaran United Robotics Groups) [10], to pet-like companion robots like Spot (Boston Dynamics) and PARO (PARORobots USA), to service robots used in health and social care facilities to optimize logistics and reduce time pressure on caregivers such as Moxi Robot ([10]) and Relay (Swisslog Healthcare).

In contrast, assistive robots, as they are colloquially known, are used for rehabilitation or assistance in everyday life. They do not have the ability to speak and are mostly based on robotic arms, but still need adaptive capabilities. Rehabilitation robots have the ability to guide the user's extremities to assist in the rehabilitation process [10]. The robots' force feedback allows progress to be tracked and adapted to each individual patient. The gamification of such processes is being explored through the implementation of computer games or virtual reality [10]. Assistive Robotic Arms (ARA) can be attached to wheelchairs to assist in performing daily tasks through a joystick, head tracking, eye tracking, or brain computer interface (BCI). The control of such systems can be facilitated by shared control that automates task execution for example by using intention recognition [22].

With the ability to recognize emotions and personality and store memories of previous interactions, such robotic systems can benefit care facilities and hospitals [1]. Adaptive behavioral models, cognitive architectures design, and the ability for perceived empathy in human machine interaction are recent challenges. The application of artificial intelligence accelerates the solution finding [15].

These challenges and requirements must be evaluated separately for each individual robot and target group (e.g., customers, elderly people or people with disabilities). In this work, the user requirements have been identified in cooperation with two related organizations. A case study is presented in which requirements and solutions for assisted work were identified and the robot Pepper was evaluated in an assisted living facility. A comparative overview with existing literature was performed to validate the results.

2   Methods

This work is based on a cooperation with Lebenshilfe im Kinzig- und Elztal e.V. and the company WfB Haslach gGmbH. Lebenshilfe im Kinzig- und Elztal e.V. is an organization that provides services for people with disabilities to ensure participation in all areas of life and to maintain a good quality of life. Services include counseling for family members, assistance in finding housing, suitable employment, and leisure activities [11]. Lebenshilfe im Kinzig- und Elztal e.V. is the shareholder of WfB Haslach gGmbH. This company offers people with disabilities a wide range of work and employment opportunities that are adapted to each individual's abilities [21].

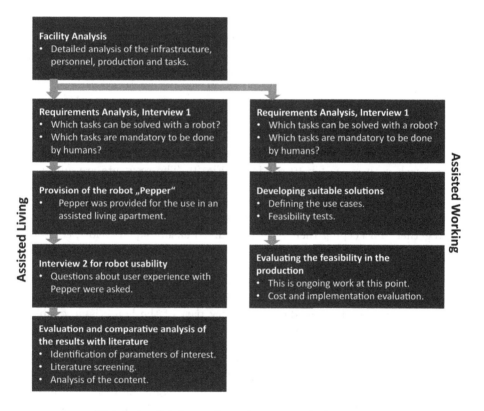

Fig. 1. Application and requirement analysis process.

In a multi-step process, as shown in Fig. 1, the authors identified the needs of both institutions with respect to robotic applications. This work took place between March and October 2023. After an on-site analysis, the evaluation was split into two requirements analyses due to the different needs and work areas. While in the assisted work a demand was found to relieve the caregivers by reducing quality control in production, in assisted living entertainment and assistance in cleaning and logistics were identified, which will be presented in detail in the following sections.

The robot Pepper was used to evaluate the user experience in assisted living (Fig. 3). The robot is 1.20 m tall and weighs 28 kg. The battery life is 12 h according to the manufacturer. Pepper was equipped with applications such as games (memory, guessing, and animal imitation) and interaction options. The interactions were limited to hugging people, speaking fixed scripts, posing for selfies, and recording greetings to relatives or friends. The robot is able to locate itself and move to predetermined locations. A speech recognition tool provides the robot with an additional input modality to the touch screen.

3   Results

The analysis began with a visit to both facilities. An on-site evaluation of both facilities was conducted. The authors spoke with management and staff. A standardized questionnaire was not used. The requirements of each facility are presented below.

3.1   Assisted Living

The community housing complex visited is divided into three communal apartments, each with space for 10 residents (Fig. 2). The residents are adults and pensioners. Most of the residents work at WfB Haslach gGmbH. People with different disabilities live together. They are cared for around the clock by experienced caregivers, but they do most of the daily tasks themselves, such as cooking together and keeping their rooms clean. To ensure equality for people with disabilities the German government supports housings like this. One goal is to allow people with disabilities to choose the type of housing that best suits their abilities and preferences. Another goal is to decentralize the complexes, which will lead to less segregated and closed living environments [14]. The modern complex represents this agenda. Since every part of the house is accessible, it provides perfect conditions for social, mobile robots such as the humanoid robot Pepper.

The requirement analysis was conducted in the steps mentioned in Fig. 1. First, caregivers and managers were asked, what features they thought the optimal robot should have and what tasks the robot would be most helpful with. The results showed that the robot would be most helpful if it could guide people to their rooms, retrieve objects from other rooms, such as a glass of water from the kitchen, assist with eating, or provide general entertainment functions such as playing a game. Due to the Covid-19 pandemic, a feature that regularly cleans surfaces and door handles according to the assisted living standards would be helpful. When walking, some residents may need assistance due to the risk of falling, or they may walk slower than the rest of the group, in which the robot could overtake the role of an additional caregiver to walk with them at their own pace. On the other hand, the robot should not access areas such as the bathroom due to privacy restrictions, nor should it distribute medication. Although the robot could help by monitoring and assisting in cleaning each resident's room, the management pointed out that this would raise further ethical questions about privacy regulations. Due to the infrastructure of the building, it was also requested that the robot could be able to ride the elevator or go outside the building to assist with walking activities.

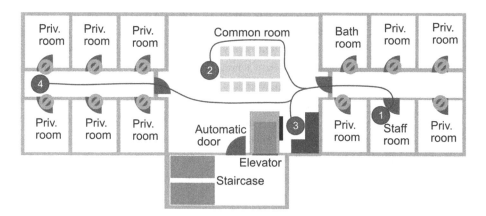

Fig. 2. Example of the shared apartment. It is accessible through an automatic door that can be opened by pressing a button. 1. Robot is loaded in this place overnight. 2. and 3. Robot location for entertainment interaction. 4. Trajectory to guide residents back to their rooms. Private rooms and the bathroom should not be accessed by the robot.

In a second step, the organization was interested in testing such a robot in the shared apartment. The Pepper robot was assigned to one floor of the building, but the residents could visit each other as the rooms are connected to a common room on each floor. Over a period of eleven days, the robot was active for at least one hour every day except for two days. Caregivers monitored the interactions and ensured a safe and privacy-compliant environment for the residents. Therefore, the robot was only used in the common areas.

In the third step, the caregivers collected the residents' experiences with Pepper and participated in the second interview. Table 1 lists the resulting feedback key points from residents and caregivers.

The setup took approximately two hours at the facility. During this time, the robot's functions were explained and a path was taught from the loading cable in the nurses' staff room to the common room, as shown in Fig. 2. Although all doors are accessible and at least 1 m wide, Pepper had difficulty maneuvering between rooms due to inaccurate positioning. While teaching the robot, instructions were displayed on the touchscreen. Especially in path-teaching mode, the robot had no artificial feedback, such as a sound or flashing LEDs, to signal that it had completed its task. Since the robot was moving away from the user

Table 1. Feedback key points regarding the usage of Pepper.

No.	Key point	Perception
1	Mood-lifting effect	positive
2	Speech abilities: discreet with personal information	positive
3	Setup time of about two hours	positive
4	Limited localization accuracy	negative
5	No artificial feedback to signal the end of a task	negative
6	Not usable in a loud group	negative
7	Switched languages unintentionally	negative
8	No pressure sensors to detect contact	negative
9	Permanent upper body movements	negative
10	Maximum volume setting was too low	negative
11	Proposal for a more stable standing position	–
12	Proposal for smart home device control	–
13	Speech model and hardware was perceived as outdated	negative
14	Proposal that robot should be able to play music	–

most of the time, this was inconvenient. Therefore, the staff suggested the possibility of a split-screen application in future systems, showing the instructions on a handheld device. Automatic movement to a loading station and additional automatic loading was requested.

In general, the residents liked Pepper. It had an uplifting effect on the residents. The reason for this, besides the entertainment functions, was that the residents could tell the robot everything without being laughed at or gossiped about with others. In the beginning, one person was against the system, but as the days went by, the acceptance improved. One reason for this was the observation of the system over several days, to see that the robot stops in front of other people when it moves and does not hurt anyone.

The usability has been criticized in some points. The robot is not usable in large groups where multiple users want to interact with the system, as it is designed to interact with one person at a time. The test environment was naturally noisy and busy which is why the robot was often unable to recognize commands. This issue is signaled by Pepper when it turns its head from one person to the next. At that moment, the attention switches to the person whose face the recognition algorithm recognizes best. In some applications, the robot unintentionally switched to English because no German version of the function was provided. Most of the residents do not speak English. Many wanted to try the handshake function. The robot has touch sensors on the back of its hands, but they were not used in the handshake function. After an internal countdown, it began shaking hands, which caused irritation. When using the Memory app, Pepper kept moving the upper body where the touchscreen is mounted. This made it difficult to select cards. Furthermore, residents wished Pepper had more conversational capabilities. With the ongoing development in smart devices technology, residents and caregivers are accustomed to newer technology. In particular, the robot's responsiveness was outdated due to their daily use of voice assistants such as Alexa (Amazon), Siri (Apple), or Google Assistant (Google). Finally, residents mentioned that Pepper is naked. They wanted to dress him up for fun. The customization of parts could increase the acceptance of the humanoid robots, and is provided by additional manufacturers for Pepper [16].

In terms of accessibility, hearing-impaired people were unable to interact with the robot because it was too quiet. In addition, the robot needs to be build more stable because it could be kicked unintentionally since some disabilities, such as spasticity, do not allow the constant control of body movement. Another issue is coverage by private and government health care systems. The cost of a robot that exceeds a certain price cannot be covered by the institutions. This is also true if the system is to be used by individuals living on their own.

Finally, feedback was sought on specific tasks that the robot should help with after trying the robot for eleven days. Household tasks should not be done by the robot, since in this type of all-day care, the residents are responsible for themselves. Instead, the robot should perform pick and place tasks. Implementing Bluetooth and smartphone applications could help, as one of the most common requests was to play music. This would also allow bedridden residents to call the robot. One suggestion to improve the robot was that it should be able to push a wheelchair, making it easier for wheelchair dependent residents to get around.

3.2   Assisted Work

The workshops of WfB Haslach gGmbH are located in Elzach, Hausach and Steinach in Baden-Württemberg, Germany. These facilities produce various products such as printed pens and shirts, wallets, key chains, sewing kits and parts for automotive and other industries. The work can be described as predominantly assembly work. Three main differences from general production lines have been identified.

- **Plant utilization:** Utilization capacities fluctuate due to varying order situations caused by delivery bottlenecks and peak production times specified by the customer.
- **Daily quantities:** Due to the disabilities of the workers, the quantity may vary from day to day. Unlike most industries and due to the low level of automation, absent workers cannot be replaced by other workers. Most workers have a limited range of varied work in which they specialize.
- **Various production lines:** The aim of this company is to promote the employment of disabled people. Therefore, most of the work is done manually. Many machines were presented to assist the workers in production, such as sewing machines, printing machines (screen and pad printing), and devices for aligning parts and assisting with assembly tasks.

The role of the workers varies. Most of manual work is done by people with disabilities. Trained supervisors monitor the work process, assist workers in the process, but also in activities such as assisting with toileting, and perform tasks such as quality control. Due to the lack of digitalization and the limited use of high-tech control and measuring equipment, quality control is also done manually. As a result, this task has to be interrupted when another person needs assistance, resulting in a restricted workflow. This is one of the company's biggest challenges. Based on these statements, the requirements for the robotic system were identified and presented in Table 2.

The second step was to develop and evaluate solutions. Using a state-of-the-art camera, solutions were provided for quality control of various products. For example, a computer vision system was set up to verify that a print on a pen was applied correctly by color and position. The algorithms used for measurement were successful. Whether the systems have the desired effect is part of future work.

4   Discussion

4.1   Comparative Overview

Pepper was released in 2014, which could be a reason for some of the negative key points such as its outdated language model. Talking with the robot about personal problems was also discussed in other studies [23]. Due to the continuous progress in speech recognition, new applications have been developed for Pepper, presenting an interface for using ChatGPT [4]. In the work of Sheridan [19] and Papadopoulos et al. [17], it was found that robots gave ore unsatisfactory answers when asked simple questions. For more complex questions, the robot's performance was similar to that of a human. As noted in [19], collision avoidance receives special attention. This requires not only localization, but also that the robot is aware of its environment. As seen with Pepper, collision avoidance worked well, but localization of the robot in the building had shortcomings. In recent years, Simultaneous Localization and Mapping (SLAM) algorithms have greatly improved through the use of more accurate sensors and more robust algorithms, providing suitable solutions to this problem.

Interaction is affected by further parameters. For example, gestures and reactions are crucial for human-robot interaction [19]. Improving the touch sensors in the hands could not only improve interaction scenarios like shaking hands, but also provide a basis for giving the robot information needed for grasping tasks. Using Pepper as an example, the grasping strength must be enhanced in order to grasp objects. The necessary trajectory planning along with velocity profiles, orientation, and prediction of the target location influence user experience and acceptance of the system [19]. In addition, it has been shown that physical interaction with a soft and warm robot is preferred [19]. In visual design, perception depends on the individual [17]. Insights gained by studying the "uncanny valley" effect showed that a mixture of abstract and human-like appearance can help. As an enabler for social humanoid robots, personalization by adapting to the

Table 2. Functional and non-functional requirements for robot application in assisted work.

No.	Requirement	Description	Funct.	Non-Funct.
1	Collaborative	The robot must stop in the event of unintentional contact with the user [7]	x	
2	Unsupervised use	The robot must perform automated tasks	x	
3	Graphical or manual teach mode	Non-expert users must be able to reprogram the robot for new tasks without extensive training	x	
4	Implementation of Quality control	Cameras and sensor systems must be deployable	x	
5	Mobile setup	The robot must be movable to be used at different work stations	x	
6	Automatic calibration	When the robot is moved to a new task/position, it should be able to recalibrate its base and teach the new position by itself	x	
7	Adaptability in speed	The robot must be adaptable to the speed of the user. The robot should wait in certain positions for the user to take the parts	x	
8	Non-intimidating appearance	Users should not feel threatened by the robot		x
9	Robotic Arm	Unlike other types of robots, the robotic arm provides the necessary flexibility to be used in a variety of environments	x	
10	Robust against high external forces	Rough contact with users cannot be excluded. Therefore, higher external forces must be expected	x	
11	Robust against lighting changes	Quality control cameras need to perform equally well in a variety of lighting conditions due to the mobile nature of the setup	x	

user's tastes and preferences has been mentioned as it is available for Pepper [17].

Additional features such as walking and supervisory tasks were suggested by the caregivers and residents. As the question arose whether a robot would be a suitable walking aid, Karunarathne et al. showed in a study that elderly people would prefer a humanoid robot over walking alone [9]. However, the humanoid robot used lacks a suitable design as a physical aid for people with motor impairments and its inability to move on roads or even on gravel. The question is whether the user would accept the robot over a human companion, and whether cognitively impaired people would react in the same way as the participants in this study. In general, a robot used as a walking aid should be small, equipped with handles, and preferably have a seat and wheels capable of traversing different floor surfaces.

In the workshop environment, the requirements analysis showed differences from general industrial production, which makes the ubiquitous use of robots more important. The focus is on maintaining manual jobs to ensure prospects for people with disabilities. In particular, the robot must be used at different workstations, assisting people without fixed cycle times, but also adapting to the abilities of each user, resulting in a high level of social behavior. Since the people living in the community housing also work in the company, the crossover effects between social and collaborative robots in the user experience are of interest. No relevant literature was found regarding this topic. It is assumed that there should be no significant drawbacks when using both systems in one facility, since the robotic arm used in the assembly tasks is very different from humanoid robots such as Pepper used in this study (Fig. 3). A challenge in using both robots could be that Pepper actively avoids unintentional contact with the user and its grasp strength is low. On the other hand, industrial and collaborative robots can grasp more strongly and move faster. The user may have difficulty estimating the robot's strength. This could result in the person refusing to cooperate with the robot.

4.2 Ethical Questions

Working with people with cognitive impairments, the institution mentioned that using the robot as a walking aid would improve the quality of life for some residents. In this case, the robot should respect the human's freedom of choice, but also ensure that the environment is safe and that the human does not get into risky situations. This raises the question of whether a robot can take over this care task from a human, and what safety functions need to be integrated to meet these requirements.

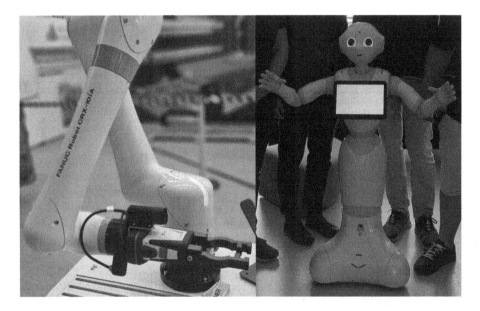

Fig. 3. Use of robots in different environments. Left: Example of a collaborative robot with a camera for assistive work, Right: Social robot Pepper used in assisted living.

In the context of privacy, this work revealed a requirement that contrasts with ethical choices. Including applications that allow the robot to be called by a bed bound person would improve the interaction possibilities between the user and the robot. However, one requirement was that the robot should not enter private rooms. Exceptions need to be defined. In addition, data security is an ongoing issue in robotics due to the use of cameras or the recording and processing of human voice in private environments, especially regarding cloud-based solutions as mentioned above. Camera imagery and LiDAR data have a huge impact due to their detailed imaging of people and environments. This raises concerns in combination with recent legal uncertainties, the increasing autonomy of robots and the replacement of human interaction [3]. With more sophisticated and faster ways to process information, the definition of human-centered data changes. For exoskeletons or adaptive robots, muscle activity detected by the robot or user behavior can be included as part of the human-centered data [3]. To find the right balance between not providing needed functionality due to a potential lack of data and creating a risk to personal security, Fosch-Villaronga et al. suggest involving real users in the process and being transparent about the data collection process so that people can decide for themselves whether they want to share the data [3]. Not only the opinion of the disabled person and caregivers should be sought, but also that of relatives, as in some cases they have a care directive. Including such a number of stakeholders the use of private information should be as clear as possible [8]. The experience from this work shows that each participant (manager, caregiver, and resident) was happy to have this opportunity to help. Nevertheless, it has to be clear that there is a huge responsibility when working with e.g. dementia patients or participants with disabilities [17].

5   Conclusion

This paper presents a case study for social robots in assisted living using the example of Pepper, which was tested for eleven days by residents and caregivers in a community housing complex for assisted living, and for general robotic application and automation in assisted work. For both cases, design and usability requirements were extracted based on user experience. The results were analyzed in comparison with existing literature. The comparison showed similar results. Ethical issues were raised by the organization and the company, which were not found in the literature.

Acknowledgements. Contacts with Lebenshilfe im Kinzig- und Elztal e.V. and WfB Haslach gGmbH were a result of ongoing research in the EU-funded Interreg project "Robot Hub Transfer". We would like to thank all participants of Lebenshilfe im Kinzig- und Elztal e.V. for their cooperation and participation in the survey and Prof. Dr. Matthias Vette-Steinkamp (Trier University of Applied Sciences) for providing the Pepper robot. The authors have no competing interests to declare that are relevant to the content of this article.

References

1. Ahmad, M., Mubin, O., Orlando, J.: A systematic review of adaptivity in human-robot interaction. In: Multimodal Technologies and Interaction, vol. 1(3), pp. 1–14. MDPI (2017). https://doi.org/10.3390/mti1030014
2. Breazeal, C.: Toward sociable robots. In: Robotics and Autonomous Systems, vol. 42(3), pp. 167–175. Elsevier Science B.V. (2003). https://doi.org/10.1016/S0921-8890(02)00373-1
3. Fosch-Villaronga, E., Lutz, C., Tamò-Larrieux, A.: Gathering expert opinions for social robots' ethical, legal, and societal concerns: findings from four international workshops. Int. J. Soc. Robot. **12**(2), pp. 441–458. Springer, Cham (2020). https://doi.org/10.1007/s12369-019-00605-z
4. Hireche, A., Belkacem, A.N., Jamil, S., Chen, C.: NewsGPT: ChatGPT Integration for Robot-Reporter (2023). https://arxiv.org/abs/2311.06640
5. International Federation of Robotics: World Robotics 2023, September 2023. https://ifr.org/img/worldrobotics/2023W_R_extended_version.pdf. Accessed 18 Jan 2024
6. ISO 8373: 2012 (EN); Robots and Robotic Devices-Vocabulary. International Organization for Standardization: Geneva, Switzerland (2012)
7. DIN ISO/TS 15066:2017-04, DIN SPEC 5306:2017-04 (DE): Robots and robotic devices - Collaborative robots (ISO/TS 15066:2016), Beuth (2017). https://doi.org/10.31030/2584636
8. Jecker, N.S.: You've got a friend in me: sociable robots for older adults in an age of global pandemics. Ethics Inf. Technol. **23**, 35–43 (2021). Springer, Cham. https://doi.org/10.1007/s10676-020-09546-y
9. Karunarathne, D., Morales, Y., Nomura, T., Kanda, T., Ishiguro, H.: Will older adults accept a humanoid robot as a walking partner. Int. J. Soc. Robot **11**, 343–348 (2019). Springer, Cham. https://doi.org/10.1007/s12369-018-0503-6
10. Kyrarini, M., et al.: A survey of robots in healthcare. Technologies **9**(8), 1–26 (2021). MDPI. https://doi.org/10.3390/technologies9010008
11. Lebenshilfe e.V.: Informieren: Beiträge der Lebenshilfe. https://www.lebenshilfe-kinzig-elztal.de/. Accessed 12 Jan 2024
12. Lee, I.: Service robots: a systematic literature review. Electronics **10**, 2658 (2021). https://doi.org/10.3390/electronics10212658
13. Liang, N., Nejat, G.: A meta-analysis on remote HRI and in-person HRI: what is a socially assistive robot to do? Sensors (Basel, Switzerland) **22**(19), 1–21 (2022). MDPI. https://doi.org/10.3390/s22197155
14. Ministerium für Soziales, Gesundheit und Integration Baden-Württemberg: Wohnen nach Maß. https://sozialministerium.baden-wuerttemberg.de/de/soziales/menschen-mit-behinderungen/wohnen. Accessed 12 Jan 2024
15. Nocentini, O., Fiorini, L., Acerbi, G., Sorrentino, A., Mancioppi, G., Cavallo, F.: A survey of behavioral models for social robots. Robotics **8**(3), 1–35 (2019). MDPI. https://doi.org/10.3390/robotics8030054
16. Nippon: Robot Runway: Pepper's Fashion Show Debut. https://www.nippon.com/en/views/b00911/. Accessed 26 Jan 2024
17. Papadopoulos, I., Koulouglioti, C., Lazzarino, R., Ali, S.: Enablers and barriers to the implementation of socially assistive humanoid robots in health and social care: a systematic review. BMJ Open **10**, 1–13 (2020). BMJ. https://doi.org/10.1136/bmjopen-2019-033096

18. Samarakoon, S.M., Bhagya P., Muthugala, M.A., Viraj, J., Jayasekara, A.G., Buddhika, P.: A review on human-robot proxemics. Electronics **11**(16), 1–21 (2022). MDPI. https://doi.org/10.3390/electronics11162490

19. Sheridan, T.B.: A review of recent research in social robotics. Curr. Opinion Psychol. **36**, 7–12 (2020). Elsevier. https://doi.org/10.1016/j.copsyc.2020.01.003

20. Søraa, R.A., Tøndel, G., Kharas, M.W., Serrano, J.A.: What do older adults want from social robots? A qualitative research approach to human-robot interaction (HRI) studies. Int. J. Soc. Robot. **15**(3), 411–424 (2023). Springer. https://doi.org/10.1007/s12369-022-00914-w

21. WfB Haslach gGmbH: Informieren: Beiträge der Lebenshilfe Der Trägerverein Lebenshilfe.   https://www.wfb-werbeartikel.de/de/ueber-uns/Traegerverein-Lebenshilfe. Accessed 12 Jan 2024

22. Yang, B., Huang, J., Sun, M., Huo, J., Li, X., Xiong, C.: Head-free, human gaze-driven assistive robotic system for reaching and grasping. In: 40th Chinese Control Conference (CCC), Shanghai, China, pp. 4138–4143 (2021). https://doi.org/10.23919/CCC52363.2021.9549800

23. Yoshikawa, Y., Kumazake, H., Matsumoto, Y., Miyao, M., Ishiguru, H., Shimaya, J.: Communication support via a tele-operated robot for easier talking: case/laboratory study of individuals with/without autism spectrum disorder. Int. J. Soc. Robot. **11**, 171–184 (2019)

Professional Digital Well-Being: An In Situ Investigation into the Impact of Using Screen Time Regulation at Work

Katrin Gratzer, Stephan Schlögl$^{(\boxtimes)}$ ⓘ, and Aleksander Groth ⓘ

Department of Management, Communication and IT, MCI — The Entrepreneurial School, Innsbruck, Austria
stephan.schloegl@mci.edu
https://www.mci.edu

Abstract. The smartphone has not only become our everyday companion but increasingly also led to concerns about digital well-being. Screen time regulation applications aim to tackle these concerns through promoting mindful smartphone use. Despite being predominantly focused on private contexts, they can also help at work in that they enhance task focus and cognitive capacity management. The development of such applications, however, faces certain challenges, particularly when it comes to finding the right balance between what users want in terms of technological functionality, and what tools need to offer to actually change smartphone interaction behavior. Thus, the goal of the preliminary research presented in this article was to investigate the effect of and peoples' perceptions towards the use of screen time regulation at work. Our analysis focused on a single case study, which ensured a sample frame where all study participants work under similar conditions and within a comparable working environment. We used a tripartite empirical study methodology, which started with a questionnaire survey ($n = 74$) exploring peoples' smartphone interaction behavior at work. This was followed by a field experiment in which $n = 20$ employees were asked to log their smartphone interactions with and without screen time regulation. Finally, we asked the same participants to provide feedback on their intervention experiences. Results show that the use of screen time regulation does indeed reduce people's time spent with the smartphone. We also found that participants find it useful, yet that they are afraid of a company-driven introduction or the implementation of respective company-wide policies.

Keywords: Smartphone Distraction · Digital Well-being · Screen Time Regulation · In Situ Exploration · Technology Acceptance

1 Introduction

Over the last decade, digital devices, particularly smartphones, have become integral to our daily life, offering users versatile, time-independent access to various tools and services [12,13,19,23]. In many professional settings, it has furthermore become commonplace for employees to have smartphones on their desks, or to take them to meetings, particularly if these devices have been provided by their employer with the goal to enhance their reachability by email or phone [23]. Yet, while on the one hand smartphones help foster close relationships, excessive interactions can also lead to adverse effects such as reduced cognitive capacity, inattention, and decreased productivity [7,16,19,33]. Consequently, professional life is impacted, by (1) private smartphone use at work creating pressure for immediate responses [21], (2) constant smartphone notifications disrupting work behavior and potentially disturbing work environments [7,11,23], and (3) a general fear of missing out increasing employees' stress levels [25,26],

To mitigate these effects, digital well-being applications seek to balance mobile connectivity, disconnectivity, and user well-being [31]. The digital well-being framework hereby emphasizes a supportive approach [5], which has been taken up by various technology providers such as *Apple* and *Google*, and led to the implementation of respective features into their mobile operating systems. The lack of a more standardized definition for digital well-being, however, has resulted in a myriad of solution approaches, ranging from screen time tracking to distraction prevention [20,31,34]. This diversity in solutions highlights the importance of considering users' actual needs and potential rejection factors inhibiting the successful implementation of digital well-being both in private and work contexts [3,29].

Taking a first step towards better understanding these needs and consequently aiming to strike a balance between productive smartphone use and digital well-being, we conducted an multi-step empirical investigation. We used the following overarching question to guide our research efforts:

> *What impact do digital well-being applications have on employees, and what are potential sociodemographic differences which need to be considered when looking at a company-wide implementation?*

The description of our work starts with a discussion of relevant background theory in Sect. 2. After this, it will elaborate on our study methodology in Sect. 3. Next, it will report on some key results in Sect. 4 and provide more general findings in Sect. 5. Finally, Sect. 6 will conclude our insights and their limitations and point out potential future research directions.

2 Theoretical Background

In the last two decades, we have seen a shift to a technology-based society in which HCI research has been increasingly trying to find solutions to problems that came with the adverse side effects of going all-digital [28]. The ongoing smartphone overuse [7] in particular, which is often triggered by an ever-increasing number of distracting notifications [16], calls for concepts which help foster one's digital well-being [4]. One such attempt is found in the theoretical concept of slow technology, first mentioned by Hallnäs & Redström [13]. It deals with the consequences of ubiquitous computing and proposes a conscious use of technology rather than the continuous search for efficiency improvements. In doing so, it sets the focus on designing long-term relationships between technology and their users, which on the one hand enables performance improvements but on the other hand also sets aside time for reflection [35]. In a sense, it aims to change existing behavioral patterns and wants to support the formation of new, more healthy technology-interaction habits. Exploring various tools for slowing down technology, Eichner [8] proposes a classification of respective tool features, distinguishing between usage tracking, distraction avoidance, feedback presentation, app blocking, time limitation and prompts, and do not disturb solutions.

Previous work on these types of digital detox [27] and well-being interventions [5] underlines the positive effects they can have on perceived user experience as well as satisfaction. Yet, an investigation of 42 digital well-being applications also highlights that restrictions with some of these tools as well as bugs can hinder their acceptance [20]. And although gamification, reward systems and community features have been shown to support tool adoption [17], respective solution providers still face the challenge of finding and implementing the right functionalities. That is, on the one hand features need to have positive effects on one's digital well-being but on the other hand they must also be acceptable to users without infringing too much on their subjectively perceived need for technology-interaction [1,3,17]. Finding this sweet spot often requires the collection and analysis of users' behavioral data [24], which increasingly leads to discussions concerning respective data protection and privacy [22]. One solution for this problem may be found in speculative design solutions, which ask for the sharing of data only if users want to have more personalized experiences. Another approach may focus on creating a better overall understanding about people's digital well-being needs and how they may best be fulfilled. This is where our study aims to add a small piece to the puzzle in that it starts to investigate the impact the use of smartphone screen time regulation has in a work context and collect employees' perceptions towards this type of digital well-being application.

3 Methodology

In order to assess the impact of and perceptions towards smartphone screen time regulation at work, we focused on white-collar employees of a national bank institute. Our respective empirical investigation used a three-stage study procedure, which started with a broader initial survey (Stage 1), followed by a smaller in situ field experiment (Stage 2) including a subsequent post-experiment questionnaire (Stage 3).

For the initial survey, which focused on people's smartphone interaction at work, self-assessment of notifications, and knowledge of digital well-being tools, we used adapted versions of the *Mobile Attachment Scale (MAS)* and the *Mobile Usage Scale (MUS)* [15], the *Mobile Phone Affinity Scale (MPAS)* [2], and the *Smartphone Distraction Scale (SDS)* [30]. We also collected data on people's demographics, hardware use and attitude towards the use of digital well-being tools.

The subsequent in situ field experiment was inspired by Vanden Abeele's research [31] and aimed at exploring the actual use of digital well-being tools at work. Using a within-subject experimental setup, a small sub-group of the previously surveyed employees went through two consecutive intervention phases. Phase 1 (one week) focused on tracking individual smartphone use, where *Android* users were asked to use the *StayFree*[1] application to track their interactions and *iPhone* users instructed to manually log their activities. During the subsequent Phase 2 (one week) *Android* users were then asked to use the smartphone's built-in *Digital Wellbeing* function whereas *iPhone* users had to turn on *Screen Time* restrictions to inhibit on their common smartphone interaction behavior. Again, we used *StayFree* and manual interaction logging to collect activity data.

The final stage of our study remained with the field experiment participants, focusing on their perceptions regarding the felt practicality of these digital well-being functions, particularly collecting feedback on usability and acceptance. Following the literature, we chose to use an already tested and validated 10-item technology acceptance questionnaire for this assessment [32].

3.1 Hypotheses

Based on the results produced by previous work, the following hypotheses are underpinning our investigation:

First, we assume that there is a difference in smartphone activities between younger employees belonging to Generation Y and Z [9] (<41 years of age) and older employees (i.e., >= 41 years of age). To this end we may argue that

- H1a: Traditional phone calls are more relevant to employees who are 41 years of age and older than to younger employees.
- H1b: Employees who are younger than 41 years of age use their smartphones for social media activity more often than older employees.

Second, since employees under 41 years of age are used to mobile technology since their childhood or at least since their youth [10], we expect for them to be more open to digital well-being applications than employees who are older. Thus, we may hypothesize that

- H2a: The term digital well-being is more familiar to employees who are younger than 41 years of age than to older employees.
- H2b: Employees who are younger than 41 years of age have more experience with digital well-being applications than older employees.
- H2c: Employees who are younger than 41 years of age consider digital well-being applications more valuable than older employees.

Third, since females' overall information security awareness seems to be lower than males' [18] we assume to find gender differences with respect to ethical concerns attached to employer-provided tools. Consequently, we also assume that ethical concerns about using employer-provided screen regulation applications are stronger among male than female employees. Thus, we may argue that

---

[1] Online: https://stayfreeapps.com/ accessed: February 2nd 2024.

- H3a: Male employees are more concerned about the potential analysis of data when using a digital well-being applications than female employees.
- H3b: The misuse of digital well-being applications for employee monitoring is of greater concern to male than it is to female employees.
- H3c: Female employees are more likely to use a company-provided digital well-being application than male employees.

Fourth, while smartphone use at work fosters connectivity with social contacts, it can also put pressure on users as they may feel the need to be available and respond to incoming messages or calls at any time [14]. We assume for this effect to be stronger for parents as they may feel responsible for being reachable by their children. Therefore, we hypothesize that the availability by phone is of higher importance to employees with children than to those without children. Consequently, we may argue that

- H4a: Parents think about missed calls more often than childless employees.
- H4b: Employees with children are more anxious if they cannot answer incoming calls immediately compared to childless employees.

Fifth, we assume that using digital well-being applications at the workplace affects employees' smartphone interaction in that

- H5a: Using digital well-being tools at the workplace reduces the frequency of smartphone checks.
- H5b: Using digital well-being tools at the workplace reduces the overall time spent on the smartphone.

And finally sixth, we hypothesize that

- H6: Employees' perceived usefulness of a digital well-being intervention is directly related to the likelihood of them using such tools consistently in the future.

3.2   Participants

For our initial survey we reached out to a total of $N = 142$ employees of the aforementioned bank institute. While we addressed different job roles we made sure that all of the contacted people work under the same company policies, aiming for a homogeneous dataset. For the subsequent field experiment (Stage 2) and the final questionnaire evaluation (Stage 3) we only involved employees who held a comparable job position and worked in an open-plan office without direct client contact or management roles.

The complete three-stage study design as well its sampling procedure received approval by MCI's Ethics Commission evaluating all research studies subject to human participation. In addition, all of the employees who agreed to participate both in the initial survey as well as in the subsequent field experiment, were informed that their participation was entirely voluntary and that all collected data would be processed anonymously. Particularly in case of the experiment, participants were informed in advance that quitting the experiment would be possible at all times and that this would not hold any negative consequences for them.

3.3   Data Analysis

All the data which was collected during the initial survey was screened for missing or invalid data points. To investigate the earlier presented hypotheses, we worked with independent groups. Consequently, the sample was split by age (H1a, H1b, H2a, H2b, H2c), gender (H3a, H3b, H3c) and whether participants indicated to have children or not (H4a, H4b). In cases where the collected data showed normal distribution, we used parametric test methods for analysis, while data that did not fulfill this criterion was subject to non-parametric testing.

For the field experiment, all participating employees submitted their individual log diaries in Excel, including logs from both experiment phases. Before proceeding with the analysis, plausibility checks were performed on each of the logs to ensure data accuracy and validity. To reduce distortions, both minimum and maximum time values for each of the participants were removed before the rest of the data was used to investigate hypotheses H5a and H5b.

Finally, we applied Spearman's Rank Correlation to the data collected by the concluding questionnaire to investigate H6.

4   Results

A total of $n = 74$ (52.11%) of the invited employees completed the initial survey (Stage 1), of whom $n = 49$ categorized themselves as male and $n = 24$ as female. One respondent preferred to not select a distinct gender classification. The sample covers representatives from various age groups ranging from 18 to 62 years. As for our hypotheses H1a, H1b, H2a, H2b, and H2c we were thus able to consider $n = 31$ respondents belonging to the assumed younger generation (<41 years) and $n = 43$ respondents belonging to the assumed older generation ($>= 41$ years). Regarding their parental status, $n = 35$ indicated to have at least one child.

Out of this pool we were able to recruit $n = 20$ employees ($n = 13$ male; $n = 7$ female) for the field experiment (Stage 2) and subsequent questionnaire (Stage 3). Here $n = 12$ belonged to the assumed younger generation and $n = 8$ to the assumed older generation, all of whom completed both stages of the investigation.

4.1   Initial Survey

The results from $n = 74$ employees who completed the initial survey show that most of their smartphone activities relate to various kinds of instant messaging tools, such as *WhatsApp* and *Signal*. Looking at age differences, the data also shows that younger employees (i.e., younger than 41 years of age = GA1) indicate significantly higher social media activities than older employees (41 years of age and older = GA2): GA1 ($M = 2.71; SD = 1.37; Median = 2$) vs. GA2 ($M = 1.72; SD = 1.16; Median = 1$); $U(n1 = 31, n2 = 43) = 978, p =< .001$. No such difference was found with regards to the use of the smartphone's call function. Consequently, we may argue that our data confirms H1b yet lacks support for H1a.

Expecting more openness towards the use of digital well-being applications with GA1, the data confirms this assumption for employees younger than 41 years of age (GA1 ($M = 3.19, SD = 1.64, Median = 3$) vs. GA2 ($M = 2.53, SD = 1.56, Median = 2$); $U(n1 = 31, n2 = 43) = 816.5, p = .046$). Yet, there seems to be no difference between the two age groups when it comes to having actual experience in using such applications. Consequently, the data also does not support the assumption of an increased usefulness perception with GA1. Thus, although the data confirms H2a it neither supports H2b nor H2c.

Looking at gender differences regarding the use of digital well-being applications at work and potentially accompanying ethical and/or privacy concerns, the collected data does not support any of the assumptions put forward. Consequently, none of the cluster 3 hypotheses, i.e., H3a, H3b and H3c, are supported. And also H4a and H4b, where we expected different smartphone connectivity needs for employees with children compared to those without such parenting duties, had to be rejected due to lack of respective evidence found in the collected data.

4.2   Field Experiment

Investigating actual smartphone use and the impact of screen regulation, the analysis of the smartphone activities logged during the first phase of the field experiment (without the use of a screen regulation application or function) showed that employees use their smartphone at work mainly for chats (95% of participating employees) and other social media activities (45% of participating employees), browsing the Internet (50% of participating employees), phone calls (25% of participating employees), reading the news (20% of participating employees) and online shopping (10% of participating employees). They reported to have checked their smartphone on average 3.2 times per day during working hours and spent on average 9.36 min (561 s) using it ($SD = 534$ s, $Median = 374$ s). Comparing this data with the data collected during the second phase of the field experiment, the number of smartphone checks did not dramatically change, yet the amount of time spent interacting was reduced significantly ($M = 209$ s, $SD = 388$ s, $Median = 90$ s; $p =< .001$) as depicted in Fig. 1. Thus, we may argue that although our data does not confirm H5a it does provide support for H5b.

4.3   Final Questionnaire

The goal of the final questionnaire was to gain insights into the experiences participants gained from taking part in the field experiment and collect attitudinal feedback towards the use of digital well-being tools at the workplace. The data shows that 17 of the 20 participants perceived the tested screen time regulation application as easy to use (rating $>= 4$ on a 7-point positively increasing Likert scale). Also, the overall usefulness of digital well-being tools at work has

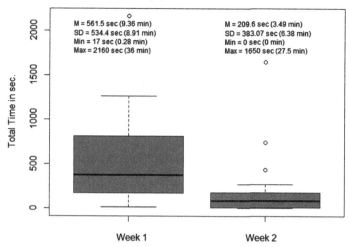

Fig. 1. Smartphone activity with and without screen time regulation.

been positively perceived ($M = 4.20$, $SD = 1.51$ on a 7-point positively increasing Likert scale). Investigating whether this perceived usefulness correlates with employee's intention to continue using such an application during work, the data points to a strong positive effect: $r(18) = .601, p = .003, |r| >= 0.40$. Thus, we may conclude that based on our arguably small field experiment and its concluding questionnaire, there is support for H6.

5   General Findings

Although the results from our survey did not confirm assumptions concerning sociodemographic differences with respect to reachability, use of digital well-being tools and respective ethical concerns, they provide valuable insights into how employees interact with smartphones at work. That is, it can be seen that employees use their smartphones primarily for chats and other types of social media, which helps them stay in touch with friends and family. While here the data confirms a significant difference between younger and older employees ($>= 41$ years of age) in that younger people seem to be more attached to smartphone interactions than older ones, it also points to a unique relationship that people of all age groups have built up with their smartphones [19]. To this end, it was also shown that most of our study participants (50 out of 74) find respective smartphone notifications from them and/or coworkers disruptive and that they would work more efficiently without being interrupted, which indicates that employees are aware of the negative affects smartphone notifications and consequent interactions can have on one's work performance [23]. And although one third (31%) of the participating employees find that an app blocker or respective screen regulation feature would support their concentration while working on complex tasks, only 26% indicate that they would use such an application.

With the field experiment, we have seen that screen regulation can significantly reduce the time such interruptions take. Surprisingly, however, such measures did not reduce the number of smartphone checks. Comparable results have already been reported by Monge Roffarello & De Russis [20], which suggests that screen regulation is often less effective than expected.

Finally, our study was particularly interested in whether participants find digital well-being applications practical and useful in a work context. Previous studies have shown that there is a strong link between perceived usefulness and continuous use of a technology or tool [6]. Our data confirms this link, as it shows a clear positive correlation between our participants' perception of the usefulness of screen time regulation at work and their stated intention of using such a feature in the future. Only when it comes to the introduction of a respective company-wide policy does acceptance drop, with merely 15% of participants indicating that they would support such an employer-driven initiative.

6   Conclusions, Limitations and Future Outlook

Distractions and interruptions by smartphone notifications can increase employees' stress levels and lower concentration as well as productivity [7,16,33]. Subsequently, they can negatively affect individuals' well-being at work [23]. To prevent these negative side effects, the presented study set out to analyze solutions that inspire employees to interact with their smartphones more consciously and trial tools to support this behavior. Our study results show that although the employees we engaged in our study interact very differently with their smartphones, especially when it comes to the frequency and duration of using messaging and other types of social media applications, their general attitude towards using screen time regulation at work is similar. That is, although we found that people under 41 years of age seem to be more open and positive towards using digital well-being applications, the actual use of such tools eventually comes down to their perceived practicality and is not necessarily driven by users' sociodemographic characteristics.

Although our hypothesis that employees with children would be more anxious about using screen regulation, as this could potentially inhibit their connection to home, is not supported by the collected study data, we believe that additional analyses to this end are needed to clearly rule out such an effect. For example, we did not interrogate on children's age, which may have had significant influence on our participants' judgement as to whether they would be more or less dependent on their smartphone. Generally, it is recommended for future work to engage in more long-term studies, since our field experiment was probably too short to produce reliable results as to the practicability of screen regulation applications at work. Such long-term analyses would also help in identifying other factors potentially influencing the adoption of these tools and further verify the impact of respective company-driven interventions and policies. Finally, we suggest that future work should consider other, more diverse working situations, addressing the impact of multi-device use and cross-device interactions as well as cultural differences in the adoption of digital well-being applications.

References

1. Almourad, M.B., Alrobai, A., Skinner, T., Hussain, M., Ali, R.: Digital wellbeing tools through users lens. Technol. Soc. **67**, 101778 (2021). https://doi.org/10.1016/j.techsoc.2021.101778, https://www.sciencedirect.com/science/article/pii/S0160791X21002530
2. Bock, B.C., et al.: The mobile phone affinity scale: enhancement and refinement. JMIR Mhealth Uhealth **4**(4), e134 (2016). https://doi.org/10.2196/mhealth.6705
3. Bødker, S.: When second wave HCI meets third wave challenges. In: Proceedings of the 4th Nordic Conference on Human-Computer Interaction: Changing Roles, NordiCHI 2006, pp. 1–8. Association for Computing Machinery, New York, NY, USA (2006). https://doi.org/10.1145/1182475.1182476
4. Büchi, M.: Digital well-being theory and research. New Media Soc. **26**(1), 172–189 (2024). https://doi.org/10.1177/14614448211056851
5. Cecchinato, M.E., et al.: Designing for digital wellbeing: a research & practice agenda. In: Extended Abstracts of the 2019 CHI Conference on Human Factors in Computing Systems, CHI EA 2019, pp. 1–8. Association for Computing Machinery, New York, NY, USA (2019). https://doi.org/10.1145/3290607.3298998
6. Davis, F.D., et al.: Technology acceptance model: TAM. Al-Suqri, MN, Al-Aufi, AS: Information Seeking Behavior and Technology Adoption, pp. 205–219 (1989)
7. Duke, É., Montag, C.: Smartphone addiction, daily interruptions and self-reported productivity. Addict. Behav. Rep. **6**, 90–95 (2017). https://doi.org/10.1016/j.abrep.2017.07.002, https://www.sciencedirect.com/science/article/pii/S2352853217300159
8. Eichner, A.A.: Planting trees and tracking screen time: a taxonomy of digital well-being features. In: Proceedings of PACIS 2020 Proceedings (2020)
9. Fidan, A.: The effect of attitudes by generations X, Y, Z, alpha, beta, gamma and delta on children. In: Being a Child in a Global World, pp. 17–33. Emerald Publishing Limited (2022)
10. Forgays, D.K., Hyman, I., Schreiber, J.: Texting everywhere for everything: gender and age differences in cell phone etiquette and use. Comput. Hum. Behav. **31**, 314–321 (2014). https://doi.org/10.1016/j.chb.2013.10.053, https://www.sciencedirect.com/science/article/pii/S0747563213004032
11. Fortin, P.E., Cooperstock, J.: Understanding smartphone notifications' activity disruption via in situ wrist motion monitoring. In: Extended Abstracts of the 2022 CHI Conference on Human Factors in Computing Systems. CHI EA 2022. Association for Computing Machinery, New York, NY, USA (2022). https://doi.org/10.1145/3491101.3519695
12. Gowthami, S., Kumar, S.: Impact of smartphone: a pilot study on positive and negative effects. Int. J. Sci. Eng. Appl. Sci. **2**(3), 473–478 (2016)
13. Hallnäs, L., Redström, J.: Slow technology-designing for reflection. Pers. Ubiquit. Comput. **5**, 201–212 (2001)

14. Kühnel, J., Tim Vahle-Hinz, J.d.B., Syrek, C.J.: Staying in touch while at work: relationships between personal social media use at work and work-nonwork balance and creativity. Int. J. Hum. Resour. Manage. **31**(10), 1235–1261 (2020). https://doi.org/10.1080/09585192.2017.1396551

15. Konok, V., Gigler, D., Bereczky, B.M., Miklósi, Á.: Humans' attachment to their mobile phones and its relationship with interpersonal attachment style. Comput. Hum. Behav. **61**, 537–547 (2016). https://doi.org/10.1016/j.chb.2016.03.062, https://www.sciencedirect.com/science/article/pii/S0747563216302333

16. Kushlev, K., Proulx, J., Dunn, E.W.: "Silence Your Phones": smartphone notifications increase inattention and hyperactivity symptoms. In: Proceedings of the 2016 CHI Conference on Human Factors in Computing Systems, CHI 2016, pp. 1011–1020. Association for Computing Machinery, New York, NY, USA (2016). https://doi.org/10.1145/2858036.2858359

17. Lyngs, U., et al.: Self-control in cyberspace: applying dual systems theory to a review of digital self-control tools. In: Proceedings of the 2019 CHI Conference on Human Factors in Computing Systems, CHI 2019, pp. 1–18. Association for Computing Machinery, New York, NY, USA (2019). https://doi.org/10.1145/3290605.3300361

18. McGill, T., Thompson, N.: Gender differences in information security perceptions and behaviour. In: 29th Australasian Conference on Information Systems (2018)

19. Melumad, S., Pham, M.T.: The smartphone as a pacifying technology. J. Consum. Res. **47**(2), 237–255 (2020). https://doi.org/10.1093/jcr/ucaa005

20. Monge Roffarello, A., De Russis, L.: The race towards digital wellbeing: issues and opportunities. In: Proceedings of the 2019 CHI Conference on Human Factors in Computing Systems, CHI 2019, pp. 1–14. Association for Computing Machinery, New York, NY, USA (2019). https://doi.org/10.1145/3290605.3300616

21. Patterer, A.S., Yanagida, T., Kühnel, J., Korunka, C.: Staying in touch, yet expected to be? A diary study on the relationship between personal smartphone use at work and work-nonwork interaction. J. Occup. Organ. Psychol. **94**(3), 735–761 (2021). https://doi.org/10.1111/joop.12348, https://bpspsychub.onlinelibrary.wiley.com/doi/abs/10.1111/joop.12348

22. Pierce, J.: Smart home security cameras and shifting lines of creepiness: a design-led inquiry. In: Proceedings of the 2019 CHI Conference on Human Factors in Computing Systems, CHI 2019, pp. 1–14. Association for Computing Machinery, New York, NY, USA (2019). https://doi.org/10.1145/3290605.3300275

23. Pitichat, T.: Smartphones in the workplace: changing organizational behavior, transforming the future. LUX: J. Transdisciplinary Writ. Res. Claremont Grad. Univ. **3**(1), 13 (2013)

24. Polanco-Diges, L., Saura, J.R., Pinto, P.: Setting the relationship between human-centered approaches and users' digital well-being: a review. J. Tourism Sustain. Well-Being **10**(3), 148–171 (2022)

25. Przybylski, A.K., Murayama, K., DeHaan, C.R., Gladwell, V.: Motivational, emotional, and behavioral correlates of fear of missing out. Comput. Hum. Behav. **29**(4), 1841–1848 (2013). https://doi.org/10.1016/j.chb.2013.02.014, https://www.sciencedirect.com/science/article/pii/S0747563213000800

26. Rodríguez-García, A.M., Moreno-Guerrero, A.J., López Belmonte, J.: Nomophobia: an individual's growing fear of being without a smartphone-a systematic literature review. Int. J. Environ. Res. Public Health **17**(2), 580 (2020). https://doi.org/10.3390/ijerph17020580, https://www.mdpi.com/1660-4601/17/2/580

27. Schmuck, D.: Does digital detox work? Exploring the role of digital detox apps for problematic smartphone use and well-being of young adults. In: 65. DGPuK-Jahrestagung, Munich, Germany (2020)
28. Shepherd, J.: What is the digital era? In: Social and Economic Transformation in the Digital Era, pp. 1–18. IGI Global (2004)
29. Stephanidis, C., et al.: Seven HCI grand challenges. Int. J. Hum.-Comput. Interact. **35**(14), 1229–1269 (2019)
30. Throuvala, M.A., Pontes, H.M., Tsaousis, I., Griffiths, M.D., Rennoldson, M., Kuss, D.J.: Exploring the dimensions of smartphone distraction: development, validation, measurement invariance, and latent mean differences of the smartphone distraction scale (SDS). Front. Psychiatry **12**, 642634 (2021). https://doi.org/10.3389/fpsyt.2021.642634, https://www.frontiersin.org/journals/psychiatry/articles/10.3389/fpsyt.2021.642634
31. Vanden Abeele, M.M.P.: Digital wellbeing as a dynamic construct. Commun. Theory **31**(4), 932–955 (2020). https://doi.org/10.1093/ct/qtaa024
32. Venkatesh, V., Morris, M.G.: Why don't men ever stop to ask for directions? Gender, social influence, and their role in technology acceptance and usage behavior. MIS Q. **24**(1), 115–139 (2000). http://www.jstor.org/stable/3250981
33. Ward, A.F., Duke, K., Gneezy, A., Bos, M.W.: Brain drain: the mere presence of one's own smartphone reduces available cognitive capacity. J. Assoc. Consum. Res. **2**(2), 140–154 (2017). https://doi.org/10.1086/691462
34. Widdicks, K.: When the good turns ugly: speculating next steps for digital wellbeing tools. In: Proceedings of the 11th Nordic Conference on Human-Computer Interaction: Shaping Experiences, Shaping Society, NordiCHI 2020. Association for Computing Machinery, New York, NY, USA (2020). https://doi.org/10.1145/3419249.3420117
35. William Odom, E.S., Chen, A.Y.S.: Extending a theory of slow technology for design through artifact analysis. Hum.-Comput. Interact. **37**(2), 150–179 (2022). https://doi.org/10.1080/07370024.2021.1913416

Developing Gamification Strategies to Reduce Handling Damages in High Precision Production Environments

David Kessing[1]([✉]) and Manuel Löwer[2]

[1] Institute for Product-Innovations, University of Wuppertal, Wuppertal, Germany
david.kessing@uni-wuppertal.de
[2] Department for Product Safety and Quality Engineering, University of Wuppertal, Wuppertal, Germany

Abstract. The challenge of handling damages in high-tech industries, where precision and quality standards are paramount, has been addressed through innovative gamification strategies which are rooted in psychology. Focusing on a German aviation company, this research aims to minimize handling damages caused by workers through a systematic development of gamification strategies. The study involved the application of the *How to design gamification* process, incorporating a *Hexad user types* survey and Marczewski's *Periodic Table of Gamification Elements* as an ideation base.

The preparation phase defined objectives, emphasizing general handling damage reduction and early reporting. Success metrics included the detection of damages, increased reporting, reduced reporting time, and positive worker feedback. The analysis phase utilized the Hexad typology, revealing a concentration on the user types philanthropists, players, and achievers. In the ideation phase, game elements were meticulously selected, and in the design phase, strategies were devised, leading to an emergency button prototype.

Implemented at a quality inspection workstation, the emergency button successfully reduced handling damages, with a noticeable increase in reporting and positive worker feedback. Unforeseen benefits included workers reporting defects in raw parts, surpassing the initial quality improvement goal. The study argues for the systematic involvement of affected groups in developing gamification solutions, citing the *How to design gamification* process as an effective framework. Future plans include a detailed technical implementation, broad prototype deployment, and integration with complementary concepts to foster an improved error culture. Overall, this research demonstrates the potential of gamification to address complex, human-centric challenges in industrial settings.

Keywords: Gamification · Handling damages · Production environment · User Types Hexad

F. F.-H. Nah and K. L. Siau (Eds.): HCII 2024, LNCS 14720, pp. 314–333, 2024.
https://doi.org/10.1007/978-3-031-61315-9_22

1 Introduction

In the rapidly evolving landscape of the high-tech industry, precision and quality standards are a key requirement. Optimizing manufacturing in this regard, particularly in the aviation sector, is fundamental, with companies struggling not only with the complexities of advanced technologies, but also with the inherent challenges posed by human factors [1]. One of these major challenges is the occurrence of handling damage in the production of high-precision components - a costly problem that requires innovative solutions [2].

The aerospace industry, as an industry with the highest quality standards, relies on the meticulous manual work of employees to ensure these quality standards. Despite the integration of cutting-edge technologies and automated processes, there are still certain manual steps, especially in the complicated quality analysis and approval of components for subsequent work steps. Herein lies a paradox: the combination of human judgment and flexibility, often considered essential for these critical tasks, carries the risk of handling errors [3].

In contrast to machine-related errors or external factors, handling errors are individual mistakes made by employees during the processing of components. These errors, which range from the incorrect use of tools to unintentional damage caused by carelessness, not only represent a financial burden, but also disrupt the smooth running of production. In the high-tech industry, where the smallest deviations from quality standards can lead to safety-critical defects, the repair of handling damage is of particular importance [4].

This study takes place against the backdrop of a German aerospace company, which provides a representative environment for the general challenges faced by industries with similarly stringent quality requirements. As technology advances, the question arises: how can companies reconcile essential human contact with the demands of precision manufacturing? One possible solution lies in a multidisciplinary approach that uses insights from psychology and innovative methods such as gamification.

Gamification, a concept that is becoming increasingly popular in various fields, is a promising way to overcome human-centered challenges. By incorporating elements of game design into real-world processes, gamification aims to influence human behavior, motivation and decision-making. The application of gamification strategies has the potential not only to mitigate the damage caused by handling damages, but also to promote a positive change in employee attitudes and engagement.

This research looks at the development and implementation of gamification strategies tailored to the specific challenges of handling damage reduction. The aim of this study is to shed light on the transformative potential of gamification in addressing high-stakes human challenges in precision manufacturing.

2 Problem Description

In the modern automated industrial production of components, manual processing by humans is still necessary at some points despite digitalization and the use of high technology [1]. Particularly in the quality analysis and release of components for subsequent work steps, human judgment and flexibility are relied upon. During these manual work steps, various causes, such as carelessness and boredom due to monotony, lack of appreciation of work or reduced motivation due to lack of transparency of the influence of the work in the company can lead to so-called handling damages by the processing humans. Due to their human-related causes, these damages cannot be sustainably remedied by process or company-related changes [2, 3].

Handling damages are part of the so-called abnormal losses. Abnormal losses refer to losses that arise for reasons beyond the control of the company, such as theft, damage or waste. Normal losses, on the other hand, describe losses that result from normal business operations, such as wear and tear of machinery or natural depletion of raw materials Abnormal losses, like normal losses, can have a significant impact on a company's profitability. They can lead to a decrease in revenue, an increase in expenses or both [4].

Abnormal losses occur in five different forms:

- Natural losses: occur due to the nature of the materials used and cannot be avoided. These can be considered in the production process.
- Accidental losses: occur due to unforeseen events such as power failures or tool damage.
- Pilferage and theft: influence the profitability of a company not only at the material level, but also through the loss of intellectual property and exclusive production knowledge.
- Process losses: occur for example due to overproduction, defective products or raw material waste.
- Shrinkage: occurs due to incorrect storage, incorrect handling or transportation of the product. Can occur as physical damage and affect the quality of a product.

Causes of abnormal losses can be:

- Poor Maintenance: machines are not maintained correctly.
- Improper material handling: raw material gets damaged leading to wastage.
- Machine breakdown: machines break down due to wear, lack of maintenance or overuse.
- Human factors: lack of training, fatigue and stress can lead to abnormal losses.
- Inefficient production processes: non-optimized processes lead to wastage and loss.
- Quality control issues: if the quality control processes are not strict enough, faulty products are produced, leading to rejects.

Possible consequences of abnormal losses can be described as follows:

- Financial damage to the company due to reduced revenue, increased losses and possibly also reduced cash flow.
- Operational impact due to paused production and resulting delays.
- Reduced employee morale due to a feeling of insecurity and the resulting drop in productivity.
- Loss of reputation of the company due to low-quality products [4].

According to these definitions, handling damages can be classified as abnormal losses of the *shrinkage* type with the causes *improper material handling* and *human factors*. The consequences described can all occur due to handling damage.

Classic solutions are described as machine maintenance, technical improvements, worker training and quality control measures. However, since handling damages are human-related causes, psychological approaches are particularly suitable for influencing motivation and behavior.

Gamification, a contemporary methodology in motivational and interaction design, has received considerable attention in the past years. Gamification's core strength lies in its ability to analyse human behaviour and motivation and thus, gives a promising approach for the application in production environments with human-related problems [5].

3 Theoretical Background and Related Research

3.1 Gamification

The term gamification first appeared in 2002 in a paper by management consultant Nick Pelling and has gained more and more attention in many areas over time. In 2011, the first scientific conference on the topic was held [6]. Gamification is a relatively young approach that started in the field of software development. In the meantime, gamification methods are already successfully used in many companies [5].

Gamification describes the integration of *game-design elements in non-game contexts* [7]. A more objective-orientated definition is provided by Huotari and Hamari as *enhancing a service with affordances for gameful experiences in order to support users' overall value creation* [8].

The concept of gamification attempts to use the potential of video games in a meaningful and targeted way. The goal is to enhance people's motivation by offering new incentives to increase interest in activities and make overcoming challenges more attractive. By systematically designing gamification strategies, the positive motivational potential is tapped and intended behavior is triggered. The theoretical background is provided by approaches from motivational psychology, such as Csikszentmihalyi's Flow theory, which describes the optimal state of concentration between boredom due to underchallenge and stress due to overchallenge [9]. Deci and Ryan's self-determination theory describes the basic drives of human action represented by autonomy, competence, and relatedness [10].

3.2 User Types Hexad

Andrzej Marczewski is a gamification pioneer and author. He makes his collected findings available in his book *Even Ninja Monkeys like to play* [11] and on the gamified.uk website [12]. Marczewski has developed a comprehensive framework for the design and implementation of gamification.

Marczewski's basic assumption is that people can be categorized according to their motivation preferences. He then developed the six HEXAD user types, based on Bartle's player types [13] and self-determination theory [10]. The intrinsically motivated types are Achiever, Socializer, Philanthropist and Free Spirit. They are motivated by Relatedness, Autonomy, Mastery and Purpose. The other two types, motivated by Change and Reward, are the Disruptor and the Player. The HEXAD model enables the classification of people in gamified applications.

The Socializers group is driven by the desire to foster relationships and promote a sense of community identity. Philanthropists, on the other hand, are motivated by their selfless intention to help without expecting anything in return. Conversely, Players derive their motivation from the desired reward for a completed task. The Free Spirits user group is motivated by the act of creating and exploring. The pursuit of self-realization, creativity and autonomy drives the beliefs and practices of this group.

The Achiever group is attracted to challenges and puzzles and is attracted towards achievements. Acquiring knowledge and gaining experience as well as striving to develop their own skills are the main drivers.

The sixth user group, the Disruptors, are motivated by the desire to disrupt the system. Rewards are not of interest to this group. The intention is to bring about a constructive or negative change through a disruption, which is initiated either by the Disruptor themselves or by another user.

Tondello et al. developed a questionnaire with 24 items to determine the typology according to the user types Hexad [14]. Krath et al. examined a shortened version of the questionnaire with 12 items, which showed a higher average reliability [15].

Figure 1 shows the user types with their main motivational drivers.

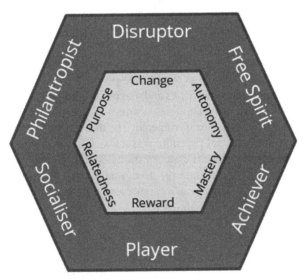

Fig. 1. User Types *HEXAD* acc. to Marczewski [11]

Marczewski assigns his identified 54 game elements to the six *User Types Hexad* and two additional categories *General* and *Schedules*. These are arranged in the so-called *Periodic Table of Gamification Elements* [12].

3.3 How to Design Gamification

In the journal article *How to design gamification? A method for engineering gamified software* Morschheuser et al. describe a design process that they developed based on a comprehensive literature analysis of gamification design methods [16]. They then interviewed 25 gamification experts on design principles for the development of gamification. The result was an integrated process for the development of gamified software, as well as a basic procedure for gamification design. The developed process was evaluated with ten gamification experts and tested on an example project. Due to the comprehensive and scientific approach, the "How to design Gamification" process according to Morschheuser et al. is a plausible basis for scientifically correct gamification design.

The process consists of seven phases that are similar to design thinking.

1. Preparation

The preparation phase serves to define and verify the objectives and requirements, as well as to answer the question: Is gamification useful within the given context? The result of this phase is the "Go Decision" and the list of requirements.

2. Analysis

The analysis includes both a context analysis and a user analysis. The context must be understood and then provided with success metrics. The user analysis consists of defining the target users with identified usage needs and motivations. The result is a typology of the target user.

3. Ideation

The ideation phase consists of brainstorming and the subsequent consolidation of ideas for the gamification of the context. For brainstorming, reference is made to various methods and frameworks, such as design thinking or Octalysis [17].

4. Design

The design phase corresponds to iterative rapid prototyping. Designing, developing and evaluating are repeated until the concept goal is achieved. The result is the development plan for the product, including specifications and budget.

5. Implementation

The implementation of the final product also takes place in iterative phases. After the decision as to whether the product will be developed internally or outsourced, a loop of design implementation and playtesting with feedback is created. Once the goal has been achieved, a pilot of the final gamified product is created. A gamification expert accompanies the process to ensure correct implementation.

6. Evaluation

Based on the success metrics defined in the preparation phase, the success of the gamified context is evaluated quantitatively. In addition, reference is made to other evaluation methods from the literature and interviews with test users.

7. Observation

After the release of the product, targeted observations can be made at the request of the client with a continuous improvement of the gamification concept [16].

3.4 Gamification Application in Production Environments

The application of gamification in production environments has already been presented in several scientific publications and also analyzed in systematic literature reviews. However, the results of the individual studies are highly specific, which is why the findings of the literature reviews will be presented here.

In their 2020 article in the Journal of Business Research *Gamification of production and logistics operations: Status quo and future directions*, Warmelink et al. analyzed 18 scientific studies on gamification applications in production and logistics environments [18]. It is concluded that research on this topic is currently in the pilot phase, mainly due to methodological and theoretical aspects of the studies. Most of the studies reviewed conducted only limited testing, often with small samples and unvalidated measurement instruments, and practically never achieved statistical significance.

The analysis shows that only one publication measured both psychological outcomes and behavioral/organizational effects in the same study. The empirical research designs were generally not very comprehensive, which is considered a common limitation in gamification research. The need to use more theory-based research designs to link psychological outcomes to behavioral or organizational effects is emphasized.

Also, it is emphasized that the gamification of production and logistics is a fascinating but largely unexplored area for organizational researchers. The authors suggest going beyond the pursuit of performance and efficiency, aiming for customized gamification and using technology to adapt gamified work contexts to individual backgrounds and needs. Another point is the definition of gamification in the context of production and logistics and the distinction between motivational aspects and important organizational design decisions. Research shows that in this area gamification is often viewed in a simplistic, deterministic and instrumental way. The authors call for more research that goes beyond pure performance or efficiency enhancement and focuses on areas such as process or product innovation [18].

In the Journal of Manufacturing Systems in 2022, Keepers et al. analyze 35 studies in their article *Current state of research & outlook of gamification for manufacturing* [19]. The authors conclude a growing interest in the field of Gamification for Manufacturing (GfM), while the existing research consists mainly of experiments on the shop floor and conceptual design, which builds a strong base for future research. The analyzed applications of gamification were not limited to a specific industry sector, but most of the times consisted of a technical, computer-based system. The benefits were split into psychological and production-orientated categories, while the psychological outcomes were discussed most. The most mentioned limitation is finding appropriate applicability of gamification in the given use case. The authors recommend to refer to gamification definition of Deterding as it is the most mentioned one and applies to various scenarios. They also recommend to use the term game elements beside other vocabularies such as game mechanics or game techniques.

The conclusion for future research in the field of GfM is to perform additional empirical research centered around the true benefits of gamification for manufacturing, specifically, how different game elements impact the users. Also, the applicability of gamification should be examined as designers struggle to identify the most fitting game elements for their individual use case.

A step-by-step guideline for implementing is suggested to support gamification design for manufacturing, focusing on the impact of specific elements in individual use cases [19].

4 Methodology

The systematic development of gamification strategies is based on the *How to design gamification*-process according to Morschheuser et al. [16].

1. The problem definition defines in the preparation phase whether gamification is a possible solution. The exact objective of the gamification strategies to be developed and the success metrics to be measured are defined.
2. The analysis phase includes an environment and user analysis according to the methodology, which is collected using qualitative or quantitative methods. Due to the scientifically proven validity, the establishment in the scientific discourse and the clear scope of the necessary questions compared to other analysis frameworks, the decision was made at this point in favor of the survey of user types Hexad profiles according to Marczewski in the Hexad-12 scale according to Krath et al. [15].
3. The interpretation of the analysis results in the form of prioritized user types in the Hexad typology [11], as well as the subsequent selection of suitable game elements in the ideation phase, take place in a workshop setting with the direct participation of all stakeholders in order to achieve a correct and in-depth understanding of the problem and the underlying psychological mechanisms of the people directly involved. *The Periodic Table of Gamification Elements* according to Marczweski provides the basis for the game element selection [12], as it enables a direct assignment to the prioritized user types. To expand the solution space and provide additional creativity support, elements from the related frameworks *Laws of UX* (User Experience Design) [20], *Design with Intent* (Design, Architecture and Ergonomics) [21] and *The Art of Game Design* (Game Design) [22] were also considered. These pattern elements cannot be directly assigned to the analysis results and therefore serve as a free source of inspiration for solution approaches. The game elements are selected by means of a systematic reduction. During the workshop, elements are evaluated in group discussions as suitable or unsuitable for the use case.

At the beginning of the design phase, the positively evaluated game elements are combined into various strategies, again within the workshop framework, with the participation of all participants. The framework conditions here are the possible integration of the elements into the use case, as well as the possible interaction of the elements with each other. A strategy consists of a clear number of elements of the positively evaluated patterns of all available frameworks. The next step in the design phase is to translate the strategies into concrete implementation ideas for the use case. This is achieved within the workshop using the rotating brainwriting method, in which implementation ideas are briefly recorded in writing and can then be supplemented by team members with their own ideas. This ensures that all stakeholders have an influence on the individual ideas and that no role remains unconsidered. At the same time, equal participation supports the acceptance of the ideas developed. The ideas for the implementation of the gamification strategies are then put into concrete terms. This is done using the *cover story* method, in which details of the ideas have to be formulated using predefined fields. A cover story is a fictitious newspaper report on the idea, in which core statements, thoughts and rough visualizations of the ideas are presented. These factors are entered in a predefined template [23].

The detailed ideas are then evaluated by the participants in the workshop by means of a dot rating with regard to their potential suitability for solving the problem in a practical application. All participants receive a certain number of dot stickers, which they can distribute to the most promising ideas. The concepts with the highest scores have priority for practical implementation [23].

4. The first step in the implementation phase is to translate the cover stories into concept boards. Concept boards define technical and organizational factors that are necessary for implementation. These are, for example, *Resources & people, activities, KPIs* or a *visualization* (see Fig. 2).

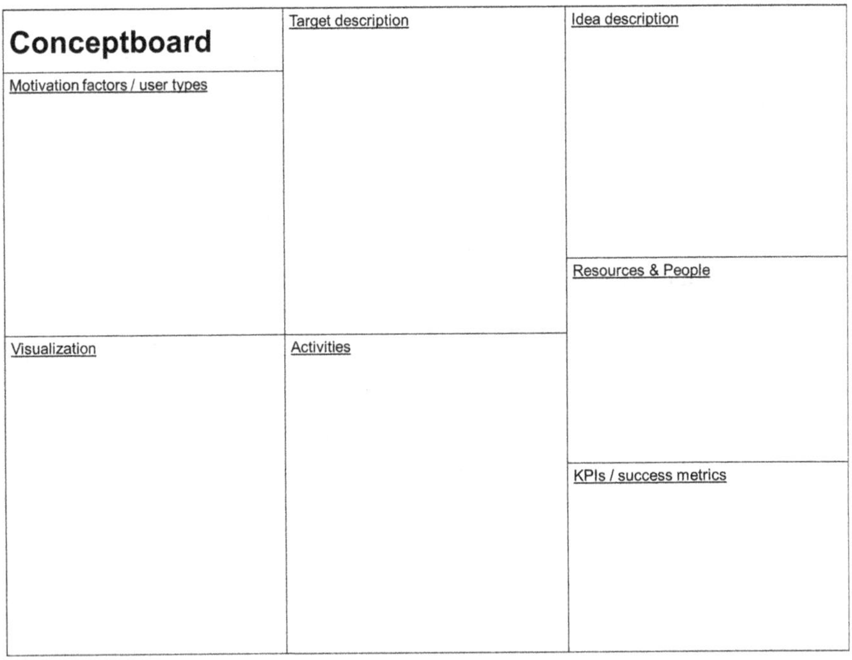

Fig. 2. Conceptboard template (own illustration)

The concept boards are now being technically implemented in a prototype status and integrated into the application context.

5. The functioning prototype is actively used in the work context over a test period. Data on the defined success metrics is collected in order to be able to carry out an evaluation.

6. The prototype is iterated based on the evaluation outcome. Further, it is implemented in the long term and transferred to various analog use cases. Continuous monitoring of KPIs and success metrics will be set up.

The complete process is illustrated in Fig. 3.

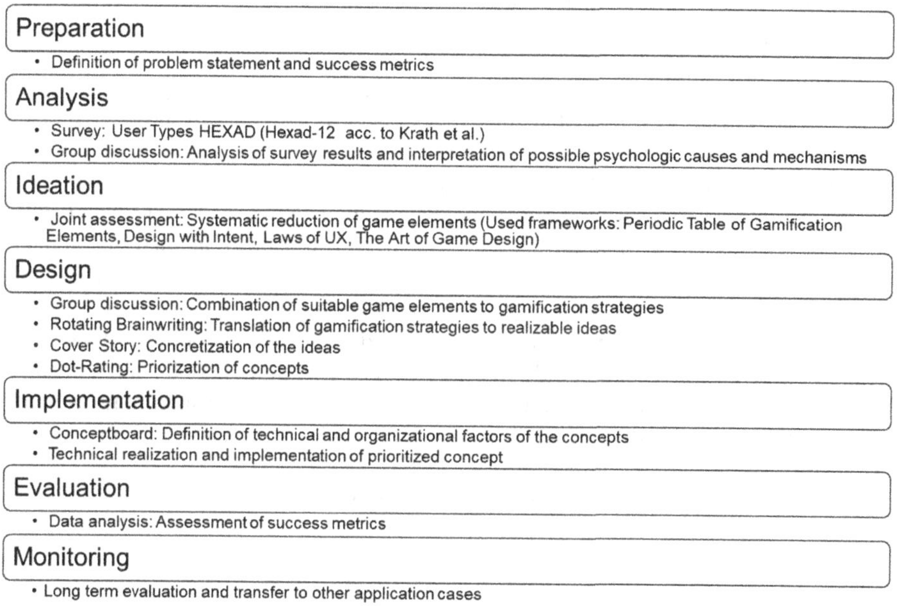

Preparation
- Definition of problem statement and success metrics

Analysis
- Survey: User Types HEXAD (Hexad-12 acc. to Krath et al.)
- Group discussion: Analysis of survey results and interpretation of possible psychologic causes and mechanisms

Ideation
- Joint assessment: Systematic reduction of game elements (Used frameworks: Periodic Table of Gamification Elements, Design with Intent, Laws of UX, The Art of Game Design)

Design
- Group discussion: Combination of suitable game elements to gamification strategies
- Rotating Brainwriting: Translation of gamification strategies to realizable ideas
- Cover Story: Concretization of the ideas
- Dot-Rating: Priorization of concepts

Implementation
- Conceptboard: Definition of technical and organizational factors of the concepts
- Technical realization and implementation of prioritized concept

Evaluation
- Data analysis: Assessment of success metrics

Monitoring
- Long term evaluation and transfer to other application cases

---

Fig. 3. Gamification design with applied methods based on the How to design gamification-process acc. to Morschheuser et al. [16]

5   Results

5.1   Research Context

The company that provides the research context for the development of gamification strategies is a German company in the aviation industry that manufactures high-precision parts with the highest quality requirements. The smallest deviations from quality standards can lead to safety-critical defects in the product, which is why quality assurance has the highest priority. Possible quality deviations can be caused by raw part errors, processing errors or handling damages by workers in production.

Handling damages describe individual errors made by people during the processing of components. These errors include, for example, the incorrect use of processing tools or unintentional damage to components due to carelessness, for example. When detected, handling damage leads to additional reworking and quality control or rejection of the machined workpiece. According to the company, the costs incurred by handling damage are in the high single-digit millions of euros per year, particularly due to the high-quality requirements placed on the products.

The company felt forced to take strategic action against handling damages and considered the use of gamification a promising modern method to find solutions to the problem.

In the following, the results of the development carried out are listed according to the individual phases of the methodology described above.

5.2 Preparation Phase

In the preparation phase, the problem, the objective of the development and the associated success metrics are defined. In addition, a decision is made as to whether gamification is a sensible approach to solving the problem.

The problem definition of the handling damages includes a human-related initial situation. In internal investigations by the company, it was found that handling damages could be reduced if workers were constantly monitored. It was concluded that the occurrence of handling damage was a psychological effect during the workers' daily work routine. Ideas for restrictive measures were rejected, as this does not correspond to the company philosophy, which includes a high regard for working people, and a negative influence on morale should be avoided. Positive methods for increasing motivation and influencing behavior should be investigated, which is why gamification is considered as a possible solution.

Two different aspects should be considered within the development:

1. the general avoidance of handling damage
2. the early reporting in case of occurring handling damages

A systematic development of targeted gamification solutions to reduce and report handling damage in production at an early stage is therefore defined as the objective of the study. The associated success metrics were defined as (Table 1):

Table 1. Defined success metrics, descriptions and targeted changing directions

Suc. Metr.	Description	Changing direction
SM1	Amount of handling damages detected	↘
SM2	Amount of reportings	↗
SM3	Time period until reporting after detection	↘
SM4	Positive feedback from the workers	↗

Possible interdependencies between the metrics are not initially considered.

5.3 Analysis Phase

For the survey of a motivation typology, the existing questionnaire on the Hexad user types with 12 items according to Krath et al. [15] was slightly adapted in consultation with the company's works council. This was necessary from the works council's point of view in order to avoid reservations in the form of insinuations on the part of the respondents. For example, questions such as "I see myself as rebel" could not be used verbatim due to their unsuitability for a work context. Similarly, it should not be assumed that workers only perform at their best in the work context if they are rewarded, as suggested by the question "If the reward is sufficient, I will put in the effort", for example. The questions were adapted in consultation with the works council in order to protect the rights of the workers on the one hand and not to distort the original intention of the question on the other. The answer options corresponded to a 5-point Likert scale: "agree, I somewhat agree, neither, I somewhat disagree, I disagree". The questionnaire was prepared in German at the company. 89 questionnaires were collected, of which 86 can be fully analyzed. The adapted questions are translated into English in the following table (Table 2):

The results of the data analysis are shown in the following table. The highest values have been highlighted (Table 3).

The results show a consistent focus on Philanthropist and Player user type. Socializer and Free Spirit share inconsistent responses to the questions, while Disruptor consistently receives low responses. The Achiever is slightly below Philanthropist and Player, but also consistent. Based on these results, the focus in the following phases of gamification development is placed on the Philanthropist, Player and Achiever user types.

The following phases up to the design were carried out as part of a workshop. In order to be able to develop the most suitable strategies, the presence of the directly affected group of people, the workers, was assumed. Furthermore, the workshop team consisted of heterogeneous participants from different departments of the company. There were 19 people present.

The results of the analysis were interpreted in a group discussion after their presentation. The focus on the Philanthropist user type was interpreted as a generally positive, selfless attitude towards solving the problem of handling damages. At the same time, the Player and Achiever focus reflects a self-critical and success-driven attitude. Within the discussion of possible psychological causes and effects, unclear processes and the resulting excessive demands when handling damages occur and the subsequent decision-making process for further processing were mentioned. There is a rough guideline on what should happen in the event of handling damage and an individual workaround for dealing with handling damage when it occurs - there was no precise definition of what constitutes handling damage to be reported and at what level of severity it must be reported. Accordingly, the decision and responsibility lay with the workers in individual cases, who acted on the basis of their subjective judgment and experience. A second reason was the lack of feedback on independent decision-making. If handling damages were reported or rectified, there was no feedback in the follow-up process as to whether the part had ended up in the scrap bin or was further processed, or whether the rectification was sufficient. The workers therefore carried out work on their own responsibility, based on their experience with the product, and did not know how the decisions made

Table 2. Adapted questionnaire based on the Hexad-12 scale from Krath et al. [15]

Hexad type	Item acc. to [15]	Original question acc. [14]	Adapted question
Philanthropist	p1	It makes me happy if I am able to help others	It makes me happy when I am able to help my colleagues
	p4	The well-being of others is important to me	I want my colleagues to do well
Socializer	s2	I like being part of a team	I like working in a team
	s4	I enjoy group activities	I enjoy group tasks
Achiever	a2	I like mastering difficult tasks	I like mastering difficult tasks
	a4	I enjoy emerging victorious out of difficult circumstances	I like overcoming challenges
Player	r2	Rewards are a great way to motivate me	Experiencing success motivates me
	r4	If the reward is sufficient, I will put in the effort	I try harder when I experience success
Free Spirit	f1	It is important to me to follow my own path	I would like to fulfill myself in my job
	f3	Being independent is important to me	Independence is important to me at work
Disruptor	d3	I see myself as a rebel	I would do/design things differently in my working environment
	d4	I dislike following rules	Too many rules restrict me

were to be evaluated. This lack of transparency led to demotivation when carrying out this work.

5.4 Ideation Phase

Suitable game elements were then selected during the same workshop. The elements of Marczewski's *Periodic Table of Gamification Elements* [12] provided the basis of the available elements. From these, the corresponding elements of the identified focus user types (Philanthropist, Achiever, Player), as well as the general mechanics (Reward Schedule, General) were used. These were supplemented by the patterns of the frameworks *Design with Intent* [24], *Laws of UX* [20] and *The Art of Game Design* [22]to create an additional source of inspiration. The total output was around 90 elements. The elements were available in physical form in the workshop and were explained to the participants in detail.

Table 3. Results of the questionnaire

Item	Mean value (n = 86)	Standard deviation
p1	**3.50**	.85
p4	**3.44**	.78
s2	**3.57**	.72
s4	2.95	.92
a2	3.31	.80
a4	3.22	.84
r2	**3.44**	.89
r4	3.27	.92
f1	2.97	.89
f3	**3.40**	.70
d3	2.82	.98
d4	3.03	.91

The reduction took place as part of a group discussion. Each element was briefly discussed individually and approved or rejected in terms of its basic suitability for the application. This reduced the selection of available elements to around 30.

5.5 Design Phase

The subsequent part of the workshop included the design of gamification ideas for the application case.

The first part involved combining game elements into gamification strategies. This also took place in a group discussion. Elements that showed synergies and were potentially compatible in the application case were combined to form a strategy. For each defined goal from the preparation phase (reducing handling damage, early reporting of handling damage), three strategies were developed, consisting of a number of three to eleven game elements. The number of elements was not limited so as not to restrict the creativity of the participants.

One strategy focused on Player User Type elements, such as *feedback-related activities*, the second on Philanthropist-associated elements, such as *Sharing knowledge*, and the third on Achievers, with *Learning* mechanics, for example.

The next step was the creative translation of gamification strategies into implementation ideas to solve the initial problem. The method chosen for this was brainwriting. Each participant had five minutes to write down their implementation idea for a gamification strategy. The ideas were then passed on to another participant, who added their ideas for implementation. This not only ensured that all participants were aware of all the ideas, but also increased acceptance of the ideas developed through their own participation.

The next step was to concretize the still relatively abstract ideas. The cover story was chosen as the method for this. The template was handed out to the group and filled in for the individual ideas. The cover story helps to visualize and identify critical aspects of the ideas and queries the necessary factors for the subsequent technical implementation.

The results of the creative phase are shown as an example in Fig. 4.

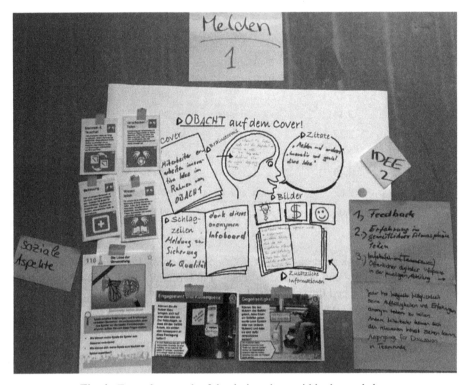

Fig. 4. Exemplary result of the design phase within the workshop

The final step of the workshop and the design phase was the evaluation and prioritization of the developed concepts in order to identify the most suitable concept for the application case. The evaluation was carried out using a dot rating. All participants were given three sticker dots, which they were allowed to distribute as they wished among the existing concepts. The three concepts with the most votes were considered for technical implementation in the next phase.

The three concepts consisted of

1. an emergency button that could be pressed if handling damage occurred in order to inform the shift quality manager. The error was then recorded. (From the Player user type related strategy)
2. an error catalog in which workers could enter regularly occurring errors and damages and their solution in order to inform their colleagues. The catalog would be an integral part of the daily meeting before the start of the shift, but could also be viewed individually at any time. (From the Philanthropist user type related strategy)

3. an escape room learning concept with monthly changing tasks on regularly occurring errors and damages which might also be used in on-boarding for new employees. (From the Achiever user type related strategy)

As the concept of the emergency button received the most points in the dot evaluation, the initial focus was on the realization of this concept.

However, as the two other concepts offer very good synergies and combination possibilities (the handling damages identified in the emergency button as input for the error catalog, current elements of the error catalog as the basis for the escape room), these are to be considered in later projects. Regarding the two objectives from the preparation phase, the emergency button addresses the early reporting of handling damages, while the other two concept focus on general reduction of handling damages.

5.6 Implementation Phase

In the following implementation, the technical realization of the emergency button is carried out as a prototype and the subsequent integration at a workstation in the application environment. A purely digital solution did not appear to be suitable for everyday use in production due to the lack of accessibility. The decision was therefore made to use a buzzer button. Due to the company's high security standards, the solution could not be integrated into the company's own intranet at the prototype stage. The button was therefore connected to a mobile laptop, which was placed in the immediate vicinity of the workers' workstations.

When handling damage occurs, the worker can now press the button if he needs additional help from the quality manager if anything is unclear. Paper form templates for recording defect details for the company's database have been placed next to the laptop, which should be filled in after pressing the button. Motivational messages then appear on the laptop as a form of feedback, such as *Thank you for always keeping an eye out for us.*

This solution addresses all identified focus user types from the analysis phase:

- Philanthropist: Supporting the company by reducing handling damages and thus increasing sales and profits
- Achiever: Further qualification by recording and documenting handling damages
- Player: Form and functionality of the solution, as well as the existing feedback

The paper form template is also to be digitized when the prototype is further developed for regular operation. There are also plans to replace the laptop (Fig. 5).

5.7 Evaluation Phase

The emergency button was integrated into a quality inspection workstation during a 3-week test phase. In general, the concept was very positively received by the workers, as a new option was introduced to request support in the event of uncertainties. The acceptance of the concept was additionally supported by the participation of the workers in the workshop, as the concept was already known at the time of implementation.

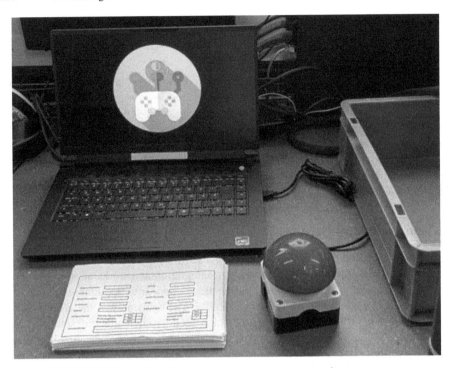

Fig. 5. Implemented working prototype of the emergency button concept

The three-week test phase took place during the summer vacation in Germany, where the throughput of products in production was slightly reduced. The emergency button was activated six times during the test phase.

While the absolute number of activations does not seem particularly high for the three-week period, the success metrics developed extremely positively.

The success metric number of existing handling damages (to be reduced) has fallen to zero. During the test period, no more handling damages were registered outside of the reports. In this test scenario, the gamification strategy therefore generated the best possible impact.

Conversely, the number of complaints (to be increased) increased to the maximum possible value, as all handling damages to be identified were discovered or reported.

The time to report after discovery (to be reduced) could not be measured in comparison to the time before the integration of the error emergency call, but it is assumed that this has been reduced due to the simplification of the process and the high willingness of the workers to accept it.

The positive feedback from the workers (to increase) was not systematically recorded, but as described above, a high level of acceptance was achieved through the participation of the workers and the simplification and clear definition of processes (Table 4).

An interesting development was made by the workers themselves. They not only reported the handling damages, as originally planned, but also discovered defects in the raw parts. As a result, in addition to the actual goal, an additional goal was achieved that

Table 4. Evaluation of the success metrics

Suc. Metr.	Description	Changing direction	Obtained?
SM1	Amount of handling damages detected	↘	✓
SM2	Amount of reportings	↗	✓
SM3	Time period until reporting after detection	↘	?
SM4	Positive feedback from the workers	↗	✓

was not part of the original development. As a result, the targeted quality improvement in production was exceeded by the avoidance of handling damages.

6 Discussion and Findings

Handling damage caused by workers is a cost-intensive problem, especially for high-tech companies with high quality standards. Due to the human cause, solutions from the field of psychology are particularly promising.

Through the systematic development of gamification strategies and the subsequent technical implementation of these, it has been possible in this case to significantly reduce handling damages.

While the survey based on the user types Hexad method revealed a focus on philanthropists, players and achievers, possible causes for the occurrence of handling damages in production that can be transferred to similar use cases were identified, in particular the lack of clarity in processes when handling damages occur, the resulting possible excessive demands in the decision-making process for further processing and a lack of transparency and feedback regarding one's own work.

In the present research, the aspect of the unclear process chain was addressed in particular in the technical implementation. A simple solution that fits into the production environment not only solved the problem within the process chain, but also provided it with a positively motivating reinforcement.

The impact of the installed concept was considerable. The success metrics defined at the beginning were positively changed. The number of handling damage reports was increased by introducing a clear process, while the time span from detection to reporting was probably reduced. Positive feedback from workers was achieved. A statistically evaluable analysis of the data is still pending and will be part of future publications.

In the future, the concept will be technically detailed from the proof-of-concept phase and transferred to a broad prototype phase at several locations in the production environment of the company in order to be able to generate statistically relevant results on the reduction of handling damages. At the same time, the combination with the other concept from the design phase is aimed at, which includes a database storage and regular discussion of the reported defects on the shopfloor level and an escape room on-boarding and education concept. The goal is to increase the awareness for currently frequently occurring errors and a positively lived error culture.

When developing gamification strategies in similar environments, the involvement of the affected groups of people is particularly recommended in order to gain a comprehensive understanding of the origin and scope of the problem on the one hand and to increase acceptance of the developed solution strategies among the employees on the other.

The *How to design gamification*-process according to Morschheuser et al. [16] in combination with the *User Types Hexad* and the *Periodic Table of Gamification Elements* according to Marczewski [11, 12] offers a systematic approach that can generate a multitude of creative ideas within a practice-oriented development.

The selected methods in the design phase could be replaced by similar ones. It is essential to support the participants through the systematic stimulation of creativity, the targeted concretization and visualization to reduce the level of abstraction and the subsequent prioritization to meaningfully reduce the concepts through the applied methods.

The presented research thus provides an exemplary application of a systematic approach, as demanded in existing academic literature, thus expanding the scientific state of the art. The empirical analysis following the acquisition of all data will be part of future publications.

References

1. Turner, C.J., Ma, R., Chen, J., Oyekan, J.: Human in the loop: industry 4.0 technologies and scenarios for worker mediation of automated manufacturing, pp. 103950–103966 (2021)
2. Sgarbossa, F., Grosse, E.H., Neumann, W.P., Battini, D., Glock, C.H.: Human factors in production and logistics systems of the future, pp. 295–305 (2020)
3. Lee, J.D., Seppelt, B.D.: Human factors in automation design. In: Nof, S.Y. (ed.) Springer Handbook of Automation: With 149 Tables, pp. 417–436. Springer, Heidelberg (2009)
4. FasterCapital: Abnormal loss: Beyond the Ordinary: Analyzing Abnormal Losses (2023). https://fastercapital.com/content/Abnormal-loss--Beyond-the-Ordinary--Analyzing-Abnormal-Losses.html#Introduction-to-Abnormal-Losses. Accessed 22 Jan 2024
5. Reiners, T., Wood, L.C. (eds.) Gamification in Education and Business. Springer, Cham (2015). https://doi.org/10.1007/978-3-319-10208-5
6. Fleisch, H., Mecking, C., Steinsdörfer, E.: Gamification4Good: Gemeinwohl spielerisch stärken (2018). https://ebookcentral.proquest.com/lib/gbv/detail.action?docID=5427441. Accessed 15 June 2023
7. Deterding, S., Dixon, D., Khaled, R., Nacke, L.E.: From Game Design Elements to Gamefulness: Defining "Gamification" (2011). http://dl.acm.org/citation.cfm?id=2181037
8. Huotari, K., Hamari, J.: A definition for gamification: anchoring gamification in the service marketing literature, pp. 21–31 (2017). https://link.springer.com/article/10.1007/s12525-015-0212-z
9. Csikszentmihalyi, M.: Flow: the psychology of optimal experience. Harper and Row, New York (1990)
10. Deci, E.L., Ryan, R.M.: Self-determination theory: when mind mediates behavior. J. Mind Behav. 1, 33–43 (1980). https://www.jstor.org/stable/43852807. Accessed 15 June 2023
11. Marczewski, A.: Even Ninja Monkeys Like to Play: Unicorn Edition (2018). 978-1724017109
12. Marczewski, A.: Gamified UK: Gamification & Life in general (2023). https://www.gamified.uk. Accessed 12 Jan 2023
13. Bartle, R.: Hearts, clubs, diamonds, spades: players who suit MUDs. J. MUD Res. (1996)

14. Tondello, G.F., Wehbe, R.R., Diamond, L., Busch, M., Marczewski, A., Nacke, L.E.: The gamification user types Hexad scale. In: Cox, A., Toups Dugas, P.O., Mandryk, R.L., Cairns, P. (eds.) Proceedings of the 2016 Annual Symposium on Computer-Human Interaction in Play, pp. 229–243, New York, NY, USA. ACM (2016)
15. Krath, J., Altmeyer, M., Tondello, G.F., Nacke, L.E.: Hexad-12: developing and validating a short version of the gamification user types Hexad scale. In: Proceedings of the 2023 CHI Conference on Human Factors in Computing Systems, New York, NY, USA. ACM (2023)
16. Morschheuser, B., Hassan, L., Werder, K., Hamari, J.: How to design gamification? A method for engineering gamified software. Inf. Technol. **95**, 219–237 (2018). https://doi.org/10.1016/j.infsof.2017.10.015
17. Chou, Y.: Actionable Gamification. Octalysis Media (2015)
18. Warmelink, H., Koivisto, J., Mayer, I., Vesa, M., Hamari, J.: Gamification of production and logistics operations: status quo and future directions. J. Bus. Res. **106**, 331–340 (2020). https://doi.org/10.1016/j.jbusres.2018.09.011
19. Keepers, M., Nesbit, I., Romero, D., Wuest, T.: Current state of research & outlook of gamification for manufacturing. J. Manuf. Syst. **64**, 303–315 (2022). https://doi.org/10.1016/j.jmsy.2022.07.001
20. Yablonski, J.: Laws of UX. O'Reilly Media, Inc. (2024)
21. Lockton, D., Harrison, D., Stanton, N.A.: The Design with Intent Method: a design tool for influencing user behaviour. Appl. Ergon. **41**, 382–392 (2010). https://doi.org/10.1016/j.apergo.2009.09.001
22. Schell, J.: The Art of Game Design: A Book of Lenses. CRC Press (2008)
23. Gray, D., Brown, S., Macanufo, J.: Gamestorming: A Playbook for Innovators, Rulebreakers, and Changemakers. O'Reilly Media, Inc. (2010)
24. Lockton, D., Harrison, D., Stanton, N.A.: Design with Intent: 101 patterns for influencing behaviour through design. Equifine, Berkshire, UK (2010)

Research on Solving the Problem of Children's Doctor-Patient Relationship from the Perspective of System Theory - The Example of Children's Medical IP "Courageous Planet"

Cheng Peng[1] and Yajie Wang[2(✉)]

[1] Tongji University, Shanghai, China
[2] Wuhan University of Technology, Wuhan, China
411349548@qq.com

Abstract. Due to children's limited cognitive ability, children's patients often suffer from fear and anxiety during treatment, which leads to resistance to medical treatment, and parents don't know how to comfort their children, so the tension between doctors and patients in children's hospitals has always been a thorny issue. Shanghai Children's Medical Centre (SCMC), one of the best children's medical institutions in Shanghai, was the first to set the goal of becoming a "no-cry children's hospital", and since 2014, the School of Design and Creativity in Tongji University has been establishing a design-driven collaboration with SCMC, aiming to improve children's healthcare services and experience.

Hospitals are a systemic problem, made up of different roles, and different environments, and essentially a communication problem in connecting doctors, parents, and children. This paper is based on Richard Buchanan's Definitions of system: The four modes of definitions of system: set, arrangement, group, and condition, of which set, and arrangement belong to the design thinking mode of problem-solving, which is also usually the way designers think about problems, but this design mode is not a good solution to the problem of communication between children, parents, and hospitals. This paper favors the thinking mode of CONDITION, which is the harmonious and orderly interaction of the whole, without the smallest part, close to the process of seeking truth, a mode of Assimilation, the process of mutual understanding and communication through natural life and extending to the social group. IP is such a bridge that it can communicate across media and at the same time assimilate each medium into a systematic ecology for its use. Through the children's healthcare IP to solve the communication problem, to build a new concern of the three, so that people with different identities can be placed in it and understand each other.

Through researching 123 doctors and 121 families with children, we produced the "Courageous Planet" medical IP brand design based on children's physiological and psychological characteristics and research in the medical field and then verified it to produce a business strategy model. The "Planet Courage" medical IP design aims to help doctors, parents, and children deliver messages and build emotional ties more effectively in the relatively special setting of a children's hospital through the introduction of IP images. The name "Planet Courage" is centered on the word "courage", hoping that every child is a little warrior who is brave enough to face up to illness and grow up healthy and happy.

Keywords: System Theory · Children's doctor-patient relationship · Children's medical IP design · Cross-media Narrative · Design method

1   Case Background and Reflection

Since 2014, Tongji University's School of Design and Creativity (D&I) has established a design-driven partnership with Shanghai Children's Medical Center (SCMC) to jointly improve children's health and medical service experience, while providing a reference sample for every researcher and practitioner involved in this interdisciplinary work. On October 30, 2018, the "Medical Care with Temperature – 'No Crying Hospital Plan' series Design Exhibition" jointly organized by Shanghai Children's Medical Center and Tongji University Design and Creativity College officially opened in the atrium of the 4th floor of the Blood-Tumor Building of Shanghai Children's Medical Center. This exhibition is the first design exhibition held in a Class III hospital in China.

In this exhibition, a large number of children's medical product design achievements were displayed, which left a deep impression on the hospital. For example, this child atomization treatment product design, which is a "small elephant" design of a simulative state, based on existing atomization treatment products, by changing the connection form of the atomizer, using a freely bending and telescopic connection tube to separate the atomizer and the mask, optimizing the product layout, making the atomization mask easier to wear comfortably and not easy to break free. At the same time, a cute and context-compatible elephant shape was adopted for the atomizer after separation, which weakened the negative experience of children forced to accept treatment and alleviated their negative impression and fear of medical equipment, so that children's atomization treatment could obtain better experience and higher efficiency as shown in Fig. 1.

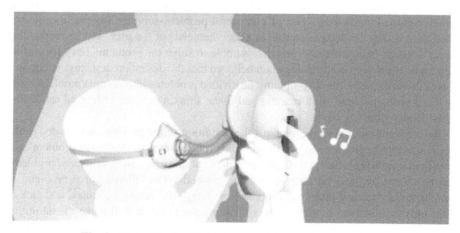

Fig. 1. "Baby Elephant" children atomizing treatment products

However, these design achievements are difficult to implement in the daily treatment of hospitals. The reason for this is the lack of investment or corporate involvement in the design results, and the fact that these designs are unlikely to bring about the immediate and visible improvement in the hospital as expected. Reflecting on the above phenomenon, the students spent a lot of time and energy in the hospital research and believe that design-thinking as the leading design research mode can bring improvement and influence on the field of children's medical care, to solve a problem in a certain link of children's medical care. To understand the "unmet needs" of doctors, parents, and child patients from the perspective of designers, to use design thinking methods such as user journey diagrams to "reasonably plan" user behavior paths, and to define and solve problems based on this. Just like the "baby elephant" children's atomization treatment product, although the use of skeuomorphic modeling design can effectively alleviate children's fear when using treatment products, a single product line and the overall unfamiliar environment of the hospital cannot make children accept a single friendly product in a short time but often ignore the overall environment and the fear brought by strangers.

Therefore, based on the problem-solving theory [2], it is easy for designers to focus on specific scenes or local tasks in the hospital when solving the problem of doctor-patient communication, hoping to improve the user experience by solving the problems in these situations, but the final effect has certain limitations.

2   Problem-Solving Theory and System Theory

2.1   Conceptual Definition of Problem-Solving Theory

"Problem-solving" [2] is usually defined as a series of purposeful and directional cognitive operation processes, which is essentially a kind of thinking activity. Simon [3], as one of the founders of the first generation of design methodology, defined design as the process of "problem-solving", and first proposed the view of "artificial science", a systematized "problem", thought that a complex problem was a system composed of several sub-problems, and established the linear model of analysis-synthesis-evaluation, which regarded method as a part of purposeful problem-solving. The premise of such a solution is that the designer needs to know the needs of the user, and corresponding needs analysis (sub-problems) and disassemble to solve the problem. Terry [4] further developed the design methodology and believed that the design process has uncertainty, especially in the definition of problems, he divided problems into "simple problems" and "wicked problem", and from rational and linear thinking modes to limited rationality and dialogue thinking mode.

In the solution of design problems, the definition of the problem is regarded as the key to solving the problem. Simon divided the problem into "ill-structured problem" and "well-structured problem" [5]. He further points out that only "well-structured problem" can be solved by "general problem-solving method". The "ill-structured problem" is similar to the "wicked problem" proposed by Terry. The scope, variables, and lack of clear purpose involved in the process of solving the problem will make the definition of the problem ambiguous and uncertain. This shows that when designing the method of "problem-solving", it is very important to define the properties of the problem: the deterministic description of the problem, the singularity of the initial problem, the clear correspondence of the solution, the certainty of the problem solver for the relevant knowledge, the universality of the problem-solving principle, etc.

2.2 Limitations of Problem-Solving Theory and Design Thinking

Richard Buchan [6] pointed out in his application of wicked problems to design thinking that every "wicked problem" is a symptom of another higher-level problem, which is essentially a social system problem with chaotic information and difficulty in framing. Richard Buchan [7] argues that in the era of artificial intelligence, decision-making is increasingly limited for these complex problems, interrelated systems of human and social behavior, and the old system-based design methods are not adapted to the new needs. After its root, the creativity of design thinking belongs to cognitive behavior, and its thought and practice are often separated in the design process so that the design in the process of dealing with the boundary of the problem, the problem of dealing with complex social organization, reflects the design in the process of solving problems there is a certain ambiguity. This exactly corresponds to the "ill-structured problem" referred to by Simon and the "wicked problem" proposed by Terry.

While "problem-solving theory" is a good basis for describing design, it also requires a detailed description of the "problem solver," including an account of the early knowledge that the "problem solver" may bring to the situation [5]. Donald Norman [8] argues that designers (engineers) are not aware of the extensive experimental and theoretical literature, nor are they well versed in statistical variability. Make plausible explanations for behaviors that have little basis in fact. This creates uncertainty about the boundaries of their approach to the problem, creating a wicked problem. The designers (engineers) he refers to here are "problem solvers".

This means that in problem-solving theory, the control of variables must be based on science and logic. As an interdisciplinary [5] subject, design includes the problems of environment, society, culture, nature, and other natural sciences and humanities. As a result, there are no fixed conditions and restrictions for design problems, forming problems that are difficult to define, that is, wicked problems. In addition, design itself is not a tool to study a single problem, and it will change with different philosophical concepts and practical situations. Therefore, the problems that designers think about are rarely limited to any single knowledge field above but will intricately and complexly consider the knowledge of multiple fields, forming the culture, logic, and method that the human world itself has, to achieve the purpose of solving the problem.

2.3 The Value of Systems Theory

In response to the uncertain process practice in design problems, Donald Schon [9] proposed practical epistemology, that is, to acquire practical knowledge through reflection in action. Based on Dewey's [10] pragmatism, the ill-structured problem is viewed from a new Angle, and the design problem is solved through the experience of design [11] practitioners and the schemes previously learned. This design process model based on inspiration experience makes up for the fact that designers (engineers) do not strictly define the problem in the early stage of design.

To describe how to solve the "ill-structured problem" in depth, Maher [12] et al. proposed the design model of "co-evolution", that is, design does not first fix the problem and then search for a satisfactory solution concept. Creative design is more a matter of the formation of a problem and the development and refinement of the idea of a solution, with the process of analysis, synthesis, and evaluation constantly repeated between two conceptual design "Spaces" - the problem space and the solution space. Building on Maher, Cross [12] proposes that creative activity in design is to build a "bridge" between the problem space and the solution space by identifying key concepts. There is a period of exploration when the problem and solution space is constantly evolving and unstable until it is taken over by an emergency bridge that identifies the problem-solution pairing. This eureka moment: Schon [14] calls it "problem framing."

Both "problem framing" and "empirical problem-solving" require designers to have a strong early understanding of each situation. Richard Buchan [7] points out that one of the greatest strengths of design over the past century has been to focus on specific things in human experience, namely the problems that humans face in their lives. And what products humans can create and manufacture to overcome those problems or problems. However, at a time when old systems and empirical cognition are showing signs of strain, designers are being asked to smooth the edges, as new technological systems powered by artificial intelligence provide opportunities for innovation in our human relationships and our relationship with artifacts and the natural world, and as networks of social-technological systems are evident all around us, in the need to seek more effective ways to serve human society.

From the beginning of the design discipline, the concept of systems has been a part of design theory and practice. A system is defined as a relationship of parts that work together in an organized way to achieve a common goal [7]. Its existence begins with our perception of the relationships between the parts and the whole in our surroundings. A designer's mode of thinking usually integrates a discrete, uncertain situation into a potentially unified system, making it a relationship between parts and whole, and defines or postulates the problem through "problem-solving theory" and divides it into several parts or postulates for practical and theoretical reflection. But when the designer studies the part and the whole, there are ambiguities about the nature of the part and the whole, such as whether the state of the part is static or dynamic, physical or immaterial, ontological or epist emological [15], and whether the part and the part, the part and the whole work together, what the relationship is, etc. This directly leads to the ambiguity and dynamics that cannot be avoided in the process of defining problems.

The definition and division of the whole and part of the system is the beginning of the system discussion. Only with a more precise definition can the system be applied to the inquiry of various disciplines and the phenomenon problems that are widely dealt with, and can it be used to distinguish the important meaning of practical action and theoretical reflection. Richard Buchanan divides system definitions into four main categories: Arrangement, Set, Group, and Condition. Each category is based on a different mode [7] of thinking, and Richard McKeon distinguishes these four modes of thinking: the process of putting parts together, the process of seeking truth based on natural science, the process of solving problems, and the process of interpreting arbitrary statements [16], and argues that these four modes of thinking are mutually exclusive. This exactly corresponds to the different design methods in the process of problem-solving and can be used as a benchmark to deal with different problem definitions, to solve the "ill-structured problem" and "wicked problem".

3 Change the Definition of Children's Medical Problems from Problem-Solving Theory to System Theory

3.1 The Doctor-Patient Relationship of Children is Reshaped Under the System Theory

Under the problem-solving theory, the hospital can generally be regarded as a system composed of multiple parts. If a single component, such as the injection or vaping process, can be effectively improved, it is possible to improve the medical experience of children. Therefore, the design process usually starts with identifying the problems in each part and delivering the results in a way that solves the problem pattern. To some extent, this is considered to be a necessary step to complete the design, but the results seem reasonable but often not satisfactory.

This shows that problem-solving theory often lacks objective and comprehensive description and cognition in the face of increasingly complex design subjects, behavior, technology, and value problems [7]. Compared with other types of hospitals, the doctor-patient relationship in children's hospitals is more complex, and the diagnosis and treatment process involves child patients, doctors, and parents. The doctor-patient relationship in children is exactly this kind of "Wicked Problem" with ambiguity and uncertainty, so it needs to stand on a higher dimension to think. Under the system theory, based on Richard Buchanan's definition of the system, the system is divided into four categories: Arrangement, Set, Group, and Condition, as shown in Fig. 2.

A system is an ARRANGEMENT of interacting parts or bodies combined under the influence of related forces [7]. Arrangement represents the mode of construction, which emphasizes rational and scientific problem-solving by recognizing physical or material objects and obeying the empirical laws of nature. The set is determined by human agency, and selected to meet a human intent or purpose in the interpretation of phenomena [7]. Set stands for mode of discernment, a set that is chosen to satisfy a human intention or purpose. Explain problems by having reflective conversations with others about thoughts, emotions, behaviors, etc.

Group stands for being broken down into constituent elements to be studied and then integrated into a new system through logic, art, or design to achieve a new or redefined purpose. In a sense, it extends design thinking from natural relationships to social groups, emphasizing the synergy of people, objects, and the environment, and emphasizing meeting human needs through situations. This model assumes that there are problems and properties in our environment that can be systematically addressed through research, action and practice. Condition represents the mode of Assimilation. Assimilation is a process of approximating the truths or principles that organize phenomena [7]. It is based on the assumption that there are no least parts and that there is an ontological unifying principle.

If we take the Set and Group perspective as an example, the designer will view the hospital as a system composed of many parts, and the discriminating and solving mode of thinking will focus on the specific situation of the hospital or the problem of individual parts while ignoring the whole system. In the scenario of children's hospital, the relationship between child patients, parents, and doctors is more inclined to a circular whole, which needs to link the communication problem of the needs of different attributes. Therefore, the perspective should be more oriented to the Condition. There is neither the largest nor the smallest part in the overall treatment process of children patients in the hospital, but a harmonious and orderly whole. It represents a mode of assimilation and is based on an ontological principle of unity. The problem of communication is complex and pluralistic, and it cannot be simply divided into a specific part of the problem. It needs to graft a new bridge to build a new relationship between the three, so that people of different identities can be placed in it and understand each other, which is a process of mutual understanding and communication between natural life and extended to social groups, a process close to the search for truth.

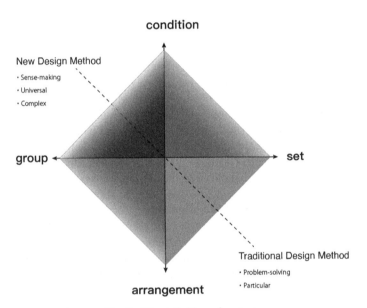

Fig. 2. The definition of systems

3.2   Multidimensional Information Integration of Children's Medical Treatment

To further understand the communication problems of children in the medical treatment process, the team interviewed 123 doctors and 121 families with children in China. Through the investigation, this paper structured the design information of children before, and scratch in a multi-dimensional design information module, and extracted the communication problems of children's patients, doctors, and parents, as shown in Fig. 3. Due to the limited cognitive ability of children, children cannot effectively communicate with doctors in the process of seeking medical treatment [17]. Preschool children between 2 and 6 years old are very susceptible to illness due to their weak body resistance. Before seeking medical treatment, children usually resist seeking medical treatment most obviously when they see a doctor and get an injection, and parents do not know how to comfort children; In medical treatment, the biggest headache for doctors is that children's crying and hyperactivity affect treatment, and parents often feel powerless and unable to communicate correctly with children. We interviewed a nurse:

We have rich experience in clinical nursing, the injection action is soon completed, and children feel no pain in the process of injection, which will not make children produce strong resistance emotions, the key is that children are not willing to extend their hands to cooperate with the treatment. Including in the atomization treatment, children will not produce pain but need children to wear a mask to cooperate with the treatment. Many times the mask and strange environment will make children feel frightened and uncomfortable, at this time it is easy to make children reject the emotion, unwilling to cooperate with the treatment.

At the completion stage of medical treatment, children usually show resistance to taking medicine, and parents lack relevant medical knowledge, which makes it difficult to communicate with children. Finally, parents will force children to take medicine through forced behavior, which will be accompanied by children's crying for a long time. In the long run, the overall performance of children in the process of medical treatment is emotional instability, fear of medical treatment, and fear of unfamiliar medical environment and other problems.

3.3 Analysis of Children's Psychological Cognition Under Medical Treatment Environment

To further seek the truth, combined with the theoretical literature research and analysis of the psychological development of preschool children [23], the reasons why children aged 2–6 do not cooperate with medical treatment can be explained from the following points:

1. The behavior is strongly emotional: the behavior is often dominated by emotions, not by reason, emotional instability, behavior is difficult to manage, parents of young children find it difficult to control children through preaching, the younger the performance is more obvious. It is easy to be affected by the external environment.
2. Love imitation: preschool children have poor independence and love to imitate others, so when they see other children crying in the hospital, they will cry inexplicably.
3. Intuitive action: thinking depends on action, do it first, think while doing, lack of behavioral planning, the predictability of the results of action, will not be comprehensive analysis, so it is difficult for children to be convinced by the predictive role of education and medical care.
4. Concrete image thinking: children usually understand adult language according to their own specific life experience, once it exceeds the experience level of children, it will produce ineffective communication. Usually, the concept of medical terminology is more professional abstract, so the words of parents and doctors are often ineffective input for children.

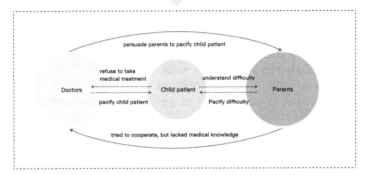

Process flows	Space places	Doctors			Child patient		Parents	
		behavior	object	tool	behavior	object	behavior	object
before medical treatment	consulting room	diagnose through interrogation	child patient &parents	stethoscope tongue spatula	cry and escape from the consulting room		pacify and control	child patient
	aisle						carry child around	child patient, belongings & medical information
be receiving medical treatment atomization as example	treatment room	wear mist mask	face	mist mask	scratch	venipuncture sites	pacify and control	child patient's body
		pacify child patient	child patient	talk				
		persuade parents	parents	talk				
after medical treatment take the medicine for example	aisle						carry child around	child patient, belongings & medical information
	pharmacy	check prescription	prescription	pharmacy system	wait		submit a prescription	prescription
		give medicine	medicine				listen to the medicine	doctor
		explain medicine use	parents	talk				
	after getting home				refuse to take medicine	medicine	pacify and control	child patient
							force children to take medicine	medicine

Fig. 3. Children's medical information structure model

Through the actual field investigation and theoretical literature research, the author reconsiders the communication problems among children's patients, parents, and doctors. Through the analysis of CONDITION design mode, it is concluded that it is necessary to help children understand the medical procedure and positive medical consciousness through a medium, and this medium can be used by the designer. Storytelling is one of the most effective ways for children to understand the world, and it is especially suitable for disseminating obscure medical terms and basic knowledge. Finally, the D&I team provides children with effective information input in the form of storytelling according to their specific image thinking, intuitive action, and love of imitation.

4 Practice of Children's Medical IP Design Based on Cross-Media Narrative

4.1 Necessity of Cross-Media Narrative IP Design in Children's Medical Treatment

The literal translation of IP (Intellectual Property) is intellectual property. This paper takes IP out of the structure of intellectual property and regards it as a symbol with strong characteristics, which can be a story or a character. Cross-media narrative belongs to the category theory of communication. In 2003, Henry Jenkins [17] first proposed the cross-media narrative theory to conduct related research on IP operation and development. Transmedia storytelling represents a process in which the components of a novel are systematically dispersed across multiple channels of communication to create a unified and coordinated entertainment experience. In an ideal world, each medium would make its own unique contribution to the unfolding of the story. In the ideal form of transmedia, each medium is doing what it does best, so the story can be introduced into film, expanded through television, novels, and comics, and its world can be explored and experienced through gameplay. Each franchise entry needs to be self-contained enough to enable autonomous consumption. That is, you don't need to have seen the film to enjoy the game and vice-versa. [18].

For example, the kid-beloved Pokémon universe, a poster child not only for kids' IP design but also for cross-media narrative IP design. Several hundred different Pokémon exist, each with multiple evolutionary forms and a complex set of rivalries and attachments. There is no one text for information about these various species. Rather, the child assembles information from various media, with the result that each child knows something his or her friends do not. As a result, the child can share his or her expertise with others [19].

As mentioned above, the essence of solving children's medical problems from the perspective of systems theory is to solve the communication problem among child patients, parents, and doctors, and the key to solving this problem is a media carrier that helps children understand medical procedures and positive medical awareness. And the cross-media narrative IP is just the entry point of such a media. Through storytelling, it can spread IP content to different media in the hospital environment, such as interior design, guide cards, doctor's cards, TV, books, medical equipment, etc., to tell a story that needs to fit the psychological cognition of preschool children and is related to medical education and entertainment, to achieve the purpose of positive and effective communication.

4.2 Application of Cross-Media Narrative Theory: Practice of Children's Medical IP Design

A high-quality IP must have a distinctive character design or a novel and unique story setting. Under the premise of maintaining the consistency of "author attributes", the story system can be disseminated through branding to form a complete cross-media narrative [20]. Therefore, the team created the "Planet Courage" medical IP brand project, aiming to create a warm and comfortable environment through IP, with a variety of playmates and interesting entertainment activities. At the same time, add social games and set up a story of the medical treatment process, so that preschool children between 2 and 6 years old can see the medical treatment process as a brave adventure. "Courage Planet" medical IP brand takes the organ image as the story to reduce the fear of children, but also spreads the scientific knowledge of human health, and becomes the bridge of communication between children's patients, parents, and doctors.

Role Setting. In terms of role setting, to establish a good communication system, according to Piaget's cognitive development theory, children aged 2–7 are in the preoperational stage [21], which has the following three significant characteristics:

1. Animism: Preschool children are not able to clearly distinguish which things are alive and which are inanimate; They will generally have the characteristics of living objects added to inanimate objects - this tendency is called "animistic", so the image of the character IP needs to be anthropomorphized
2. Self-centeredness: Piaget used his classic "Three Mountains experiment" to explain this phenomenon. Children cannot divert from their own perspective and realize that others can view the same scenery in different ways, and they can only view the world from their own point of view. Therefore, the role setting should conform to the children's cognitive range and avoid unfamiliar professional terms.
3. Strong imagination: children can use symbols (e.g. words, and images in their heads) to understand the world. In the presence of pretend play, children can clearly distinguish between reality and fantasy. If the child likes, an imaginary world can be very different from reality, and the child no longer deals with the environment directly but interacts with it through the mental representation of the environment.

To sum up, we chose five organs within the cognitive range of children, namely teeth, brain, heart, lung, and stomach, as the prototype of IP character design, giving them different IP identities and personalities, and opening the adventure of Courage planet. In addition, since preschool children have a weak ability to process three-dimensional information, they mainly use two-dimensional images. 3–4-year-old children cannot distinguish all the colors in the spectrum, and distinguish the brightness of the color, so the overall use of bright, high saturation colors, to avoid brown, and gray, as shown in Fig. 4.

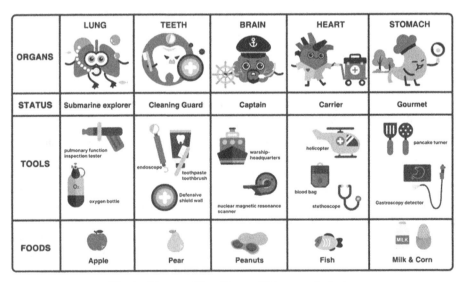

Fig. 4. "Courage Planet" organ IP image role design

Setting of Story Plot. Because preschool children have the characteristic of "self-centered" cognition, the story should convey information through the picture at the experience level of children. The plot should be abstract so that children can roughly understand the plot without looking at the text. They can only see the world from their point of view, so the interactive media lets children think that they are the hero of the whole game, through the adventure rescue story immersive experience, lets them feel fun and interesting. Ultimately, the theme of the story is adventure, and the story begins when a virus invades Planet Courage intending to destroy its Organ Spirits. In the case of dentistry TEETH, TEETH eats too much sugar and attracts the attention of the virus, which begins to destroy TEETH 's body, which slowly develops black spots and eventually turns into cavities. At this time, TEETH began to seek the help of children to get rid of the virus. Therefore, children must use toothbrushes and dental mirrors to protect the health of TEETH. A series of stories to seek the rescue of children through organ fairies, lets children understand the cause of each organ illness. At the same time, the IP character is put into reality, when the child has a similar condition of an organ and knows how to deal with it.

Interactive Media. Through the cross-media narrative theory, different media will generate new interactive forms and narrative contents of information. Each media is further combined and derived from IP with its best characteristics. Different media and languages participate in and promote the construction of the trans-media narrative world and form new cultures and values. As preschool children have the characteristic of "strong imagination" in cognition, children will automatically imagine between media to make it reasonable and interact with the mental image of the environment in a fabricated fantasy world. The IP narrative form of "Planet of Courage" precisely provides such a spatial environment. Its core narrative mechanism is based on the story plot, and the complete IP story world of "Planet of Courage" is constructed through cross-media communication of picture books, toys, and animations through medical cards, guide cards, and TVS. It enables children to experience IP stories and scenes in the strange environment of the hospital, effectively reducing children's fear.

As shown in Fig. 5, before preschool children seeing a doctor, some scenes of the hospital are integrated into the picture book so that preschool children can get familiar with the hospital environment in advance, and children patients can learn simple pathological knowledge through medical science animation and story plots; In the process of hospital treatment, doctors and parents need to cooperate with children's "adventure" actions, so that preschool children can treat the process of medical treatment as an adventure journey, which needs courage to face, and design corresponding anthropomorphic and friendly cartoon image of "Planet Courage" in the hospital to let children accept the strange scene of the hospital; After medical treatment, deepen the cognition of medical knowledge of preschool children, parents can write stories that are more easily understood and loved by children, tell their children what kind of little warrior they are playing today to defeat the virus, and give the role a new personality charm.

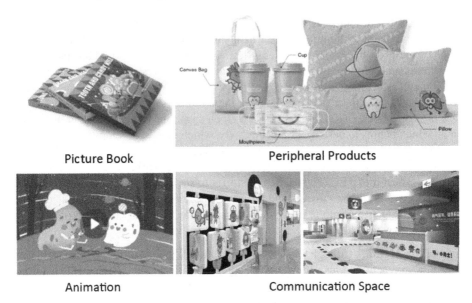

Picture Book   Peripheral Products

Animation   Communication Space

Fig. 5. "Courage Planet" Medical IP brand design

4.3 Children's Medical IP: A Communication System to Ease Children's Doctor-Patient Relationship

This paper takes medical experience as a starting point, rethinks the relationship between design and medical service, explores a truth-seeking process to understand the communication problems among children's patients, parents, and doctors, and extends to the whole link of children's medical service. As shown in Fig. 6, unlike most research and innovation projects aimed at developing health promotion interventions for children in health care is primarily based on the involvement of parents, caregivers, and other stakeholders [22], this child healthcare IP is designed based on the needs and ideas of children. Through the dissemination of narrative stories in various media, the IP enters the real world, interacts with children through storytelling, to strike a chord with children, and makes children more receptive to medical devices and the hospital environment, aiming to reconstruct the relationship between child patients, parents and doctors; As a medium, IP creates a space for communication in the process of children's medical treatment, and through multi-channel communication, strengthens children's cognition of the real world, and transmits medical and health knowledge through narrative and story ways. Therefore, we need to recognize that the medical service experience is part of children's lives and part of their growing awareness of health issues. And to rethink the relationship between design and healthcare delivery in a truth-seeking process of exploration and communication between child patients, parents, and doctors, and by extension, the whole healthcare service experience.

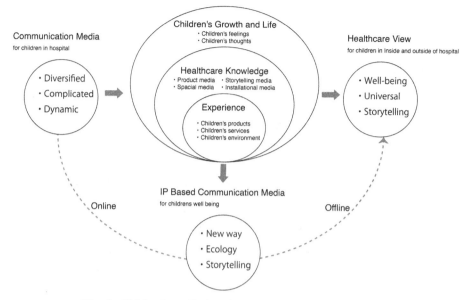

Fig. 6. Children's medical IP design mode from System Theory

5 Conclusion

The human need for health and happiness is growing, and healthcare services for children are an integral part of the design of human happiness. Children's medical treatment has always been a difficult problem, most of the previous studies were to solve the problem of specific situations or individual tasks, which is because the decision-making ability of problem-solving theory is increasingly limited in dealing with the interrelated human social system, especially in dealing with boundary problems, dealing with complex social and organizational problems, the problem-solving theory reflects a certain ambiguity. Therefore, we try to analyze the thinking mode of designers considering problems in the past, define and divide them, and gradually find a more suitable thinking mode in the process of medical treatment.

This paper provides a new perspective, this paper through the reflection, discussion, and elaboration of the problem-solving theory of the design model, to provide a framework for its re-positioning. Through field research, literature research, using the perspective of system theory, constantly explores the truth, to restore the real problems of the existing reality. The relationship between children's patients, parents, and doctors is regarded as a circular whole system, and how to deal with the communication among children's patients, parents, and doctors is regarded as the essential problem of children's medical treatment. Finally, we intervene in the IP narrative media of "Planet Courage" to create a medical experience more acceptable to children, reconstruct the relationship between children's doctors and patients, and interact with children through storytelling. Let children participate in an "adventure" action, keep brave, and make it easier for them to accept treatment. By improving the design model, we can effectively solve the complex problems related to the human social system, to improve the health and happiness of human beings.

References:

1. Geneva: World Health Organization. Constitution of the World Health Organization-Basic Documents, forty-fifth edition [DB/OL], 9 October 2006. https://www.who.int/publications/m/item/constitution-of-the-world-health-organization. 26 Nov 2023
2. Xin, Z.: A Century of problem-solving research: retrospect and prospect. J. Capital Normal Univ. (Soc. Sci. Ed.) **6**, 101–107 (2004)
3. Simon, H.A.: The Sciences of the Artificial. MIT Press, Cambridge (1969)
4. Irwin, T., Design, T.: Designing for system-level change. Decoration **10**, 12–22 (2018)
5. Simon, H.A: The structure of ill-structured problems. Artif. Intell. **1973**(4), 181–201 (1973)
6. Buchanan, R.: Wicked problems thinking in design. Des. Issues **8**(2), 5–21 (1992)
7. Buchanan, R.: Systems thinking and design thinking: the search for principles in the world we are making. She Ji: J. Des. Econ. Innov. **5**(2), 85–104 (2019)
8. Norman, D.A.: Wir brauchen neue Designer! Why Design Education Must Change [DB/OL], 16 November 2011. https://www.core77.com/posts/17993/why-design-education-must-change-17993. 26 Nov 2023
9. Schon, D.: The Practitioner of Reflection. Education Science Press, Beijing (2007)
10. Cui, G., Zhu, M.: Learning from Doing" and the interpretation of education's existentialism: an analysis of Dewey's pragmatic existentialism in learning and education. Foreign Educ. Stud. **4**, 15–19 (2005)
11. Fan, F., Dong, W., Yang, L.: Academic influence of Donald Schon's reflective practice theory on design research. J. Nanjing Univ. Arts (Fine Arts Des.) **2**, 65–70 (2022)
12. Dorst, K., Cross, N.: Creativity in the design process: co-evolution of problem–solution. Des. Stud. **22**, 434 (2001)
13. Dorst, K., Cross, N.: Creativity in the design process: co-evolution of problem–solution. Des. Stud. **22**, 435 (2001)
14. Schon, D.A.: The Reflective Practitioner: How Professionals Think in Action. Basic Books Press, London (1983)
15. Zhang, L., Sujie, W.: Research on Arturo Escobar's anthropological thought of design. Art Des. Res. **3**, 42–47 (2023)
16. McKeon, R.: Philosophic Semantics and Philosophic Inquiry. The University of Chicago Press, Chicago (1966)
17. Jenkins, H.: Convergence Culture: Where Old and New Media Collide. New York University Press, New York (2006)
18. Jenkins, H., Deuze, M.: Editorial convergence culture. Convergence: Int. J. Res. New Media Technol. **14**, 5 (2008)
19. Scolari, C.A.: Transmedia storytelling: implicit consumers, narrative worlds, and branding in contemporary media production. Int. J. Commun. **3**(3), 586–606 (2009)
20. Liang, Y.: The Innovation of IP Operating Mode in the Perspective of Transmedia Storytelling. Huazhong University of Science and Technology, Wuhan (2017)
21. Rudolph Schaffer, H.: Introducing Child Psychology. (Li Wang, Trans.) China Industrial Information Publishing Group Press (2004)
22. Nygren, J.M., Lindberg, S., Warnestal, P., Svedberg, P.: Involving children with cancer in health promotive research: a case study describing why, what, and how. JMIR Res. Protoc. **6**(2), 19 (2017)
23. Chen, G., Feng, X.X., Pang, L.J.: Developmental psychology of preschool children. Beijing Normal University Press (2013)

LingglePolish: Elevating Writing Proficiency Through Comprehensive Grammar and Lexical Refinement

Kai-Wen Tuan[1], Alison Chi[1], Hai-Lun Tu[2(✉)], Zi-Han Liao[1], and Jason S. Chang[1]

[1] Department of Computer Science, National Tsing Hua University, Hsinchu, Taiwan
{kevintuan,alisonchi,annaliao,jason}@nlplab.cc
[2] Department of Library and Information Science, Fu Jen Catholic University, Taipei, Taiwan
helen.tu@nlplab.cc

Abstract. Many existing writing enhancement tools excel at correcting syntactic errors but fall short in addressing lexical inaccuracies or providing valuable word choice suggestions for learning purposes. We present *LingglePolish*, an innovative interactive tool designed to assist language learners in rectifying both syntactic and lexical errors, thereby improving learners' overall writing proficiency. Utilizing masked language models, we augment an existing Grammar Error Correction (GEC) corpus to encapsulate comprehensive editorial insights concerning syntactic and lexical inaccuracies. We fine-tune a pre-trained generative model on this augmented dataset, culminating in the development of *LingglePolish*. Our evaluation of *LingglePolish* entails a comparative analysis with *Grammarly*, employing a corpus of authentic essays penned by language learners at the university level. Preliminary findings reveal that although *LingglePolish* demonstrates weaker performance concerning syntactic errors, it also exhibits remarkable proficiency in providing word choice suggestions. Notably, when addressing word-choice errors, our system significantly outperforms Grammarly, boasting a coverage rate that is three times higher. This enhanced capability underscores the potential of *LingglePolish* to mark a significant advancement in the domain of language learning tools.

Keywords: Grammar Error Correction · Lexical Substitution · Language Learning

1 Introduction

In light of the growing digitization of contemporary life, an increasing number of people are utilizing Computer Assisted Language Learning (CALL) tools to enhance their language writing proficiency. There are numerous studies demonstrate the efficacy and advantages of computer-assisted writing tools [20,40].

© The Author(s), under exclusive license to Springer Nature Switzerland AG 2024
F. F.-H. Nah and K. L. Siau (Eds.): HCII 2024, LNCS 14720, pp. 349–363, 2024.
https://doi.org/10.1007/978-3-031-61315-9_24

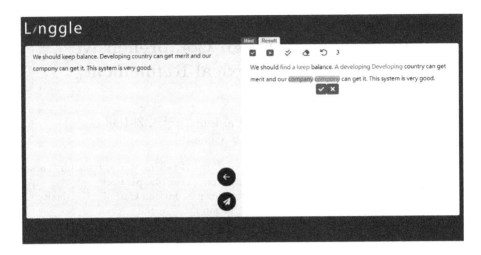

Fig. 1. The Interface of *LingglePolish*.

This pedagogical paradigm harnesses the capabilities of neural machine learning technology to facilitate automatic language acquisition, thereby affording learners a markedly personalized and interactive educational experience. Through the strategic deployment of such technological resources, learners can substantially enhance their English language skills.

Notably, prevalent writing assistance tools such as *Grammarly, Microsoft Office AutoCorrect,* and similar tools primarily concentrate on correcting syntactic errors. These tools offer corrections for issues related to function words or word forms, prepositions, articles, tense, and singular/plural distinctions.

It is also imperative to understand the significance of word choice and lexical usage in the process of language acquisition. A nuanced understanding of vocabulary and collocations is paramount for second language learners [30,42]. Despite the availability of writing tools, the persistence of word choice errors is still challenging for second language learners [32]. This can often be attributed to an inadequate mastery of vocabulary or an insufficient understanding of the proper contextual usage of words [4]. The absence of robust lexical suggestions provided by writing tools exacerbates the perpetuation of such errors in the language learners' writing.

Therefore, we present *LingglePolish*[1], a system that combines both Grammar Error Correction (GEC) and Lexical Substitution (LS) to correct and improve learners' writing skills. At run-time, users submit their writing on the left, and *LingglePolish* then displays the corrected text, processed by the underlying transformer model, on the right. The edits are color-coded: green indicates insertions, and red indicates deletions. The user can choose to accept or reject the suggestions provided by the system.

---

[1] https://w.linggle.com/.

We organize the rest of the paper as follows. We review the related works in the next section. Then, we present our method for creating our custom dataset and training our sequence-to-sequence transformer model in Sect. 3. As part of our evaluation, we asked a language learning teacher to evaluate our system on university freshman's learner writing in comparison to the popular GEC system *Grammarly* in Sect. 4. Lastly, we discuss the potential of our system and the possible future works in Sect. 5.

2 Related Works

Writing assistant systems have integrated themselves into our daily routines. Recently, state-of-the-art writing assistants utilize transformer language models to execute Grammar Error Correction (GEC) proficiently [35,39]. Our work diverges from the conventional GEC focus to address a critical aspect of writing enhancement. We integrate lexical error correction into our writing assistant framework, specifically through Lexical Substitution.

2.1 Writing Assistants

Automatic Writing Assistants have progressed alongside digital technology. Traditionally, spell checkers like GNU Aspell [2] identify misspelled words and offer corrections through a process of cross-referencing input strings with a dictionary. More recently, tools such as Writing Assistant [7] and LinggleWrite [35] address grammar errors by leveraging prevalent grammatical patterns, such as common V-N pairings, for their correction. The state-of-the-art writing assistants, including WordTune [39] and Grammarly [12], take advantage of the breakthrough capabilities of transformer models to facilitate GEC through the application of large language model.

2.2 Text Correction and Substitution

GEC. The most common type of automatic text correction is GEC [19]. Earlier methods relied on pre-written rules [5,17,21], but data-driven approaches improved performance drastically [4,38]. Today, training or fine-tuning transformer-based models tends to produce the best performance [11,27, 31,34]. Synthetically generating data has also proven useful for making larger GEC datasets [11].

Other Errors. Although great progress has been made on GEC, less work has been done on other types of error correction. Most GEC datasets only annotate "minimal" corrections, and lexical errors often go uncorrected [4]. Previous work on correcting semantic or collocation errors, even recent, often relies on traditional NLP techniques [6–8,14,16,18,23,37]. This may be partially due to the lack of large datasets that can be used for training.

Lexical Substitution. The related task of Lexical Substitution (LS) involves finding appropriate substitutes for a given word in context [24, 25, 33]. Unlike error correction, it aims to find appropriate substitutes regardless of whether the original text contains an error. Many recent LS works make use of pre-trained masked language models (MLMs) such as BERT [9] to generate state-of-the-art substitution candidates [1, 36, 41].

In this work, we aim to provide verb and adjective substitution suggestions to language learners. We have chosen to focus on verbs and adjectives because errors in verb-noun collocations, which can often be corrected by replacing the verb, represent the most common type of collocation error [32]. Additionally, adjective errors, while also prevalent, are significantly under-represented in common GEC datasets [38]. Additionally, we have decided against generating lexical substitutions for nouns, as replacing nouns is more likely to alter the sentence's intended meaning-a consequence we aim to avoid.

3   Methods

In this paper, our goal is to train a sequence-to-sequence transformer model capable of performing both Grammar Error Corrections (GECs) and providing Lexical Substitution (LS) suggestions. Rothe et al. (2021) demonstrated that a sufficiently effective GEC model can be achieved by performing a single fine-tuning stage on a cleaned parallel corpus. [31] To align with our objectives, we modify this method by generating a corpus that includes LS suggestions as well as GECs first. The data augmentation process involves identifying target words and extracting the trigrams centered around the target word. We then generate substitution candidates using a masked language model (mask-LM). We calculate the probabilities of entailment, contradiction, and neutrality between the original and candidate trigrams using a pre-trained RoBERTa model on Natural Language Inference (NLI). If the contradiction probability is low enough and entailment probability is high enough, we consider these candidates valid for Lexical Substitution and modify our database accordingly. We then fine-tune T5 base [28] with this new dataset to train a model that proficiently performs both GEC and LS tasks. Finally, we build a system around this model that provides valuable suggestions for learner writing.

3.1   Data Augmentation

Dataset. To be able to generate both GECs and LS suggestions, we have modified the cLang-8 dataset [31] to include Lexical Substitution suggestions. cLang-8 is derived from the widely used Lang-8 Learner Corpus[2], a large, multilingual, parallel corpus composed of texts written by language learners, along with user-annotated corrections of these texts. The creation of cLang-8 involved initially training a supervised GEC model, then evaluating the original target sentences

---

[2] https://lang-8.com.

from Lang-8 and retaining only the top 50% of sentence pairs based on their scores. The English version of cLang-8 that we employed contains 2,372,119 sentence pairs. In the following example, the target sentence identifies an article error in the input sentence.

Input: You know, in Japan we are confronted with power shortage.
Target: You know, in Japan we are confronted with a power shortage.

Data Modification. Our data augmentation process, as outlined in Algorithm 1, involves several steps.

(1) For each {input, target} sentence pairs, we tag the part of speech (POS) for each word using spaCy [15].

(2) For each verb and adjective, we extract trigrams, denoted as $TriOrig_w$, centered on the target word w. One example trigram we extracted from the example above is "are confronted with".

(3) We utilize the Lexical Substitution capabilities of BERT-like models [26,41], generating ten masked candidates, $TriMask_w[0-9]$, for each trigram $TriOrig_w$ using RoBERTa-base [22]. Some of the masked candidates of "are confronted with" are shown here:

(i) are dealing with
(ii) are struggling with
(iii) are faced with
(iv) are suffering with
(v) are living with

(4) To filter out unsuitable masked suggestions, we employ Natural Language Inference (NLI) probabilities for 'entailment' and 'contradiction' using the SNLI version of RoBERTa [3], as NLI probabilities have been effective in assessing meaning similarity [10,13]. For each masked trigram $i \in TriMask_w[i]$, if the contradiction probability between $TriOrig_w$ and $TriMask_w[i]$ or between $TriMask_w[i]$ and $TriOrig_w$ exceeds 10%, we consider the masked candidate as altering the original meaning and exclude it. Conversely, if the entailment probability between $TriOrig_w$ and $TriMask_w[i]$ or between $TriMask_w[i]$ and $TriOrig_w$ is above 70%, we accept the masked trigram as a viable Lexical Substitution suggestion. To continue with our example, the entailment probability between "are confronted with" and "are dealing with" is 0.864. The entailment probability between "are dealing with" and "are confronted with" is 0.896. This indicates "are dealing with" is an viable LS suggestion.

(5) After excluding masked suggestions that introduce contradictions or exhibit low entailment probabilities, we replace each original token in the sentence with its **highest probability** mask.

Algorithm 1. Algorithm for modifying each target sentence. Sentences are tokenized with spaCy. Mask candidates with the same root word as the original word are filtered out. e, c, and n stand for entailment, contradiction, and neutral, respectively.

---

for tok in $sent$ **do**
    if tok is VERB or ADJ **then**
        $tokTri \leftarrow getTokTrigram(tok)$
        $tokMasks \leftarrow genTop10Masks(tokTri)$
        for $masks, maskTri$ in $tokMasks$ **do**
            $eProb, cProb, nProb \leftarrow getNLIprobs(tokTri, maskTri)$
            $eProbFlip, cProbFlip, nProbFlip \leftarrow getNLIprobs(maskTri, tokTri)$
            if $cProb >= 10\%$ or $cProbFlip >= 10\%$ **then**
                $mask \leftarrow removed$
            end if
            if $eProb >= 70\%$ and $eProbFlip >= 70\%$ **then**
                $mask \leftarrow kept$
            end if
        end for
        if $masks$ exist **then**
            $newTok \leftarrow getTopProbMask(masks)$
            $sent \leftarrow replace(sent, tok, newTok)$
        end if
    end if
end for

This method could in theory create multiple synthetic sentences per original, but we opt to only use the top probability masks in order to maximize accuracy. Consequently, this process yields a dataset that is the **same size** as the original. Of the 2,372,118 sentence pairs in cLang-8, 1,138,010 or 47.96% of the dataset is modified with viable LS suggestions. The example sentence pair then becomes the following:

Input: You know, in Japan we are confronted with power shortage.
Target: You know, in Japan we are dealing with a power shortage.

3.2   Model

For training, we employ a 80-10-10 train-validation-test split. Similar to Rothe et al. (2021) [31], we fine-tune the pre-trained transformer T5 [29]. Due to GPU limitations, we use T5-base (600M parameters), a batch size of 32 after gradient accumulation, and maximum decoding length of 300 tokens. We train for 3 epochs and automatically select the checkpoint with the lowest validation loss.

3.3   Interface

LingglePolish (https://w.linggle.com/) is a writing assistant that offers suggestions for both grammar and word choice improvements. To begin the process, users submit their writing on the left side of the interface and click on the arrow button located in the lower middle section to initiate correction. *LingglePolish* then displays the corrected text, processed by the underlying transformer model, on the right.

LingglePolish highlights spelling, punctuation, grammar, and word choice issues with a simple color-coded system: green indicates insertions, and red indicates deletions. Users can choose to accept or reject the suggestions by clicking the tick or cross, respectively. A simple example is shown in Fig. 1. In this example, the user wrote the following passage, and *LingglePolish* provided the following corrections.

Input: We should keep balance. Developing country can get merit and our compony can get it. This system is very good.

Correction: We should find a keep balance. A developing Developing country can get merit and our company compony can get it. This system is very good.

In this instance, *LingglePolish* offers three suggestions to the user. Firstly, the system indicates "find a balance" as a better phrase than "keep balance". Secondly, the user made an article error by omitting "A" in the phrase "A developing country" instead of "Developing country." Lastly, *LingglePolish* identified the user's spelling mistake regarding the word "company."

4   Results

Due to the dual-purpose nature of our system, employing standard evaluation sets for both Grammatical Error Correction (GEC) and Lexical Substitution (LS) would be insufficient to yield definitive results. To this end, we enlisted the expertise of Dr. Y. Harada from Waseda University, Japan, to conduct a focused expert evaluation of our system in comparison to *Grammarly*. This evaluation was performed on 23 students' writing samples from Dr. Harada's freshman English class. In order to assess the efficacy of each edit suggested by the systems, we tasked Dr. Harada with categorizing the edits into word choice, idiom,

Table 1. Evaluation Metrics

Rating	Metrics
0	erroneous edit/original contains no error
1	reasonable suggestion/no improvement over the quality of writing
2	reasonable suggestion/improve the writing slightly
3	good suggestion/educational
4	very good suggestion/correct grammar error
5	excellent suggestion/improve the writing quality

grammar error, spelling and capitalization, punctuation, and incorrect sugges-
tion. Several points to note, preposition errors are classified as word choice in
Dr. Harada's assessment and idioms include common phrases such as "as an
example" or "in other word." For each edit made by the systems, we asked Dr.
Y. Harada to rate the edit from 0 to 5 according to the metrics shown in Table 1.
The evaluation results are shown in Table 2.

Table 2. Edit Type Distributions of *Grammarly* and *LingglePolish* on Learner
Essays

System	Word Choice	Idiom	Grammar Error	Spelling & Capitalization	Punctuation	Incorrect Suggestion	Total Edits
Grammarly	113	25	423	17	150	95	823
LingglePolish	375	16	320	19	2	137	869

An overview of the correction distinctions between *LingglePolish* and *Grammarly* reveals that both systems produce a similar number of edits. However,
the two systems have a significant discrepancy in the distribution of edit types.
From Table 2, it is evident that *LingglePolish* generates substantially more word
choice suggestions, accounting for 43.2% of all *LingglePolish* edits, compared
to 13.7% done by *Grammarly*. On the other hand, *Grammarly* identifies over
30% more grammar errors than *LingglePolish*, with more than 50% of its corrections targeting grammar errors. An interesting observation is that 18% of
Grammarly's edits address punctuation errors, such as adding a comma after
'and,' whereas *LingglePolish* makes almost no punctuation edits. Furthermore,
Grammarly demonstrates better accuracy with their edits, having an 11.5% error
rate compared to the 15.8% error rate of *LingglePolish*. Edits in idioms, spelling,
and capitalization categories are relatively rare for both systems. It is noteworthy that some grammar errors corrected by *Grammarly* are also addressed by
LingglePolish through lexical substitution, effectively resolving the collocation
errors. We take a more granular look into each category below.

Figure 2 displays the score distribution for word choice suggestions between
the two systems. In addition to offering a greater number of word choice suggestions, the majority of edits provided by *LingglePolish* receive a score of 4
or higher. This outcome is by design, as *LingglePolish* seeks to provide more
context-specific wording. An illustration of this approach is the modification
from 'candidate to realize the importance of planning' to 'candidate to understand the importance of planning.' Conversely, *Grammarly* often suggests synonyms in its word choice corrections, such as changing 'a very good experience'
to 'an excellent experience.' Overall, compared to other categories, the score
distribution for word choice suggestions from both systems is more varied, indicating a less consistent outcome than other categories. To some extent, this
variability is expected, as word choice suggestions are more about enhancing
the writing rather than correcting it, thus reflecting a broader range of possible
improvements.

Figure 3 illustrates the score distribution for idiom suggestions between the
two systems. Interestingly, all but one of the idiom corrections received a score

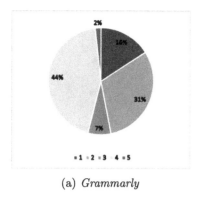

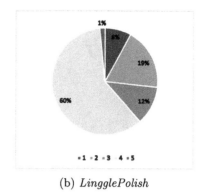

(a) *Grammarly*   (b) *LingglePolish*

Fig. 2. Score distribution graph of *Grammarly* and *LingglePolish* in Word Choice

of 4. This consistent scoring is attributed to the fact that idiom suggestions typically emerge when language learners misuse or misremember the idiom they intended to use, and the corrections are uniform between the two systems. The sole exception is an edit made by *LingglePolish*, which changed 'take anybody to anywhere' to 'take anybody anywhere.' Dr. Harada rated this edit as 1, indicating that both phrasings are acceptable and the distinction between them does not significantly impact meaning or grammar.

Figure 4 illustrates the score distribution for grammar error corrections between the two systems. Not surprisingly, both systems demonstrate higher precision in this category. *Grammarly* has specifically targeted this aspect of writing, even incorporating the word 'Grammar' into their brand name, which underscores their focus on grammatical accuracy. For *LingglePolish*, the training dataset was initially a parallel dataset designed to provide comparisons between grammar errors and their corrections. The observed deviation in the number of edits provided by *LingglePolish* compared to *Grammarly* likely originates from

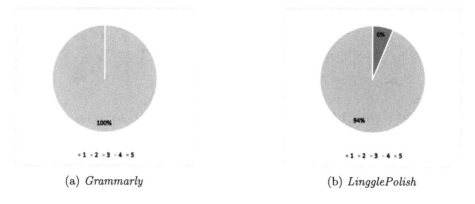

(a) *Grammarly*   (b) *LingglePolish*

Fig. 3. Score distribution graph of *Grammarly* and *LingglePolish* in Idiom

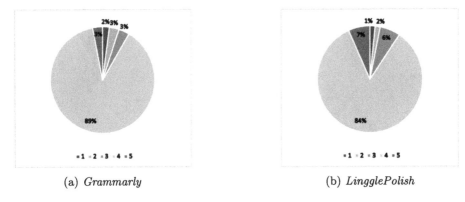

(a) *Grammarly*   (b) *LingglePolish*

Fig. 4. Score distribution graph of *Grammarly* and *LingglePolish* in Grammar Error

the fact that we modified approximately 40% of the training data to include lexical substitution, representing a trade-off in our approach. Nevertheless, *LingglePolish* still provides high quality edits in this category.

Figure 5 shows the score distribution for spelling and capitalization corrections between the two systems. It is evident that *LingglePolish* does not perform as consistently in this category as in others. Upon analyzing the data, we identified a vulnerability in *LingglePolish*: it routinely marks compound words as erroneous and suggests that users separate them. For instance, it incorrectly flags 'cannot' and 'smartphones' as incorrect, recommending a change to 'can not' and 'smart phones,' respectively. This behavior stems from the tokenization process used to identify parts of speech (POS) tags for each word. However, this issue is relatively simple to address with system post-processing to eliminate such mistakes in the future.

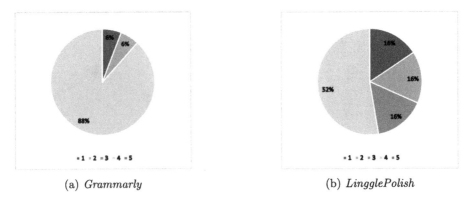

(a) *Grammarly*   (b) *LingglePolish*

Fig. 5. Score distribution graph of *Grammarly* and *LingglePolish* in Spelling & Capitalization

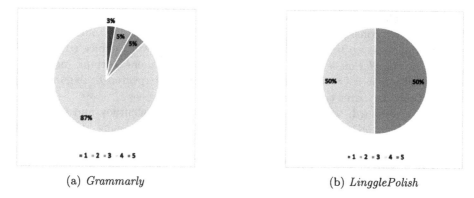

(a) *Grammarly*   (b) *LingglePolish*

Fig. 6. Score distribution graph of *Grammarly* and *LingglePolish* in Punctuation

In regard to punctuation, as shown in Fig. 6, *LingglePolish* performs almost no edits in this category. We observed that *Grammarly* appears to have a rule hard-coded to insert a comma after the word 'and,' which constitutes the majority of the punctuation edits it made in this evaluation set. Notably, *LingglePolish* frequently suggests adding a space after a quotation mark or deleting a space before a paragraph starts. Additionally, *LingglePolish* struggles to recognize different symbols representing the same punctuation. For instance, it would suggest users' correct full-width parentheses "" to the Unicode standard quotes". These types of edits were not included in our calculation of the system's effectiveness, as we deemed these issues did not alter the text in a meaningful way.

5   Conclusion

In this paper, we introduced *LingglePolish*, a writing assistant that combines Grammar Error Correction (GEC) and Lexical Substitution (LS) to enhance the quality of users' writing. Unlike conventional writing assistants, which primarily address GEC issues, our system additionally offers word choice suggestions. This feature encourages learners to reflect on and refine their lexical choices, consequently facilitating the organic expansion of their vocabulary.

We conducted an evaluation of our system against *Grammarly* using essays from university-level students. While *Grammarly* proved more adept at identifying grammar errors, *LingglePolish* generated more than threefold the number of word choice suggestions. Our evaluation further revealed that the majority of corrections provided by *LingglePolish* have a positive impact on the overall quality of writing.

During the evaluation, we identified limitations stemming from the training data and tokenization process: *LingglePolish* struggles with recognizing indentation common in structured writing as acceptable and incorrectly parses certain punctuation, leading to issues such as breaking up contractions like "we're" into "we' re".

Overall, *Grammarly* delivered more consistent edits across various categories. However, when considering *LingglePolish* primarily as a learning tool rather than an editorial tool, we firmly believe that the extensive word choice suggestions offered by *LingglePolish* would challenge language learners to think more critically about their vocabulary usage. This, in turn, is expected to expand their active vocabulary over time.

Looking forward, we aim to enhance the system's accuracy and efficiency. Currently, *LingglePolish* is not optimized for parallel processing, which results in a slightly extended runtime. We are dedicated to overcoming this challenge and improving the system's speed in future updates.

Acknowledgements. We would like to thank Professor Yasunari Harada for testing our system on his student's writing and helping to evaluate the performance of our system. This research is sponsored by the National Science and Technology Council, Taiwan, under Grant no. NSTC 112-2221-E-007-008- and NSTC 112-2410-H-030-021 -.

References

1. Arefyev, N., Sheludko, B., Podolskiy, A., Panchenko, A.: Always keep your target in mind: studying semantics and improving performance of neural lexical substitution. In: Proceedings of the 28th International Conference on Computational Linguistics, pp. 1242–1255. International Committee on Computational Linguistics, Barcelona, Spain (Online), December 2020. https://doi.org/10.18653/v1/2020.coling-main.107, https://aclanthology.org/2020.coling-main.107
2. Atkinson, K.: GNU Aspell (2003). http://aspellsourceforge.net/
3. Bowman, S.R., Angeli, G., Potts, C., Manning, C.D.: A large annotated corpus for learning natural language inference. In: Proceedings of the 2015 Conference on Empirical Methods in Natural Language Processing, pp. 632–642. Association for Computational Linguistics, Lisbon, Portugal, September 2015. https://doi.org/10.18653/v1/D15-1075, https://aclanthology.org/D15-1075
4. Bryant, C., Yuan, Z., Qorib, M.R., Cao, H., Ng, H.T., Briscoe, T.: Grammatical error correction: a survey of the state of the art. Comput. Linguist. **49**, 1–59 (2022)
5. Burstein, J., Chodorow, M., Leacock, C.: CriterionSM online essay evaluation: an application for automated evaluation of student essays. In: IAAI, pp. 3–10. Citeseer (2003)
6. Carlini, R., Codina-Filba, J., Wanner, L.: Improving collocation correction by ranking suggestions using linguistic knowledge. In: Proceedings of the Third Workshop on NLP for Computer-assisted Language Learning, pp. 1–12. LiU Electronic Press, Uppsala, Sweden, November 2014. https://aclanthology.org/W14-3501
7. Chang, Y.C., Chang, J.S., Chen, H.J., Liou, H.C.: An automatic collocation writing assistant for Taiwanese EFL learners: a case of corpus-based NLP technology. Comput. Assist. Lang. Learn. **21**(3), 283–299 (2008)
8. Dahlmeier, D., Ng, H.T.: Correcting semantic collocation errors with L1-induced paraphrases. In: Proceedings of the 2011 Conference on Empirical Methods in Natural Language Processing, pp. 107–117. Association for Computational Linguistics, Edinburgh, Scotland, UK, July 2011. https://aclanthology.org/D11-1010

9. Devlin, J., Chang, M.W., Lee, K., Toutanova, K.: BERT: pre-training of deep bidirectional transformers for language understanding. In: Proceedings of the 2019 Conference of the North American Chapter of the Association for Computational Linguistics: Human Language Technologies, vol. 1 (Long and Short Papers), pp. 4171–4186. Association for Computational Linguistics, Minneapolis, Minnesota, June 2019. https://doi.org/10.18653/v1/N19-1423, https://aclanthology.org/N19-1423

10. Dušek, O., Kasner, Z.: Evaluating semantic accuracy of data-to-text generation with natural language inference. In: Proceedings of the 13th International Conference on Natural Language Generation, pp. 131–137. Association for Computational Linguistics, Dublin, Ireland, December 2020. https://aclanthology.org/2020.inlg-1.19

11. Flachs, S., Stahlberg, F., Kumar, S.: Data strategies for low-resource grammatical error correction. In: Proceedings of the 16th Workshop on Innovative Use of NLP for Building Educational Applications, pp. 117–122. Association for Computational Linguistics, Online, April 2021. https://aclanthology.org/2021.bea-1.12

12. Ghufron, M.A., Rosyida, F.: The role of grammarly in assessing English as a foreign language (EFL) writing. Lingua Cultura **12**, 395 (2018). https://doi.org/10.21512/lc.v12i4.4582

13. Goyal, T., Durrett, G.: Evaluating factuality in generation with dependency-level entailment. In: Findings of the Association for Computational Linguistics: EMNLP 2020, pp. 3592–3603. Association for Computational Linguistics, Online, November 2020. https://doi.org/10.18653/v1/2020.findings-emnlp.322, https://aclanthology.org/2020.findings-emnlp.322

14. Herbelot, A., Kochmar, E.: 'Calling on the classical phone': a distributional model of adjective-noun errors in learners' English. In: Proceedings of COLING 2016, the 26th International Conference on Computational Linguistics: Technical Papers, pp. 976–986. The COLING 2016 Organizing Committee, Osaka, Japan, December 2016. https://aclanthology.org/C16-1093

15. Honnibal, M., Montani, I.: spaCy 2: natural language understanding with Bloom embeddings, convolutional neural networks and incremental parsing (2017, to appear)

16. Huang, P.Y., Tsao, N.L.: Using collocation clusters to detect and correct English l2 learners' collocation errors. Comput. Assist. Lang. Learn. **34**(3), 270–296 (2021)

17. Jensen, K., Heidorn, G.E., Miller, L.A., Ravin, Y.: Parse fitting and prose fixing: getting a hold on ill-formedness. Am. J. Comput. Linguist. **9**(3–4), 147–160 (1983). https://aclanthology.org/J83-3002

18. Kochmar, E., Briscoe, T.: Detecting learner errors in the choice of content words using compositional distributional semantics. In: Proceedings of COLING 2014, the 25th International Conference on Computational Linguistics: Technical Papers, pp. 1740–1751. Dublin City University and Association for Computational Linguistics, Dublin, Ireland, August 2014. https://aclanthology.org/C14-1164

19. Kwasny, S.C., Sondheimer, N.K.: Relaxation techniques for parsing grammatically ill-formed input in natural language understanding systems. Am. J. Comput. Linguist. **7**(2), 99–108 (1981). https://aclanthology.org/J81-2002

20. Lai, S.L., Chang, J.S.: Toward a pattern-based referencing tool: learner interactions and perceptions. ReCALL **32**(3), 272–290 (2020)

21. Leacock, C., Gamon, M., Brockett, C.: User input and interactions on Microsoft research ESL assistant. In: Proceedings of the Fourth Workshop on Innovative Use of NLP for Building Educational Applications, pp. 73–81 (2009)

22. Liu, Y., et al.: RoBERTa: a robustly optimized BERT pretraining approach. arXiv preprint arXiv:1907.11692 (2019)
23. Makarenkov, V., Rokach, L., Shapira, B.: Choosing the right word: using bidirectional LSTM tagger for writing support systems. Eng. Appl. Artif. Intell. **84**, 1–10 (2019)
24. McCarthy, D., Navigli, R.: The English lexical substitution task. Lang. Resour. Eval. **43**, 139–159 (2009)
25. Melamud, O., Goldberger, J., Dagan, I.: context2vec: learning generic context embedding with bidirectional LSTM. In: Proceedings of the 20th SIGNLL Conference on Computational Natural Language Learning, pp. 51–61. Association for Computational Linguistics, Berlin, Germany, August 2016. https://doi.org/10.18653/v1/K16-1006, https://aclanthology.org/K16-1006
26. Michalopoulos, G., McKillop, I., Wong, A., Chen, H.: LexSubCon: integrating knowledge from lexical resources into contextual embeddings for lexical substitution. In: Proceedings of the 60th Annual Meeting of the Association for Computational Linguistics (Volume 1: Long Papers), pp. 1226–1236. Association for Computational Linguistics, Dublin, Ireland, May 2022. https://doi.org/10.18653/v1/2022.acl-long.87, https://aclanthology.org/2022.acl-long.87
27. Qorib, M., Na, S.H., Ng, H.T.: Frustratingly easy system combination for grammatical error correction. In: Proceedings of the 2022 Conference of the North American Chapter of the Association for Computational Linguistics: Human Language Technologies, pp. 1964–1974. Association for Computational Linguistics, Seattle, United States, July 2022. https://doi.org/10.18653/v1/2022.naacl-main.143, https://aclanthology.org/2022.naacl-main.143
28. Raffel, C., et al.: Exploring the limits of transfer learning with a unified text-to-text transformer. J. Mach. Learn. Res. **21**(1), 5485–5551 (2020)
29. Raffel, C., et al.: Exploring the limits of transfer learning with a unified text-to-text transformer. J. Mach. Learn. Res. **21**(140), 1–67 (2020). http://jmlr.org/papers/v21/20-074.html
30. Richards, J.C., Rodgers, T.S.: Approaches and Methods in Language Teaching. Cambridge University Press, Cambridge (2014)
31. Rothe, S., Mallinson, J., Malmi, E., Krause, S., Severyn, A.: A simple recipe for multilingual grammatical error correction. arXiv preprint arXiv:2106.03830 (2021)
32. Sari, B.N., Gulö, I.: Observing grammatical collocation in students' writings. Teknosastik **17**(2), 25–31 (2019)
33. Seneviratne, S., Daskalaki, E., Lenskiy, A., Suominen, H.: CILex: an investigation of context information for lexical substitution methods. In: Proceedings of the 29th International Conference on Computational Linguistics, pp. 4124–4135. International Committee on Computational Linguistics, Gyeongju, Republic of Korea, October 2022. https://aclanthology.org/2022.coling-1.362
34. Tarnavskyi, M., Chernodub, A., Omelianchuk, K.: Ensembling and knowledge distilling of large sequence taggers for grammatical error correction (2022)
35. Tsai, C.T., Chen, J.J., Yang, C.Y., Chang, J.S.: LinggleWrite: a coaching system for essay writing. In: Proceedings of the 58th Annual Meeting of the Association for Computational Linguistics: System Demonstrations, pp. 127–133. Association for Computational Linguistics, Online, July 2020. https://doi.org/10.18653/v1/2020.acl-demos.17, https://aclanthology.org/2020.acl-demos.17
36. Wada, T., Baldwin, T., Matsumoto, Y., Lau, J.H.: Unsupervised lexical substitution with decontextualised embeddings (2022)

37. Wanner, L., Verlinde, S., Alonso Ramos, M.: Writing assistants and automatic lexical error correction: word combinatorics. In: Electronic Lexicography in the 21st Century: Thinking Outside the Paper, pp. 472–487 (2013)

38. Yannakoudakis, H., Briscoe, T., Medlock, B.: A new dataset and method for automatically grading ESOL texts. In: Proceedings of the 49th Annual Meeting of the Association for Computational Linguistics: Human Language Technologies, pp. 180–189. Association for Computational Linguistics, Portland, Oregon, USA, June 2011. https://aclanthology.org/P11-1019

39. Zhao, X.: Leveraging artificial intelligence (AI) technology for English writing: introducing wordtune as a digital writing assistant for EFL writers. RELC J. **54**, 00336882221094089 (2022)

40. Zheltukhina, M.R., Kislitsyna, N.N., Panov, E.G., Atabekova, A., Shoustikova, T., Kryukova, N.I.: Language learning and technology: a conceptual analysis of the role assigned to technology. Online J. Commun. Media Technol. **13**(1), e202303 (2023)

41. Zhou, W., Ge, T., Xu, K., Wei, F., Zhou, M.: BERT-based lexical substitution. In: Proceedings of the 57th Annual Meeting of the Association for Computational Linguistics, pp. 3368–3373. Association for Computational Linguistics, Florence, Italy, July 2019. https://doi.org/10.18653/v1/P19-1328, https://aclanthology.org/P19-1328

42. Zimmerman, C.B.: Historical trends in second language vocabulary instruction. Second Lang. Vocabulary Acquist. **6**(1), 5–19 (1997)

Author Index

A

Aiba, Tomoya I-66
Alhadreti, Obead II-133
Alharasis, Esraa Esam II-149
Alkhwaldi, Abeer F. II-149
Alston-Stepnitz, Eli II-233
Arai, Hiromi I-66
Asici, Jasmin I-108
Assila, Ahlem II-79
Auinger, Andreas I-46, I-157

B

Baldauf, Matthias II-38
Barros, Carlos II-221
Bartscherer, Thomas I-289
Bekiri, Valmir II-38
Beladjine, Djaoued II-79
Blöchle, Jennifer J. I-289
Brandtner, Patrick II-253

C

Chakraborty, Debapriya II-233
Chang, Jason S. I-349
Chen, Langtao I-145, II-13
Chen, Tzuhsuan I-3
Chi, Alison I-349
Chu, Tsai-Hsin II-200
Chung, Youri I-254
Cimene, Angelika II-233
Cruz-Cárdenas, Jorge II-3

D

de Vries, Marné II-96
Demaeght, Annebeth I-200, II-174, II-209
Deng, Kailing II-13
Deyneka, Olga II-3

E

Eibl, Stefan I-157
Eschenbrenner, Brenda I-145
Evans, Megan Rebecca II-27

F

Favetti, Matthew II-233
Fina, Robert A. I-157
Fischer-Janzen, Anke I-289
Fu, Xiaoliang I-23

G

Gapp, Markus I-289
García-Tirado, Johny II-221
Gerster, Sabine I-33
Götten, Marcus I-289
Grassauer, Franziska I-46
Gratzer, Katrin I-302
Greschuchna, Larissa II-209
Groth, Aleksander I-302
Gücük, Gian-Luca I-270
Guler, Martin II-38
Guo, Nickolas I-254
Gupta, Samrat I-237

H

Heuer, Marvin II-50
Hirschfelt, Kate II-233
Hsieh, Pei-Hsuan I-176
Hu, Youhong I-145
Huang, Hui-I. I-187
Huang, Yinghsiu I-3

I

Israel, Kai II-63

J

Jiang, Hang II-189
Jin, Yurhee I-91

F. F.-H. Nah and K. L. Siau (Eds.): HCII 2024, LNCS 14720, pp. 365–367, 2024.
https://doi.org/10.1007/978-3-031-61315-9

K

Kessing, David   I-314
Kojima, Toi   I-66
Kolbe, Diana   I-200
Kong, Nathaniel   II-233
Kordyaka, Bastian   I-79
Kučević, Emir   I-270

L

Lee, Yen-Hsien   II-200
Lee, Yi-Cheng   II-200
Leible, Stephan   I-270
Liang, Siyi   I-23
Liao, Zi-Han   I-349
Lin, Yuan-Fa   I-210
Ling, Zipei   I-23
Liu, Lili   I-23
Löwer, Manuel   I-314

M

Maeda, Soshi   I-66
Mao, Zhewei   I-23
Monla, Ziad   II-79
Moodley, Tevin   II-27
Müller, Andrea   I-200, II-174, II-209

N

Nakatsu, Robbie T.   I-223
Navarro, Evaristo   II-221
Nerb, Josef   II-174
Niehaves, Bjoern   I-79
Nishigaki, Masakatsu   I-66

O

Ohki, Tetsushi   I-66
Ostermann, Philip   II-50

P

Palacio-Fierro, Andrés   II-3
Panwar, Tarun   I-127
Pappas, Ilias   I-237
Peng, Cheng   I-334
Phong, Ambrose   I-176
Ponomarjova, Katrin-Misel   I-289

Q

Qian, Xiarui   I-23

R

Ramírez, Javier Alfonso   II-221
Ramirez, Rosmery Suarez   II-221
Ramos-Galarza, Carlos   II-3
Reiter, Joachim   II-209
Richter, Magdalena   II-253

S

Sanguinetti, Angela   II-233
Schlögl, Stephan   I-302
Schmager, Stefan   I-237
Schmidt, Felix N.   I-270
Sierra, Joaquín   II-221
Simic, Dejan   I-270
Sohn, Jung Joo   I-254
Sung, Youjin   I-91

T

Tan, Caroline Swee Lin   I-127
Tu, Hai-Lun   I-349
Tuan, Kai-Wen   I-349

V

Van Laerhoven, Kristof   I-289
Vassilakopoulou, Polyxeni   I-237
Venter, Anthea   II-96
von Brackel-Schmidt, Constantin   I-270

W

Wang, Chou-Wen   I-187, I-210
Wang, Yajie   I-334
Wang, Yongle   II-189
Wang, Zichan   II-115
Weber, Sebastian   I-79
Wendt, Thomas M.   I-289
Woerz, Barbara   I-200
Wohn, Kwang-Yun   I-91
Wu, Chin-Wen   I-187, I-210
Wu, Jiangqiu   II-189
Wyszynski, Marc   I-79

Y

Yang, Yu-chen   II-200
Yum, Joosun   I-91

Z

Zabelina, Ekaterina   II-3
Zaharia, Silvia   I-108
Zerres, Christopher   II-63
Zghal, Mourad   II-79

Zhu, Kaiyan   I-127
Zhuang, Beini   II-189
Zimmermann, Hans-Dieter   II-38
Zimmermann, Robert   II-253

Printed in the United States
by Baker & Taylor Publisher Services